Vitamin D: A Comprehensive Guide for Medical Professionals

Vishal Kumar • Sandeep Patel
Editors

Vitamin D: A Comprehensive Guide for Medical Professionals

Springer

Editors
Vishal Kumar
Department of Orthopaedics
PGIMER
Chandigarh, Chandigarh, India

Sandeep Patel
Department of Orthopaedics
PGIMER
Chandigarh, Chandigarh, India

ISBN 978-981-96-6490-0 ISBN 978-981-96-6491-7 (eBook)
https://doi.org/10.1007/978-981-96-6491-7

This Springer imprint is published by the registered company Springer Nature Singapore Pte Ltd.
The registered company address is: 152 Beach Road, #21-01/04 Gateway East, Singapore 189721, Singapore

Preface

Vitamin D deficiency is a pandemic. It is really infrequent to find an individual nowadays who has visited a doctor for any medical reason and not got his vitamin D and calcium levels done by the medical personnel. Nowadays, it is rampant to find patients visiting outpatient department with their serum vitamin D and calcium levels tested prior presuming it to be the culprit for whatsoever their health problem may be. Myths and whims about vitamin D have plagued not only the common man but the medical professionals as well. Vitamin D has in recent times been a witch on the roads rather than a witch in the woods. It is also not infrequent to find individuals taking sachets and shots of Vitamin D as a prophylactic and precautionary measure all by themselves without the advice of a registered medical practitioner. Otherwise, dietary habits, modern lifestyle, increased longevity, packaged food, and milk have contributed immensely in making the normal vitamin D levels in the general population an extinct entity. It is a rarity nowadays to find normal levels of vitamin D of an individual in clinics.

Evidence-based acquaintance with these harsh realities about the deficiency of Vitamin D inspired and aspired to perspire for this compendium on a common topic of immense significance. Vitamin D is intricately related to the well-being of an individual, family, society, and nation as a whole.

The authors of the different chapters in this academic treatise are from all across the nation and are the masters of their core specialty. The chapters are authoritative yet simple and lucid with the sole intent to curb the epidemiological menace of vitamin D deficiency and its droughty consequences.

This handy, crispy evidence-based treatise on vitamin D covers all its aspects in toto. This is a must-keep reference book by each and every paramedical and medical professional across all level of career including trainees and teachers alike, inclusive of all specialties of medicine and surgery, and in fact few of the chapters are worth a read by the common man. We wish this academic wholesome treatise serves its goal of being the definitive and guiding textbook on vitamin D. We trust the book will provide an answer to whatever query bewilders one about vitamin D.

Thank You Team Springer Nature for giving our pen an immortal place to perpetuate knowledge!

Chandigarh, Chandigarh, India Vishal Kumar

Contents

Chapter 1
Vitamin D: Prevention and Management from a Public Health Perspective

Risha Gupta and Shankar Prinja

Abstract Even though there is huge evidence on the association of vitamin D with the onset, progression, and treatment of various diseases and health maintenance, deficiency of this quintessential micronutrient, i.e. vitamin D, is very common. Originating as a key source by exposure to sunshine and partly from dietary sources, a large proportion of the population suffers from vitamin D deficiency. It is associated with the treatment and etiopathogenesis of various disorders such as hypertension, cancer, autoimmune diseases, osteoporosis, muscle weakness, rheumatoid arthritis, and type I diabetes. In addition, the widely prevalent deficiency of vitamin D delineates consideration of its prevention and management through public health strategies such as generating awareness among people and healthcare professionals, implementation of widespread policies such as fortification of staple food, supplementation, and dietary diversity.

Keywords Vitamin D insufficiency · Deficiency · Screening · Supplementation · Fortification · Awareness generation

Learning Objectives
- To understand the epidemiology of vitamin D deficiency.
- To elicit the significance of addressing the issue of vitamin D deficiency.
- To review various public health strategies for the prevention and management of vitamin D deficiency.

R. Gupta · S. Prinja (✉)
Department of Community Medicine and School of Public Health, PGIMER, Chandigarh, India

V. Kumar, S. Patel (eds.), *Vitamin D: A Comprehensive Guide for Medical Professionals*, https://doi.org/10.1007/978-981-96-6491-7_1

1.1 Background

Vitamin D is a quintessential micronutrient. Its status has changed from that of a basic vitamin to an essential hormone with numerous important effects on the human body. It is highly important for the development, expansion, and maintenance of the robust skeletal system. It helps in maintaining the calcium balance in the body from conception till death [1]. Vitamin D is a group of fat-soluble prohormones, known for its anti-rachitic activity. Vitamin D3 (cholecalciferol) and vitamin D2 (ergocalciferol) are the two major biologically inert precursors of vitamin D [2]. Sunlight is a major source of vitamin D. 80% of the daily requirement accomplishes by exposure to adequate sunlight. The dietary sources of vitamin D are usually animal sources such as dairy products, mackerel and sardines, fishes like salmon, eggs, and meat products. Plant sources provide ergocalciferol instead of cholecalciferol (vitamin D3) [3].

1.1.1 History of Vitamin D

There is a unique and rich background of vitamin D. In the early 1600s, for the first time the symptoms of rickets in children and osteomalacia in adults had been described. After approximately 250 years, physicians and biochemists evident the role of sunlight in the management of these conditions. Next, the chemical compositions of ergocalciferol (D2) and cholecalciferol (D3), the two primary forms of vitamin D, were determined. Then in 1900–1920 period, the anti-rachitic effect of cod liver oil had been identified which could be attributed to a vitamin found in cod liver oil. This vitamin was then named as “D” as vitamin A, B, and C had already been discovered earlier.

Subsequently, 1,25-dihydroxyvitamin D (1,25-$(OH)_2D$), the active form of vitamin D, was identified and its metabolism was explained in 1967. After that it has been realized that vitamin D has its own biological mechanism and it consist of transport proteins, metabolic enzymes, and vitamin D receptor (VDR) to mediate its actions in bone and many other tissues in the body leading to different physiological effects [4] (Fig. 1.1).

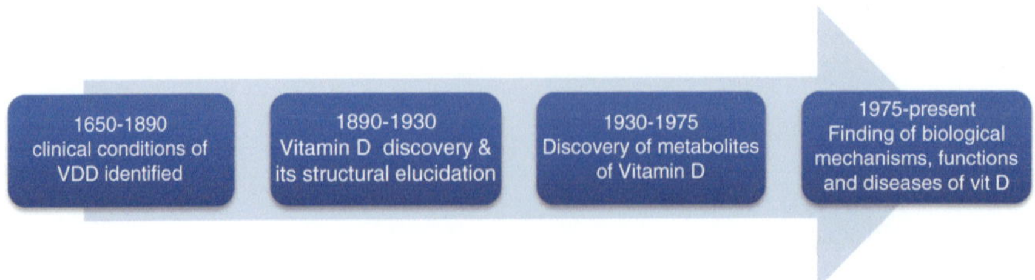

Fig. 1.1 The historical stages in vitamin D discovery

Table 1.1 The serum 25(OH)D values determining the vitamin D status in the body [5]

Serum 25(OH)D values	Vitamin D status
< 30 nmol/L (12 ng/mL)	Deficiency
30–50 nmol/L (12–20 ng/mL)	Insufficiency
≥50 nmol/L (20 ng/mL)	Sufficient

1.1.2 Vitamin D Deficiency

A serious public health concern, vitamin D deficiency (VDD), has become a pandemic in recent years. For vitamin D, the Recommended Dietary Allowance (RDA) is 200 IU daily. To prevent its insufficiency, however, 600–800 IU per day is needed if there is less exposure to sunshine [5]. During pregnancy and lactation, an intake of 600 IU is advised, and the RDA varies for different age groups. The most common method for assessing the body's vitamin D status was to measure the total serum 25(OH)D concentration in circulation. Diverse agreement guidelines exist about the definition of threshold levels for vitamin D insufficiency. However, the US Institute of Medicine (IOM) decided the cut-off points in view of the risk of various metabolic diseases as shown in Table 1.1.

1.1.3 Epidemiology of Vitamin D Deficiency

Vitamin D deficiency (VDD) is a global public health problem. Approximately 1 billion people worldwide have vitamin D deficiency, while 50% of the population are vitamin D insufficient [3].

1.1.3.1 Predisposing Risk Factors for Vitamin D Deficiency

Vitamin D deficiency can result from several causes [6]. Studies have shown that the prevalence of VDD is highest in hospitalized patients or nursing home residents, the elderly, the obese patients, and patients with chronic diseases. The various risk factors are as follows:

- Limited dietary exposure to vitamin D, which includes vegetarian diets, low dietary intake of natural sources of vitamin D such as fish, eggs, and low intake of vitamin D fortified foods.
- The presence of certain malabsorption syndromes such as inflammatory bowel disease, chronic pancreatic insufficiency, celiac disease, short bowel syndrome, gastric bypass, and cystic fibrosis that may cause vitamin D deficiency.
- Less time in direct sunshine. Sunlight exposure of about 40% of the skin for 20 min a day is necessary for the body to produce enough quantities of vitamin D [8].

  - Due to the higher melanin concentration in their skin, persons with darker complexion have been shown to have less cutaneous vitamin D production.
  - Hospitalized individuals from a very long time can have less exposure [7].
  - Individuals who use sunscreens consistently.
  - The other risk factors that contribute to decreased exposure to sunlight are higher latitudes (above 40° North), winter season, cloud cover and atmospheric pollution, and wearing a dress that covers large portion of the body.
- Reduced endogenous synthesis. Patients suffering from chronic diseases such as liver cirrhosis can have defective 25-hydroxylation, leading to deficiency of active vitamin D. Defects in 1-alpha 25-hydroxylation can also be seen in primary hyperparathyroidism, renal failure, and 1-alpha hydroxylase deficiency.
- There are certain drugs which by inducing hepatic p450 enzymes activate the degradation of vitamin D or induce the hepatic catabolism of vitamin D. For example, phenobarbital, clotrimazole, rifampin, spironolactone, carbamazepine, dexamethasone, and nifedipine [8].
- Vitamin D status of infants depends upon vitamin D status of mother during pregnancy.
- Genetic factors affecting physiology and metabolism of vitamin D. In hereditary vitamin D resistant rickets, there is end organ resistance to vitamin D.
- With age, cutaneous synthesis of vitamin D decreases.
- Obesity.

1.1.3.2 Population at Risk

In concordance with the above mentioned risk factors, there are some sub-population groups who are at increased risk of VDD [9].

- Breastfed infants as human milk is a poor source of vitamin D. These should be given a supplementation of 400 IU daily of vitamin D (as per APA guidelines).
- Older persons, as their kidneys have a reduced capacity to convert vitamin D into its active form and as their cutaneous synthesis of the vitamin decreases with age.
- People with dark skin due to higher melanin content.
- People with malabsorption syndromes and who have undergone gastric bypass surgery.
- Obese people as vitamin D gets bind to their body fat and does not circulate into the blood.
- People with chronic conditions such as kidney or liver disease.
- People on certain medicines such as anti-seizure medications, steroids, and weight-loss medicines.

1.1.3.3 Scope of the Problem: Prevalence of Vitamin D Inadequacy

In the United States, 47% of African American infants and 56% of Caucasian infants are vitamin D deficient. While in Iran, Turkey, and India, over 90% of infants have vitamin D deficiency. Among the adult population, 35% of adults in the United States and over 80% of adults in Pakistan, India, and Bangladesh have vitamin D deficiency. Additionally, 61% of senior citizens in the United States lack sufficient vitamin D, compared to 90% in Turkey, 96% in India, 72% in Pakistan, and 67% in Iran [12].

The global prevalence of vitamin D deficiency is about 15.7% and vitamin D insufficiency is about 47.9%. The decrease in prevalence has been observed from 2000–2010 to 2011–2022. The investigations also shown that there are significant differences in the prevalence of VDD in each of the six WHO regions. In the Eastern Mediterranean Region, the prevalence of VDD (serum 25 (OH) D) <30 nmol/L was 35.2%, while in the Americas, it was 5.5%. The Eastern Mediterranean Region had a prevalence of 71.8%, while the African Region had a prevalence of 18.9% for serum 25(OH)D < 50 nmol/L [7]. The Eastern Mediterranean region may have a higher prevalence because of their cultural practice of wearing veils, which reduces exposure to sunshine [11]. On the other hand, high socioeconomic position, fortified milk and other food products, more knowledge, and increased government initiatives may be responsible for the low frequency among Americans [13].

Due to inadequate sunshine, higher latitude could be a risk factor for increased vitamin D deficiency. The reviewed studies showed that 14.9% and 57.4% of people living at high latitudes had serum 25(OH)D < 30 nmol/L and < 50 nmol/L, respectively [7]. There are great differences in terms of vitamin D deficiency among world bank income groups. The prevalence is higher in Lower-middle-income countries; 26.7% for serum 25(OH)D < 30 nmol/L and 56.0% for serum 25(OH) D < 50 nmol/L. Upper-middle-income countries had significantly lower rates; 10.2% of the population has serum 25(OH)D < 30 nmol/L and 38.2% of the population has serum 25(OH)D < 50 nmol/L. The prevalence was comparatively low in high-income nations as well, with 15.1% for serum 25(OH)D < 30 nmol/L and 49.2% for serum 25(OH)D < 50 nmol/L.

Prevalence of Vitamin D Deficiency in India

Vitamin D deficiency is quite rampant in India attributing to the following reasons [10].

- Increased indoor lifestyle mainly in the industrialized areas.
- Air pollution blocking ultraviolet light exposure.
- Altered food habits leading to less intake of calcium and vitamin D from diet [11].
- High consumption of phytates and phosphates which interfere with the absorption of calcium and vitamin D [12].
- Widespread application of sunscreens.

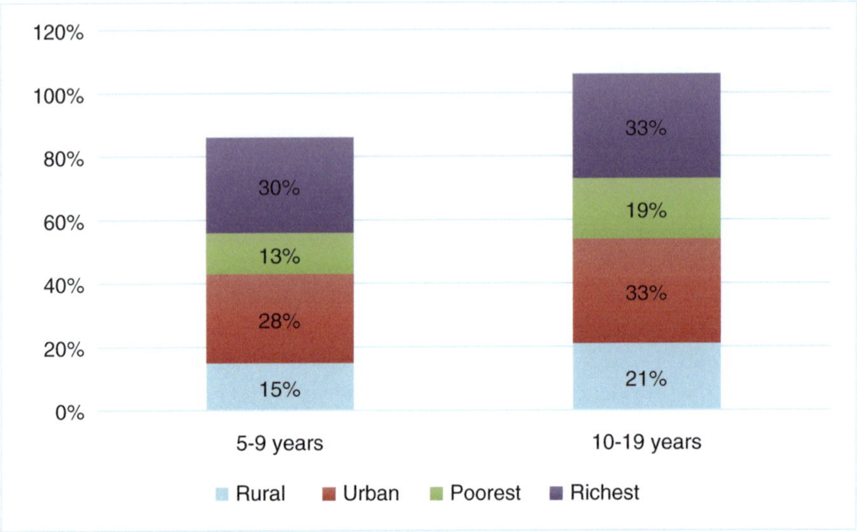

Fig. 1.2 Vitamin D deficiency prevalence among children in India, based on residence and household income index, 5–9 years and 10–19 years old, CNNS 2016–18 [14]

- Increasing rate of skin pigmentation.
- Culture of wearing *burqa* or *purdah* system [10].
- High incidence of unspaced and unwanted pregnancies leading to dietary deficit.

As per CNNS 2016–18 report, 14% of pre-school children 1–4 years of age, 18% of school going children from 5 to 9 years of age, and 24% of adolescents 10–19 years of age had vitamin D deficiency. Vitamin D deficiency was found to be more common in children aged 1–4 in urban regions (19%) compared to rural areas (12%), in wealthier households (18%) compared to poorest households (14%), and in vegetarian (18%) versus non-vegetarian (9%) children. Similar trend has been observed for children of other age groups [13] (Fig. 1.2).

Vitamin D deficiency is prevalent in other age groups also. A review on both community-based and hospital-based studies reporting VDD prevalence in the country stated that vitamin D deficiency could be seen in 37% to 99% among different categories of respondents reflecting the magnitude of the problem (Table 1.2).

Table 1.2 Studies showing vitamin D deficiency prevalence among different categories of respondents in India

Author (year of the study)	Category of respondents	Study area	Sample size	Pravelence/associated factors
Khadilkar et al. [15]	School children from 40 schools older than 7 years of age	Six Indian states covering wide geographical areas	2500	26.2% Significantly higher in girls than boys (28.4% vs 23.9%) Higher in the urban population than rural (30.2% vs 21.6%)
Kaliya et al. [16]	Medical professionals	Medical college of western Gujarat, India	509	78.39% Sunlight exposure, geographical distribution, type of food, and seasonal variations are the associated factors
Mustafa et al. [17]	Indian adolescent population	CNNS survey conducted between 2016 and 2018 in India	13,065	24.97%
Chowdhury et al. [18]	6–30 months old children	Delhi, India	960	34.5% The risk of ALRI was significantly higher among vitamin-D deficient
Suryanarayana et al. [19]	Urban elderly (≥60 years)	Hyderabad, South India	298	56.3% The mean ± SE plasma vitamin D was 19.3 ± 0.54 (ng/ml)
Alok et al. [20]	Urban and rural pregnant subjects and their newborns	Queen Mary's hospital, King George medical university, Lucknow	207 pregnant subjects and 117 newborns	Mean maternal serum 25(OH)D was 14 ± 9.3 ng/mL, and cord blood 25(OH)D was 8.4 ± 5.7 ng/mL. 84.3% of urban and 83.6% of rural women had Vit D deficiency
Kapil et al. [21]	Children (6–18 years)	Shimla, Himachal Pradesh	626	93% were found vitamin D deficient and more were females
Misra et al. [22]	Rural women (20–60 years)	Ballabhgarh	381	90.8% were Vit D deficient and 8.9% were Vit D insufficient
Rattan et al. [23]	All the healthy individuals	Cuttack, Odisha	3056	84.9% were Vit D deficient (more were females).
Srimani et al. [24]	Rural post-menopausal women (45–70 years)	West Bengal	222	22% were Vit D insufficient and 53% were deficient

1.2 Significance of Addressing the Issue of Vitamin D Inadequacy

Vitamin D deficiency has a significant impact on health and quality of life. It has a variety of skeletal and extra skeletal manifestations.

1.2.1 Health Impact and Future Consequences of Vitamin D Deficiency

Chronic vitamin D deficiency leads to decrease in the calcium and phosphorus absorption from small intestine. This could lead to hypocalcemia and secondary hyperparathyroidism. This secondary hyperparathyroidism further causes increased phosphorous excretion in urine and accelerates demineralization of bone. Rickets in children and osteomalacia and osteoporosis in adults are the common consequences of vitamin D deficiency.

A calcium and phosphorus content imbalance in the bone causes a poor mineralization of the bone that leads to skeletal deformities in rickets patients. Bone pain and muscle weakness are other symptoms. Low vitamin D consumption also causes poor calcium absorption from the diet and increased calcium resorption from the bone in adults. This lowers bone mineral density, which raises the risk of falls, causes osteoporosis and osteomalacia, and weakens the muscles [19].

It has been proposed that when vitamin D levels are at their optimal, muscle strength increases and helps to reduce falls and accidents. Numerous studies have demonstrated a link between low vitamin D levels and an increased risk of falls and fractures in the elderly. There are various extra skeletal manifestations as well. Vitamin D plays a pivotal role in maternal and child health, reducing the risk of cancer and many respiratory infections. The subjacent vitamin D levels is a potential risk factor for some chronic disabling conditions such as cancer, heart disease, autoimmune disorders, depression, etc. [25].

1.2.2 Importance of Adequate Levels of Vitamin D for Maternal and Child Health

Children and pregnant women are especially susceptible to vitamin D insufficiency. A mother's lack of vitamin D during the preconception or pregnancy stage can have serious consequences for the foetus and baby, such as congenital or nutritional rickets. Due to low vitamin D levels during pregnancy and childbirth, children are at risk for this debilitating neuromuscular condition. It may cause skeletal malformations and low stature due to abnormal chondrocyte differentiation and bone mineralization at the growth plates. Nutritional rickets has effects that go beyond the bone such as seizures, tetany, widespread weakness, cardiomyopathy, and elevated intracranial pressure [21].

A lack of vitamin D is also linked to a higher chance of unfavourable outcomes for mothers, including pre-eclampsia, gestational diabetes mellitus (GDM), and a higher chance of caesarean birth. An association has also been seen between vitamin D deficiency and recurrent loss of pregnancy and post-partum depression without establishing a causal relationship [26]. Therefore, to prevent both the biochemical and radiological signs of nutritional or congenital rickets, the global consensus on nutritional rickets suggested that all pregnant women get 600 IU/day of extra vitamin D [27].

1.2.3 Vitamin D in Cancer Prevention

As it inhibits angiogenesis, promotes apoptosis, and reduces inflammation, vitamin D has a protective impact on several tissues. Cancers of the lung, breast, colon, prostate, ovary, pancreas, and oesophagus are linked to low vitamin D reserves. A daily consumption of 2000 IU of vitamin D3 was predicted to result in a 25% decrease in the incidence of breast cancer and a 27% decrease in the incidence of colorectal cancer in North America. A regular dosage of 2000 IU/day of vitamin D was suggested as the upper limit (UL) by the National Academy of Sciences [24]. A yearly global reduction of 220,149 new instances of breast cancer and 254,105 new cases of colorectal cancer would result from serum 25(OH)D concentrations being raised to roughly 40 to 60 ng/ml, which is, in general, associated with oral intake of 2000 IU of vitamin D3 per day (Fig. 1.3).

Ten mechanisms have been reported that account for the role of vitamin D and calcium in reducing cancer incidence and mortality [29]. These mechanisms include: (a) increased epithelial cell adhesion and signalling; (b) contact inhibition of proliferation; (c) differentiation; (d) stabilization of the cell cycle; (e) promotion of apoptosis; (f) anti-neo angiogenesis; (g) down-regulation of glycogen synthase kinase 3 (GSK-3), which reduces colorectal, prostate, and pancreatic cancer proliferation in vitro; (h) downregulation of the canonical Wnt signalling pathway, which is active in colorectal and other cancers; (i) increased expression of DKK-1 protein, a tumour suppressor in colon cancer cells with mutations in the Wnt/beta-catenin pathway; (j) down-regulation of DKK-4 transcription; colorectal cancer is associated with up-regulated DKK-4, a target of the Wnt/beta-catenin pathway that increases cellular motility, autonomy, and invasiveness. By binding to the DKK-4 promoter, the VDR1,25(OH)2D complex significantly inhibits transcription of the gene.

Hence, to decrease the incidence of cancer in human body, the serum 25(OH) D concentration needs to be in the recommended limits.

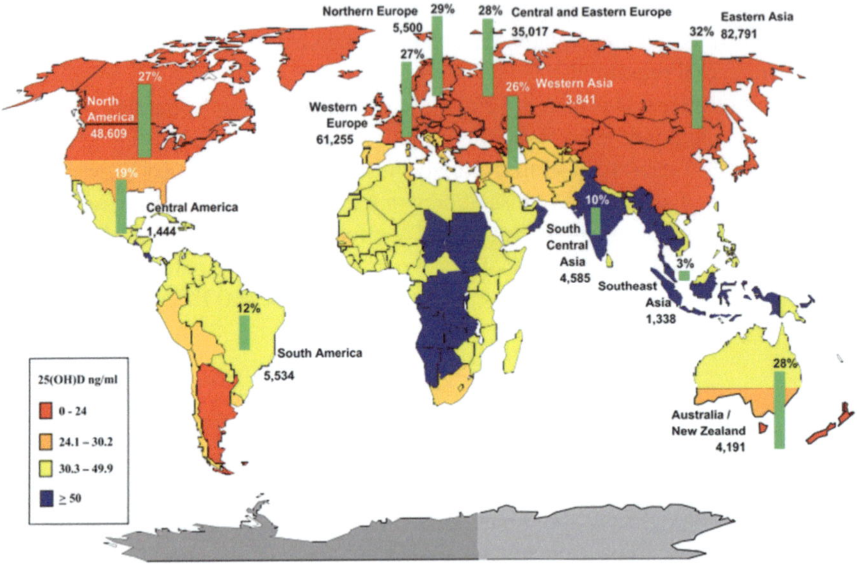

Fig. 1.3 Estimated serum levels of 25(OH)D and the anticipated yearly decrease in new instances of colorectal cancer with 2000 IU of vitamin D3 and, where practical, moderate sun exposure (10 min per day) that does not exceed the erythema threshold [28]. (Source: Garland et al. Vitamin D for cancer prevention: a global perspective. Ann Epidemiol. 2009;19(7):468–83)

1.2.4 Vitamin D for the Prevention and Management of Respiratory Infections Including COVID-19

Vitamin D deficient individuals are at an increased risk of certain infectious diseases such as tuberculosis, influenza, or other upper respiratory tract infections [30]. Studies show that severe complications of lower respiratory tract infections are more likely to develop in patients with vitamin D deficiency. When compared to individuals with appropriate levels of vitamin D, they exhibit a much higher probability of needing oxygen therapy (by 117%) and mechanical ventilation (by 217%) [27]. Additionally, the findings demonstrate a linear correlation between the cumulative incidence of respiratory infections and vitamin D status [28].

Role of vitamin D both in the prevention and management of SARS-CoV-2 infection was evident. A potential mechanism of lung tissue damage caused by SARS-CoV-2 is the induction of Th1 pro-inflammatory cytokines, which can result in cytokine storm. The development of acute respiratory distress syndrome (ARDS) is further exacerbated by this [29]. By changing the pro-inflammatory Th1 and Th17 response to the anti-inflammatory Th2 and Treg response, vitamin D prevents this cytokine storm [30].

Vitamin D may even lower the chance of contracting COVID-19 through several potential pathways. The induction of defensin and cathelicidin transcription is one of the processes. They encourage macrophage chemotaxis to the site of inflammation and lower the rate of virus replication. To rapidly replenish 25 (OH) D insufficiency, Grant et al. advised individuals who are at risk of COVID-19 to think about taking vitamin D supplements for a few weeks at 10,000 IU/day and then 5000 IU/day. The ideal range for the blood levels of vitamin D is 40–60 ng/ml [31].

In addition, vitamin D also helps to alleviate the various neurological symptoms caused due to SARS-CoV-2 infection. It controls the synthesis of neurotrophins, which are essential for nerve cell survival and differentiation [32]. It also has a protective effect on haemostatic disorders and hypercoagulability related to COVID-19 hospitalized patients leading to hypoperfusion and multiorgan failure. Its anticoagulant action may be due to a reduction in tissue thromboplastin and an increase in glutathione and thrombomodulin [33].

1.2.5 *Other Diseases or Conditions Associated with Vitamin D Deficiency*

1.2.5.1 Depression

The secosteroid hormone vitamin D can cross the blood-brain barrier (BBB). Several brain areas, including the limbic system, cortex, and cerebellum, contain its receptors [34]. It has been shown that VDD may cause an increase in the hormone parathormone, which is linked to depression. Additionally, compared to those without a vitamin D deficiency, patients with a deficiency recovered from depression far more slowly. It emphasizes how crucial it is to address hypovitaminosis D to effectively control depression [35].

1.2.5.2 Autoimmune Diseases

Vitamin D also functions as immune system modulator. It plays a role in controlling the division and proliferation of cells. According to the studies, children with type 1 diabetes had much higher VDD (91%) than children without the disease (85%). These children who took vitamin D supplements had a 30% lower chance of getting type 1 diabetes [36]. Rheumatoid arthritis has also been linked to lower levels of vitamin D [37].

1.2.5.3 Heart Diseases

According to the Framingham Heart Study, individuals with lower than 15 ng/ml vitamin D levels were 60% more likely to develop heart disease, which is mediated by the renin-angiotensin hormone system, than those with greater amounts. Patients with an acute myocardial infarction often have severe VDD, which is linked to numerous risk factors for the condition [38].

1.2.6 Economic Impact of Vitamin D Deficiency

Numerous ailments and disorders can be prevented and managed with the help of vitamin D. Various forms of cancer, cardiovascular disorders, diabetes mellitus, a number of bacterial and viral infections, and autoimmune diseases are among the illnesses for which there is strong evidence of benefit, and which have significant economic implications. Due to their high latitudes, sedentary lifestyles, lack of access to natural vitamin D-rich food sources like cold-water ocean fish, and general absence of efficient vitamin D fortification programs, Europeans typically have low vitamin D levels. It has been noted that raising the mean serum 25(OH) D level to 40 ng/ml may lessen the direct and indirect economic burden of disease. A daily dosage of 2000–3000 IU of vitamin D could accomplish this. The reduction in economic burden for 2007 was projected to be about €187,000 million annually, while the estimated cost of consuming 2000–3000 IU of vitamin D3 per day, along with related expenses like testing and education, would be roughly €10,000 million [39].

In Canada, a comparable decrease in financial strain has been noted. If we raise the serum 25 (OH) D levels to 105 ng/ml, the death rate could decrease by 37,000 deaths (22,300–52,300 deaths), or 16.1% (9.7–22.7%) of annual deaths, and the economic burden could decrease by 6.9% (3.8–10.0%), or $14.4 billion ($8.0 billion–$20.1 billion), less the program's cost [40].

1.3 Clinical Management of Vitamin D Deficiency [31]

- Vitamin D deficiency patients may benefit from an initial 8-week course of vitamin D3 therapy, either at 6000 IU daily or 50,000 IU weekly with a maintenance dose of 1000 to 2000 IU per day.
- Initial vitamin D3 supplementation at a greater dose of 10,000 IU daily may be necessary in high-risk adults following a maintenance dosage of 3000–6000 IU daily.
- Children who are vitamin D deficient may take a daily 2000 IU or weekly 50,000 IU vitamin D3 supplement for a period of 6 weeks following a maintenance dose of 1000 IU per day.
- The American Academy of Paediatrics states that 400 IU of vitamin D supplementation is necessary for breastfed newborns and kids who drink less than 1 l of fortified milk.
- Consider using calcitriol if the deficiency does not improve after taking vitamin D2 and/or D3 supplements.
- Calcidiol may be taken into consideration for people who have significant liver disease or fat malabsorption.

1.4 Public Health Strategies for the Prevention and Management of Vitamin D Deficiency

Even while we know what causes vitamin D insufficiency, we have not been able to stop it from occurring very often. India is a tropical nation and receives enough sunlight. Thus, it is anticipated that the levels of vitamin D will be sufficient among Indian population. Nevertheless, India has a significant rate of vitamin D insufficiency. This necessitates a suitable and effective public health response. The following actions can be performed to lessen the disease burden.

1.4.1 Awareness Generation

To raise public awareness of vitamin D insufficiency, educational initiatives are essential. People should be encouraged for adequate exposure to sunlight between 10 AM and 4 PM and adequate dietary intake of foods rich in vitamin D [32]. They should be aware of the other factors that limit the development of vitamin D deficiency. The educational programs could be conducted at various health and wellness centres, schools, colleges, and other community settings. Adequate time, financial, and labour investments are needed to create, implement, and maintain such initiatives.

1.4.2 Screening

It is imperative to screen for vitamin D deficiency in high-risk individuals who do not exhibit any symptoms. By doing this, future consequences will be avoided. Nursing home residents, elderly people, osteoporosis-affected women, pre- or post-menopausal women, hospitalized patients, patients with chronic liver or renal disease, and patients with malabsorption syndromes are among the high-risk groups of people [43].

1.4.3 Supplementation for Preventing Vitamin D Deficiency

There are various forms of vitamin D available commercially. It has been demonstrated that vitamin D3 (cholecalciferol) is more effective than vitamin D2 (ergocalciferol) in obtaining the ideal levels of 25-hydroxyvitamin D, suggesting that vitamin D3 should be used as the preferred therapeutic option [44]. Vitamin D supplements of good quality should be made available at PHC level for the population at risk, i.e. pregnant women, lactating women, children, and elderly. The newborn infants are also now-a-days given the vitamin D supplementation for the first year of their life.

In order to prevent deficiency, adults under 65 who do not get enough sun exposure should take 600–800 IU of vitamin D3 daily as a precaution. Older persons who are 65 years of age or older should take 800–1000 IU of vitamin D3 every day. This will lower the chance of falls and fractures [7].

1.4.4 Food-Based Strategies: Ensuring Food Quality and Diversity

The natural sources of intake of vitamin D should be considered. Balanced diet by including variety of food products should be taken. The rich dietary sources of vitamin D are pippali, curry leaves, pomegranate, oyster mushroom, gingelly seeds (brown, black), soya bean (white, brown), amaranth seed (black and pale brown), nutmeg, walnut, cloves, mace, cardamom, pork (shoulder, ham, chops, liver, kidney), beef heart, egg (poultry, country, duck), meat (goat, sheep) [33]. In addition to the food diversity, one should take proper care ensuring food quality and safety. While preparing the above-mentioned food items, the nutrition value should not be compromised and appropriate hygienic practices should be followed.

1.4.5 Fortification of Staple Food Items

Food fortification with vitamin D is the best option to address this issue. Fortification of widely consumed staple food items offers one of the simplest and most practical methods to combat micronutrient deficiencies for both poor and wealthy societies. Vitamin D was added to milk in the United States in 1930s to help prevent rickets in children [34]. Vitamin D can be added to oil and milk products, including butter, yogurt, curd, and baby formulae. It is also possible to fortify commonly consumed foods like rice flour, *atta*, and *maida*. Food products fortified with vitamin D should be inexpensively accessible to the general people. This ought to be a part of the PDS, or public distribution system. Development of a fortification program requires persistent political, administrative, and legislative commitment and support in addition to effective legislation.

1.4.5.1 Policy Related to Vitamin D Fortification in the United States and Canada

The 1930s saw the start of vitamin D fortification of milk in the United States. Originally, irradiating milk or giving the cows' irradiated yeast was the method used to fortify milk with vitamin D. The easier and more efficient procedure took the place of this one in the 1940s. In this nation, adding vitamin D to food is optional, but it is tightly regulated based on the types of food, how they work, and how much they are used to counter over fortification. Most of the milk sold in the United States has vitamin D added to it. Some businesses also include vitamin D in rice, soy milk, and breakfast cereals [46].

In Canada, fortification of milk (180 IU/250 mL), milk substitutes, and margarine (530 IU/100 g) is required by law. Like in the United States, vitamin D fortification is optional for other foods that are allowed, but the amount of fortification is controlled. In Canada, it is illegal to fortify ready-to-eat breakfast cereals with vitamin D [46].

1.4.5.2 Vitamin D Fortification in some Other European Countries

An outline of some European nations' policy regarding vitamin D fortification is provided below [35] (Table 1.3).

1.4.5.3 Vitamin D Fortification Policy in India

Fortification is not new to India either. Fortification of *Vanaspati* with vitamin D started 50 years ago and has been mandatory in the country since 1953. Every 100 g, it is often supplemented with 200 IU of vitamin D. Certain brands of milk products

Table 1.3 Fortification with vitamin D in some European countries

Country	Fortification strategy
Austria	No mandatory fortification
Belgium	Margarine and spreadable fats (6–7 μg/100 g) are mandatory; optional foods and beverages include milk, dairy products, milk substitutes, cereals, biscuits, fruit juices, lemonades, cocoa powder, and weight-loss items
Czech Republic	Minimal voluntary fortification; not mandatory
Denmark	Minimal voluntary fortification; not mandatory; rarely done
Estonia	Allowed but not mandatory
Finland	Milk, margarine, and fat spread are fortified in large quantities. Yogurt, orange juice, plant-based milk, bread, and cereals are fortified in specific brands
Ireland	Milk (~ 160 IU/200 ml) and grains (~ 118 IU/35 g) can be fortified voluntarily, but margarine must be fortified
Spain	All meals, except for newborn formulae and infant cereals, are optionally fortified. To get the initial levels of whole milk, it is common practice to fortify skimmed and semi-skimmed milks with vitamin D. Additionally, various commercial items like biscuits, yoghurt, margarine, cheese, cereals, and juices are also fortified with vitamin D
UK	Infant formula and foods meant for energy-restricted diets, bread (200 IU/100 g), orange beverage (1000 IU/240 ml), and margarine (not less than 7.05 μg and not more than 8.82 μg)

also contain vitamin D fortification. In the context of India, fortifying some foods with vitamin D to especially meet the nutritional needs of newborns and children is a feasible strategy. Ready-to-eat foods fortified with vitamin D should be included in government programs aimed at meeting the nutritional needs of children in India, particularly through midday meal programs. For instance, the protein-dense Indian food known as *sattu*, which is formed of roasted flour derived from cereals and legumes, has a great deal of promise to be used as a technologically and financially viable fortification vehicle for vitamin D [47].

The Food Safety and Standards Authority of India has formulated comprehensive regulations for fortified foods namely Food Safety and Standards (fortification of food) regulations, 2016. The regulations constitute standards for fortification of wheat flour, maida, rice, milk, oil, salt, and vanaspati. The edible oil fortification standard as established by FSSAI is to add 4.5 IU vitamin D per gram of oil [36]. The government's plan was to maximize the availability of fortified *vanaspati* either through public-funded nutrition programs such as ICDS, midday meal, and Public Distribution System or through open market channels. To increase the supply through open market channels, a great deal of public awareness regarding the importance of fortification is needed.

1.4.6 Other Strategies

- Educating kids who are in school can be advantageous. The availability of foods fortified with vitamin D during midday meals in schools, the necessity for a healthy lifestyle, and other relevant information should be known to them. It would be best to urge them to engage in regular physical activity so they can get enough exposure to sunlight.
- Vitamin D testing facilities should be made reasonably priced and available to individuals who are at a high risk of developing clinical vitamin D deficiency, such as pregnant women, children, the elderly, and others, as mass screening is not practical.
- Further investigation is required in this field, and it is imperative that the government provides assistance to research teams studying and tracking the effects of fortification techniques and supplementation programs.

1.4.7 Research Strengthening the Implementation of Public Health Strategies

The public health strategies mentioned above have been evaluated in different settings and in different combinations in many studies as shown below:

1.4.7.1 The Economic Case for Prevention of Population Vitamin D Deficiency: A Modelling Study Using Data from England and Wales [37]

An individual-level simulation model was utilized in this study to assess the following interventions:

(I) supplementing at-risk groups;
(II) fortifying wheat flour;
(III) combining flour fortification and supplementation; and,
(IV) a "no additional intervention" scenario.

The estimation of the probability of deficiency under the alternative scenarios was done using data from national nutrition surveys. The expenses borne by the public, local government bodies, the health care industry, and the government were considered. The overall cost and impact of each treatment, as well as the cost per "prevented case of VDD" and cost per "Quality Adjusted Life Year (QALY)," were used to express the results. The findings demonstrated that wheat flour fortification was economical because the savings from avoiding VDD more than compensated for its expenses. Fortification plus supplementation shown to be a cost-effective strategy (£9.5 per QALY gained). According to the model, just fortifying wheat flour would lead to a 25% decrease in VDD cases, while the combined approach would lower the number of cases by further 8%.

1.4.7.2 Cost-Effectiveness and Budget Impact of Empirical Vitamin D Therapy on Unintentional Falls in Older Adults in the UK [38]

The costs and advantages of empirical treatment with 800 IU of cholecalciferol daily were evaluated by simulating patient transitions between major fall-related outcomes using a 5-year horizon and annual cycles in a Markov health state transition model. According to our model, giving 800 IU of cholecalciferol to the UK population 60 years of age and above would, over the course of 5 years:

1. prevent more than 430,000 minor falls;
2. 190,000 catastrophic falls;
3. avert 1579 acute fatalities;
4. avert 84,000 person-years of long-term care; and,
5. avert 8300 deaths linked to higher long-term care mortality.

Those 75 years of age and beyond show the most gains. The intervention is cost-effective and results in higher quality adjusted life years for all persons 65 years of age and older, even on the basis of a reduction in falls alone. Treating every adult over 60 results in a net savings of £420 million over the course of 5 years, with an intervention cost of £2.70 billion less than fall-related expenditures of £3.12 billion.

According to this study, giving cholecalciferol 800 IU daily to the senior population in the UK could lower mortality and result in significant cost savings by preventing falls.

1.4.7.3 Preventing Vitamin D Deficiency (VDD): A Systematic Review of Economic Evaluations [39]

The full economic evaluation of VDD preventive strategies was examined in this study. Public health campaigns, supplements, and dietary fortification were all taken into consideration as interventions. For certain demographic groups, these interventions were found to be both cost-effective and cost-saving (Table 1.4).

1.5 Conclusion

Deficiency of vitamin D is prevalent worldwide. The clinical cases of VDD represent only the tip of the iceberg. We can envision the burden this silent epidemic will cause, as it has multiple consequences. A population-based strategy is required to address the widespread vitamin D insufficiency in India and its health effects. This includes dietary fortification, supplementation, and awareness-raising initiatives. Given the variety of foods and cultures, all of these strategies have been proven to be efficient and affordable, so they could be used. To raise mean daily dietary intakes of vitamin D in India, we suggest adding vitamin D to wheat flour, *maida*, rice, and rice flour. A notable improvement in general health and bone health overall can be anticipated, with minimal risk at a modest cost, because of the ability to capitalize on existing technology and distribution systems.

Table 1.4 Economic evaluations of various VDD prevention strategies

Study	Study area	Study participants	Intervention	Results
Sandman et al. [40]	Germany	Elderly above 65 years	No food fortification vs voluntary food fortification with 20 μg or 800 IU vitamin D3 + 200 mg of calcium	Annual net cost savings of €315 million. Almost 36,705 fractures prevented
Zarca et al. [41]	French population	Over 65 years of age	Four interventions: (1) no treatment, (2) supplementation without testing Vit D serum levels, (3) supplementation with Vit D serum level testing 3 months after initiation followed by treatment, and (4) population screening for vitamin D insufficiency followed by treatment	"Treat, then check" and "screen and treat" were two cost-effective strategies and dominated "treat without check" with incremental cost-effectiveness ratios of €5219/quality-adjusted life-years (QALY) and €9104/QALY, respectively
Lee et al. [42]	USA	Community-dwelling women and men aged 65 to 80.	Compared the cost-effectiveness of population screening vs universal vitamin D supplementation	Both population screening and universal supplementation are found to be cost-effective, but population screening is more cost-effective
Patil et al. [43]	Finland	Older home-dwelling Finnish women from 70 to 80 years	RCT with 4 groups: (1) no exercise + placebo, (2) no exercise + vitamin D 800 IU/day, (3) exercise + placebo, and (4) exercise + vitamin D 800 IU/day	Exercise was effective in reducing fall-related injuries at a moderate cost. Vitamin D supplementation had marginal additional benefit
Poole et al. [44]	UK	Over 65 years of age	Assessed the budgetary impact of administering 800 IU daily cholecalciferol on hip fracture	Could reduce the number of incident hip fractures from 65,400 to 45,700, saving almost 1700 associated deaths
Church et al. [45]	Australia	Older adults living in residential aged care facilities	Vitamin D supplementation, annual medication review, multifactorial intervention (risk assessment, medication review, vision assessment, and exercise), and hip protectors	Vitamin D supplementation and medication review are cost-effective interventions

References

1. Sarma D. Vitamin D – public health significance. Endocrinol Metab Int J. [Internet]. 2019 [cited 2024 Feb 29];7(5). Available from: https://medcraveonline.com/EMIJ/EMIJ-07-00258.php
2. Ross AC, Taylor CL, Yaktine AL, Del Valle HB. DRI Dietary reference intakes calcium Vitamin D Committee to review dietary reference intakes for Vitamin D and Calcium food and nutrition board. 2011 [cited 2024 Mar 28]; Available from: http://www.nap.edu
3. Nair R, Maseeh A. Vitamin D: The "sunshine" vitamin. J Pharmacol Pharmacother [Internet]. 2012 [cited 2024 Mar 13];3(2):118. Available from: /pmc/articles/PMC3356951/.
4. Jones G, Rauner M. Historical aspects of vitamin D. 2022 [cited 2024 Mar 30]; Available from: https://doi.org/10.1530/EC-21-0594.
5. Ross AC, Manson JE, Abrams SA, Aloia JF, Brannon PM, Clinton SK, et al. The 2011 Dietary reference intakes for calcium and Vitamin D: what dietetics practitioners need to know. J Am Diet Assoc [Internet]. [cited 2024 Mar 7];111(4):524–7. Available from: http://www.jandonline.org/article/S0002822311000058/fulltext
6. Sizar O. Vitamin D Deficiency – StatPearls – NCBI Bookshelf. StatPearls [Internet]. 2021.
7. Thomas MK, Lloyd-Jones DM, Thadhani RI, Shaw AC, Deraska DJ, Kitch BT, et al. Hypovitaminosis D in medical inpatients. N Engl J Med [Internet]. 1998 [cited 2024 Mar 13];338(12):777–83. Available from: https://pubmed.ncbi.nlm.nih.gov/9504937/
8. Gröber U, Kisters K. Influence of drugs on vitamin D and calcium metabolism. Dermatoendocrinol [Internet]. 2012 [cited 2024 Mar 13];4(2):158–66. Available from: https://pubmed.ncbi.nlm.nih.gov/22928072/
9. Cui A, Zhang T, Xiao P, Fan Z, Wang H, Zhuang Y. Global and regional prevalence of vitamin D deficiency in population-based studies from 2000 to 2022: a pooled analysis of 7.9 million participants. Front Nutr. 2023:10.
10. Ajmani SN, Paul M, Chauhan P, Ajmani AK, Yadav N. Prevalence of Vitamin D deficiency in burka-clad pregnant women in a 450-bedded maternity hospital of Delhi. J Obstet Gynaecol India [Internet]. 2016 [cited 2024 Mar 9];66(Suppl 1):67. Available from: /pmc/articles/PMC5016404/.
11. Babu US, Calvo MS. Modern India and the vitamin D dilemma: evidence for the need of a national food fortification program. Mol Nutr Food Res [Internet]. 2010 [cited 2024 Mar 7];54(8):1134–47. Available from: https://pubmed.ncbi.nlm.nih.gov/20440690/
12. Harinarayan C V., Ramalakshmi T, Prasad U V., Sudhakar D, Srinivasarao PVLN, Sarma KVS, et al. High prevalence of low dietary calcium, high phytate consumption, and vitamin D deficiency in healthy south Indians. Am J Clin Nutr [Internet]. 2007 [cited 2024 Mar 9];85(4):1062–7. Available from: https://pubmed.ncbi.nlm.nih.gov/17413106/
13. Comprehensive National Nutrition Survey 2016-18, Ministry of Health and Family Welfare, Govt. of India. Available at https://nhm.gov.in/WriteReadData/l892s/1405796031571201348.pdf.
14. Comprehensive National Nutrition Survey 2016-18. Available at https://nhm.gov.in/WriteReadData/l892s/1405796031571201348.pdf
15. Khadilkar A, Kajale N, Oza C, Oke R, Gondhalekar K, Patwardhan V, et al. Vitamin D status and determinants in Indian children and adolescents: a multicentre study. Sci Rep [Internet]. 2022;12(1):1–13. https://doi.org/10.1038/s41598-022-21279-0.
16. View of Prevalence of Vitamin—D Deficiency & Associated Factors Among the Medical Personnel of Medical College of Western Gujarat India [Internet]. [cited 2024 Sep 19]. Available from: https://ijirms.in/index.php/ijirms/article/view/1500/1140
17. Mustafa A, Shekhar C. Factors associated with Vitamin D deficiency and their relative importance among Indian adolescents: an application of dominance analysis. Int J Endocrinol [Internet]. 2023 [cited 2024 Sep 19];2023. Available from: /pmc/articles/PMC10597726/.
18. Chowdhury R, Taneja S, Bhandari N, Sinha B, Prakash Upadhyay R, Kishan Bhan M, et al. Vitamin-D deficiency predicts infections in young north Indian children: a secondary data analysis. 2017 [cited 2024 Sep 20]; Available from: https://figshare.com/s/

19. Suryanarayana P, Arlappa N, Sai Santhosh V, Balakrishna N, Lakshmi Rajkumar P, Prasad U, et al. Prevalence of vitamin D deficiency and its associated factors among the urban elderly population in Hyderabad metropolitan city, South India. Ann Hum Biol [Internet]. 2018 [cited 2024 Sep 12];45(2):133–9. Available from: https://pubmed.ncbi.nlm.nih.gov/29307228/
20. Sachan A, Gupta R, Das V, Agarwal A, Awasthi PK, Bhatia V. High prevalence of vitamin D deficiency among pregnant women and their newborns in northern India. Am J Clin Nutr [Internet]. 2005;81(5):1060–4. https://doi.org/10.1093/ajcn/81.5.1060.
21. Kapil U, Pandey R, Goswami R, Sharma B, Sharma N, Ramakrishnan L, et al. Prevalence of Vitamin D deficiency and associated risk factors among children residing at high altitude in Shimla district, Himachal Pradesh, India. Indian J Endocrinol Metab [Internet]. 2017 [cited 2024 Sep 22];21(1):178. Available from: /pmc/articles/PMC5240061/.
22. Misra P, Srivastava R, Misra A, Kant S, Kardam P, Vikram NK. Vitamin D status of adult females residing in Ballabgarh health and demographic surveillance system: a community-based study. Indian J Public Health [Internet]. 2017 [cited 2024 Sep 22];61(3):194–8. Available from: https://journals.lww.com/ijph/fulltext/2017/61030/vitamin_d_status_of_adult_females_residing_in.8.aspx
23. Rattan R, Sahoo D, Mahapatra S. Prevalence of vitamin D deficiency in adults in the coastal regions of Odisha, India. IOSR J Pharm Biol Sci [Internet]. 2016;11(6):49–52. Available from: https://www.iosrjournals.org/iosr-jpbs/papers/Vol11-issue6/Version-5/E1106054952.pdf
24. Srimani S, Saha I, Chaudhuri D. Prevalence and association of metabolic syndrome and vitamin D deficiency among postmenopausal women in a rural block of West Bengal, India. 2017 [cited 2024 Sep 22]; Available from: https://doi.org/10.1371/journal.pone.0188331.
25. Tuohimaa P. Vitamin D, aging, and chronic diseases. In Vitamin D oxidative stress immunity, aging. 2012; Vol. 8, issue 3, p. 385–406.
26. Tamblyn JA, Pilarski NSP, Markland AD, Marson EJ, Devall A, Hewison M, et al. Vitamin D and miscarriage: a systematic review and meta-analysis. Fertil Steril [Internet]. 2022 [cited 2024 Mar 30];118:111–22. Available from: https://doi.org/10.1016/j.fertnstert.2022.04.017.
27. Pérez-López FR, Pilz S, Chedraui P. Vitamin D supplementation during pregnancy: an overview. Curr Opin Obstet Gynecol [Internet]. 2020 [cited 2024 Mar 30];32(5):316–21. Available from: https://journals.lww.com/co-obgyn/fulltext/2020/10000/vitamin_d_supplementation_during_pregnancy__an.3.aspx
28. Garland CF, Gorham ED, Mohr SB, Garland FC. Vitamin D for cancer prevention: global perspective. Ann Epidemiol. 2009;19(7):468–83.
29. Garland CF, Garland FC, Gorham ED, Lipkin M, Newmark H, Mohr SB, et al. The role of vitamin D in cancer prevention. Am J Public Health. 2006;96(2):252–61.
30. Urashima M, Segawa T, Okazaki M, Kurihara M, Wada Y, Ida H. Randomized trial of vitamin D supplementation to prevent seasonal influenza A in schoolchildren. Am J Clin Nutr. 2010;91(5):1255–60.
31. Holick MF, Binkley NC, Bischoff-Ferrari HA, Gordon CM, Hanley DA, Heaney RP, et al. Evaluation, treatment, and prevention of vitamin D deficiency: an Endocrine Society clinical practice guideline. J Clin Endocrinol Metab [Internet]. 2011 [cited 2024 Mar 13];96(7):1911–30. Available from: https://pubmed.ncbi.nlm.nih.gov/21646368/
32. Rockwell M, Kraak V, Hulver M, Epling J. Clinical management of low vitamin D: a scoping review of physicians' practices. Nutrients. 2018;10(4)
33. Nutrition atlas [Internet]. [cited 2024 Mar 31]. Available from: http://www.nutrition-atlas.nin.res.in/nutritionatlas/vitd.php
34. Ritu G, Gupta A, Gupta A. Fortification of foods with vitamin D in India. Nutrients. 2014;6(9):3601–23.
35. Niedermaier T, Gredner T, Kuznia S, Schöttker B, Mons U, Lakerveld J, et al. Vitamin D food fortification in European countries: the underused potential to prevent cancer deaths. Eur J Epidemiol [Internet]. 2022 [cited 2024 Mar 27];37(4):309–20. Available from: https://link.springer.com/article/10.1007/s10654-022-00867-4

36. FSSAI FS and SA of I. Food Safety and Standards (Fortification of Foods) Regulations, 2018. 2018;(1):1–12. Available from: https://www.fssai.gov.in/upload/uploadfiles/files/Compendium_Food_Fortification_Regulations_30_09_2021.pdf
37. Aguiar M, Andronis L, Pallan M, Högler W, Frew E. The economic case for prevention of population vitamin D deficiency: a modelling study using data from England and Wales. Eur J Clin Nutr [Internet]. 2020 [cited 2024 Mar 31];74(5):825–33. Available from: https://pubmed.ncbi.nlm.nih.gov/31427760/
38. Poole CD, Smith J, Davies JS. Cost-effectiveness and budget impact of Empirical vitamin D therapy on unintentional falls in older adults in the UK. BMJ Open [Internet]. 2015 [cited 2024 Mar 31];5(9):e007910. Available from: https://bmjopen.bmj.com/content/5/9/e007910
39. Aguiar M, Andronis L, Pallan M, Högler W, Frew E. Preventing vitamin D deficiency (VDD): a systematic review of economic evaluations. Eur J Public Health [Internet]. 2017 [cited 2024 Mar 31];27(2):292–301. Available from: https://pubmed.ncbi.nlm.nih.gov/28204459/
40. Sandmann A, Amling M, Barvencik F, König HH, Bleibler F. Economic evaluation of vitamin D and calcium food fortification for fracture prevention in Germany. Public Health Nutr. 2017;20(10):1874–83.
41. Zarca K, Durand-Zaleski I, Roux C, Souberbielle JC, Schott AM, Thomas T, et al. Cost-effectiveness analysis of hip fracture prevention with vitamin D supplementation: a Markov micro-simulation model applied to the French population over 65 years old without previous hip fracture. Osteoporos Int. 2014;25(6):1797–806.
42. Lee RH, Weber T, Colõn-Emeric C. Comparison of Cost-Effectiveness of Vitamin D Screening with That of Universal Supplementation in Preventing Falls in Community-Dwelling Older Adults. J Am Geriatr Soc [Internet]. 2013 [cited 2024 Sep 22];61(5):707–14. Available from: https://onlinelibrary.wiley.com/doi/full/10.1111/jgs.12213
43. Patil R, Kolu P, Raitanen J, Valvanne J, Kannus P, Karinkanta S, et al. Cost-effectiveness of vitamin D supplementation and exercise in preventing injurious falls among older home-dwelling women: findings from an RCT. Osteoporos Int. 2016;27(1):193–201.
44. Poole CD, Smith JC, Davies JS. The short-term impact of vitamin D-based hip fracture prevention in older adults in the United Kingdom. J Endocrinol Investig. 2014;37(9):811–7.
45. Church JL, Haas MR, Goodall S. Cost Effectiveness of falls and injury prevention strategies for older adults living in residential aged care facilities. Pharmacoeconomics [Internet]. 2015 [cited 2024 Sep 22];33(12):1301–10. Available from: https://link.springer.com/article/10.1007/s40273-015-0313-8
46. Calvo MS, Whiting SJ, Barton CN. Vitamin D fortification in the United States and Canada: current status and data needs 1-4. Am J Clin Nutr. 2004;80:1710–6.
47. GR, Gupta A. Fortification of Foods With Vitamin D in India: Strategies Targeted at Children. J Am Coll Nutr. 2015 May 4 [cited 2024 Mar 27];34(3):263–72.

Chapter 2
Immune System Modulation and Vitamin D

Shaveta Jain, Alisha, and Sant Ram

Abstract The current chapter emphasises the role of vitamin D in various autoimmune diseases and its immunomodulatory effects. Vitamin D, primarily known for its role in calcium homeostasis, has been found to interact with the vitamin D receptor (VDR) located on various immune cells, influencing both innate and adaptive immune responses. Through its action on dendritic cells, T cells, B cells, and macrophages, vitamin D promotes an anti-inflammatory environment by inhibiting pro-inflammatory cytokines and enhancing regulatory T cell function. Experimental evidence had also suggested that vitamin D deficiency correlates with increased severity of autoimmune disorders such as multiple sclerosis, type I diabetes mellitus, and rheumatoid arthritis. Clinical studies exploring vitamin D supplementation in autoimmune patients have shown promising but varied results, indicating its potential as an adjunct therapy. Further research is needed to elucidate the relationship between vitamin D deficiency and autoimmunity, so that the current therapeutic strategies for these complex diseases may further be improved for better outcomes.

Keywords Vitamin D · Autoimmune diseases · Immunomodulation · Vitamin D receptor · Clinical trials

---

S. Jain · Alisha · S. Ram (✉)
Department of Biochemistry, PGIMER Chandigarh, Chandigarh, India

V. Kumar, S. Patel (eds.), *Vitamin D: A Comprehensive Guide for Medical Professionals*, https://doi.org/10.1007/978-981-96-6491-7_2

2.1 Introduction

Importance of vitamin D became clear after the fact was established that its deficiency causes rickets and osteomalacia in children and adults, respectively. Later, it was found to be associated with cardiovascular diseases (CVD), hypertension, type 1 and type 2 diabetes, rheumatoid arthritis (RA), cancer, multiple sclerosis (MS), cognitive decline, dementia, etc. [1]. Various extra-skeletal roles of vitamin D have also been reported apart from its conventional role in calcium homeostasis [2]. Vitamin D signals by interacting with vitamin D receptor (VDR), which is a ligand-regulated nuclear transcriptional factor. All nuclear receptors consist of a ligand binding domain (LBD) and a cavity in the lower part of this domain referred to as the ligand binding protein (LBP). LBP consists of 40 hydrophobic amino acid, well adapted to the shape of 1,25 $(OH)_2D_3$ [3]. Vitamin D receptor, other than in muscle and intestine, is also expressed by almost all cells in brain, bone marrow, breast, colon, malignant cells, and other cells of the immune system. Vitamin D binds with VDR along with retinoid-X-receptor (RXR) to form a heterodimer which is then translocated into the nucleus and then binds with vitamin D response element. This leads to the transcription of downstream genes C-MYC, CYP24A1, etc. [4].

Vitamin D inhibits proliferation of cell and stimulates its differentiation, especially monocytes and macrophages; hence, it acts as a modulator of immune system [1]. Thus, it regulates the innate immunity by inhibition of inflammation as an antimicrobial response and also reduces the capacity to initiate an adaptive immune response.

Immunomodulation by Vitamin D The active form of vitamin D, i.e. 1,25 $(OH)_2D_3$ functions by binding to the VDR and starts its signalling cascade. This VDR is present in almost every human cell including blood lymphocytes, neutrophils, monocytes, and dendritic cells. These immune cells can also be modulated by the concentration of vitamin D at the site of inflammation [5]. Therefore, these cells are susceptible to vitamin D-mediated modulation of both innate and adaptive immune systems.

Innate Immunity

Immune cells can enhance the level of enzyme 1-α-hydroxylase which converts inactive calcidiol to active calcitriol. In monocytes and macrophages, increased level of vitamin D triggers the production of antimicrobial peptides such as cathelicidin antimicrobial peptide (CAMP), defensins β2 and β4 which altogether results in enhanced chemotaxis, phagolysosomal fusion, and autophagy of immune cells [6]. Increased levels of vitamin D in keratinocytes, epithelial cells, lung cells, intestinal cells, and corneal cells augment antimicrobial effects and maintain the physical barrier or body's first line of defence [7].

2.1.1 Adaptive Immunity

Dendritic Cells Immune response is triggered when an antigen is presented by a mature dendritic cell (DC) whereas immature DCs induce tolerance. Vitamin D prevents autoimmunity by inhibiting differentiation and maturation of DCs by downregulating class II MHC molecules, TNFα, and IL-12, and thus promoting self-tolerance. The DCs differentiating from monocytes in the presence of vitamin D remain in its tolerogenic state. This tolerogenic property can be induced by vitamin D in DC by switching toward glycolysis and activating PI3K-Akt-mTOR pathway [8]. These tolerogenic dendritic cells (tDCs) are not fully capable of inducing proinflammatory T cell proliferation but trigger differentiation of T regulatory cells (Treg). Different types of DCs adopt different mechanisms for differentiation of Treg cells—for example, bone marrow-derived tDCs use herpes virus entry mediator mechanism for this, Langerhans tDCs use TGFβ, and dermal tDCs use IL-10 for Treg induction [9]. These Treg can also induce apoptosis of auto-active T cells to prevent autoimmunity and maintain immune homeostasis.

Macrophages Vitamin D plays a dual role in macrophage differentiation and activation. It maintains an anti-inflammatory cellular environment by maintaining increased production of IL-10 and decreased production of proinflammatory factors such as IL-1β, TNFα, IL-6, COX-2, NO, and RANKL [10]. Anti-inflammatory effect of vitamin D on macrophages is induced by decreasing IL-6 and TNFα expression and targeting the NFκB signalling pathway [11]. This results in a decrease in the release of proinflammatory cytokines. Thioesterase superfamily member 4 (THEM4) protein also blocks NFκB-mediated suppression of SOCS via miR155 signalling pathway. This anti-inflammatory milieu altogether induces polarization of more and more anti-inflammatory M2 macrophages while inhibiting proinflammatory M1 macrophages [12].

B Cells In the immune response, B cells play an essential role by differentiation into plasma cells and antibody production. Vitamin D directly affects B cell differentiation, proliferation, antibody production, and cytokine secretion. VDR directly binds to the IL-10 promotor region and maintain an anti-inflammatory milieu. To prevent autoimmunity, where disease onset is due to the production of auto-antibodies, vitamin D reduces the proliferation and class switching of B cells. In the case of early differentiation, vitamin D blocks the CD40 co-stimulation signalling pathways by inhibiting the translocation of NF-κB p65 thus reducing B cell proliferation [13]. It is evident that auto-antibody production is directly correlated with low levels of vitamin D, and its supplementation decreases the auto-antibody titer in autoimmune diseases [14].

T Cells Various T cells including CD8+ Tc cells, CD4+ Th cells, and TCRγδ cells are the direct immunological target of vitamin D as they express VDR when activated. T cells are an important target for vitamin D. In the absence of VDR on T cells, vitamin D is unable to prevent progression of MS in an experimental autoimmune encephalomyelitis (EAE) model [15]. Both adaptive and humoral immunity are affected by VDR-deficient Th1 and Th2 cells, respectively. Vitamin D also affects T cell differentiation and cytokine production. Cytokine production is mediated by increasing the production of IkBa (NF-kB inhibitory protein), thus blocking NF-kB p65 activation. This inhibits Th1 cell proliferation and production of proinflammatory cytokines such as Il-6, IL-12, IL-8, IFN-γ, IL-9, and TNF-α, and thus downregulates Th1-mediated immune responses [16].

CD4+ T Cells are a group of different types of cells, including T helper 1 (Th1), T helper 2 (Th2), T helper 17 (Th17), and T regulatory (Treg cells). In immune homeostasis, Th1 cells fight against intracellular pathogens, Th2 cells against helminth infections, and Th17 cells against extracellular pathogens and fungi. However, Tregs are responsible for mediating immunological tolerance against self-antigens. Whereas, in autoimmune disorders, T cells-mediated immune response is generated against the body itself. Experimentally, it was evident that vitamin D prevents the migration of CD4+ T cells in CNS and prevents the onset of EAE in the disease model [17].

(i) **Th1 cells** are characterized by the expression of IFNγ and T-bet. Some studies suggest that Th1 cells mediate autoimmune diseases because experimental mice devoid of transcriptional factor T-bet prevent the onset of EAE [18]. Similarly, another study showed that the transfer of IFNγ+ cells specific for myelin induces the progression of EAE disease. Therefore, it is clear that Th1 cells mediate the progression of autoimmunity. Experimental evidence suggests that during initial phases of differentiation, $1,25(OH)_2D_3$ inhibits the production of IFNγ, thereby preventing Th1 cell differentiation and thus preventing autoimmunity [19]. Since $1,25(OH)_2D_3$ reduces IL-2 release, therefore exogenous IL-2 is required for vitamin D-mediated inhibition of IFNγ release. Therefore, it is clear that vitamin D modulates Th1 not always but only in certain circumstances and plays a little role in the prevention of autoimmune disease.

(ii) **Th2 cells** produce transcription factor like GATA3 and various cytokines such as IL-13, IL-5, and IL-4. Th2 plays a pathogenic role in asthma and allergies, but they are also well known for prevention from Th17 cell-induced autoimmunity. In the experimental arthritis mice model, the protective role of GATA3 via Th17 suppression was proved by van Hamburg et al. [20]. IL-4 is a pleiotropic anti-inflammatory cytokine that functions mainly by suppressing the proinflammation causing immune suppression. Vitamin D plays a crucial role in Th2-mediated suppression of autoimmunity. In the EAE model, vitamin D inhibits disease progression via cytokine IL-4 secreted by Th2 cells. Vitamin D also induces the expression of GATA3 in this same disease model to suppress autoimmunity. Further, this hypothesis was strengthened in STAT6-KO mice, where no inhibition of EAE by vitamin D was seen in the absence of STAT6

[21]. In another experimental study, treatment naïve RA patients' PBMCs, where IL-4 production is reduced, supplementation of vitamin D upregulates IL-4 levels equals to that of healthy controls [22]. However, cholecalciferol supplementation in patients does not alter Th2-mediated IL-4 production. Therefore, it is suggested that Th2 differentiation and IL-4 release induced by vitamin D may play an important role in autoimmune suppression only when insufficient IL-4 level is present, not always.

(iii) **Th17 cells** are the subset of CD4+ T cell which produce IL-17 which is a strong proinflammatory cytokine essential for host defence against bacteria and fungi. Recently, they are considered to play an important role in pathogenesis of autoimmunity. Th17, when stimulated by IL-12 or TNFα, starts upregulated secretion of IFNγ and IL-17A or high secretion of IFNγ but negligible IL-17A secretion. These cells with high IFNγ but negligible IL-17A secretion are known as non-classic Th1 or Th17.1 cells as they express T-bet along with CCR6, RORC, and CXCR3 [23]. Therefore, these cells seem to be more pathogenic in the progression of autoimmune disease. It was found that vitamin D inhibits T cell pathogenicity in autoimmune disease as it decreases the IL-17A+ IFNγ and IFNγ+ IL-17A+ Th17.1 producing cells. Also, supplementation of vitamin D for 6 months in SLE patients reduces Th17.1 cells count [24].

(iv) **Regulatory T cells** express FoxP3, CD25, and inhibitory co-receptor CTLA4. They secrete anti-inflammatory cytokines like IL-10 and TGFβ. In comparison to other CD4+ T cell types, Treg also immunomodulate other cells such as DCs, T cells, and macrophages to maintain immune homeostasis. Patients with mutation in FoxP3 suffer from autoimmune disease IPEX syndrome [25]. In EAE mice, vitamin D supplementation protects from autoimmune disease as they induce FoxP3+ Tregs in the lymph node, spleen, and spinal cord. It was found that vitamin D supplementation in in vitro polarized Tregs enhances their suppressive capacity as it induces high expression of Treg markers such as PD1, CD25, and CTLA4, but no such results were seen in sorted Treg cells from peripheral blood [26]. This suggests that Treg function can be regulated by vitamin D in order to suppress autoimmunity.

Overall, in CD4+ T cells, 1,25-dihydroxyvitamin D3 stimulates Treg activity while inhibiting the proinflammatory Th cell functions. Therefore, beneficial effects of 1,25-dihydroxyvitamin D3 on autoimmunity may be exhibited by restoring the balance between Treg cells and effector T cells.

CD8+ T Cells CD8+ is another important class of T cells and induces apoptosis of infected/cancer cells. The role of CD8+ T cells is still not well known in the pathogenesis of an autoimmune disease like CD4+ T cells. It was found that myelin-specific and hsp60-specific CD8+ T cells induce EAE and intestinal inflammation respectively in mice [27]. Also, in psoriatic arthritis patients, IL-17A+ CD8+ T cells proportion is more in synovial fluid. CD8+ T cells show higher expression of VDR; hence, play a much important role in Vit D-mediated suppression of autoimmunity [28].

Therefore, it is evident that vitamin D is anti-inflammatory in nature and prevents autoimmune reaction because of its immunomodulatory action. This shows that autoimmune disorders result from an aberrant activation of the immune system. Since, vitamin D deficiency frequently occurs in patients suffering from autoimmune diseases, its supplementation might be helpful in decreasing the disease severity or may enhance the effectiveness of current treatments. Therefore, the correlation with serum level of vitamin D, vitamin D intake, UV exposure, and polymorphism in VDR with autoimmune severity is the subject of future research.

2.1.2 Relation to Vitamin D and Autoimmune Diseases: In Vivo Studies

Numerous experimental studies in the past few decades have been done to understand the role of vitamin D in the incidence and severity of autoimmune disorders. Experiment studies have discovered that in IL-10 knock-out mice model of inflammatory bowel diseases (IBDs), deficiency of vitamin D increases the disease severity [29]. Additionally, in the autoimmune encephalomyelitis model of multiple sclerosis and collagen-induced model of rheumatoid arthritis (RA), vitamin D supplementation could arrest the occurrence and diseases progression [30]. Similarly, vitamin D supplementation reduces the onset and severity of diabetes in diabetic model and SLE in MRL/1 mice [31].

2.1.3 Vitamin D Role in Human Autoimmune Disease

Despite the advantageous effect of vitamin D augmentation in experimental model setup, the therapeutic potential of vitamin D is limited in clinical practice for treatment of autoimmune diseases.

Multiple Sclerosis Multiple sclerosis (MS) is an autoimmune disorder affecting the central nervous system. In most of the patients, it begins with recurrent episodes of neuronal dysfunction which is then followed by partial or complete remission referred to as relapsing-remitting multiple sclerosis (RRMS). In some RRMS patients, the disease can transform into continuous progression of neurological deficits. This phase is known as the secondary progressive multiple sclerosis (SPMS). In others, there is a slow, progressive accumulation of neuronal dysfunction known as primary progressive multiple sclerosis (PPMS). In this disease, the myelin sheath covering the neurons is attacked by the immune system causing demyelination. Eventually it causes breakdown of the blood-brain barrier, formation of inflammatory lesions, oligodendrocyte loss, axonal degeneration, and reactive gliosis. Earlier it was proposed that CD4+ T helper cells cause the formation of inflammatory lesions by the production of IFN-gamma. Evidence suggest that T cells responsible for inflammation in this disease are characterised by CD4+ T helper cells producing IL-17. A monoclonal antibody alemtuzumab against CD52 removes lymphonuclear cells and reduction in inflammatory status indicating the role of these cells in multiple sclerosis [32].

Various epidemiological studies show that at higher latitudes the prevalence of MS is higher with reduced exposure to UV light. However, in recent decades, an increased trend of avoiding exposure to sun and spending majority of the time indoors may have led to a decline in the latitudinal finding [33]. Studies have also shown a correlation between the birth month of the individual and exposure of sunlight to the mother during pregnancy. Children born during the months of autumn have a lower risk of MS as compared to those born in the spring. The patients with high vitamin D have low chances of MS relapse whereas low vitamin D levels are associated with the severity of disease [33]. MS genome wide association studies predicted two genes, CYP27B1 (rate limiting step of vitamin D synthesis) and CYP24A1 (degradation of 1,25-dihydroxyvitamin D), that could increase the susceptibility to multiple sclerosis by decreasing active vitamin D levels [34].

Type I Diabetes Type I diabetes or diabetes mellitus is a T cell-mediated autoimmune disorder, characterised by destruction of insulin producing beta cells by mononuclear cells infiltration into the islets of Langerhans. In most of the cases during diagnosis, only a few (10–20%) of pancreatic beta cells remain functional. Various studies have proved that vitamin D is effective in increasing insulin sensitivity and secretion. In clinical trial type 1 diabetes, $1{,}25(OH)_2D_3$ or an analogue is used in clinical trials in spite of cholecalciferol. In these studies, both forms of vitamin D were able to delay but cannot prevent the progression of disease. This could be due to the fact that a small numbers of functional beta cells were remaining when vitamin D supplementation was started [35].

Crohn's Disease Crohn's disease (CD) is referred to as chronic relapsing inflammatory bowel disease affecting any part of the gastrointestinal tract. In CD, there is increased Th 1 and Th 17 cell response leading to the release of proinflammatory cytokines IL-22, IL-17A, IL-17F, and IFN-gamma. Whereas Treg response is reduced leading to a decrease in anti-inflammatory cytokine production such as TGF-beta and IL-10 [36]. It has been established that vitamin D is important for maintaining epithelial integrity, innate immunity, and T cell development. Studies have shown that in CD patients, vitamin D deficiency is usually higher as compared to the general population [37].

Rheumatoid Arthritis In rheumatoid arthritis (RA), the immune cells attack the inner lining of the joints, i.e. synovium. Although in experimental arthritis model, vitamin D supplementation was found to be beneficial, but a limited data about vitamin D's beneficial effect is present in RA patients. Till date, only three clinical trial studies investigating about vitamin supplementation in RA are present. Studies conducted by Salesi and Farajzadegan et al. confirmed the beneficial effects of vitamin D supplementation on RA disease activity [38]. Another study by Dehghan et al. shows the effect of vitamin D supplementation on RA relapse suggesting that with cholecalciferol treatment, one out of ten patients prevent relapse [39]. But none of these studies reach the statistical significance. Also, no effect of ergocalciferol treatment on disease activity was found. Therefore, whether vitamin D supplementation can be used as a therapeutic agent in RA is not clear.

Systemic Lupus Erythematosus SLE patients are sensitive to UV exposure; therefore, they prevent themselves from sunlight. Studies conducted by Lima et al. and Abou et al. confirmed low SLE activity score and auto-antibodies level when supplemented with either 2000 IU per day or with 5000 IU per week [29, 40]. Therefore, vitamin D supplementation in SLE is more beneficial than any other autoimmune diseases.

Conclusion In addition to the classical function of vitamin D, i.e. calcium homeostasis, it also plays an important role in immune regulation. Vitamin D acts by interacting with the VDR which is present on almost all immune cells. Therefore, these cells are susceptible to vitamin D-mediated adaptive and innate immune responses. It has become evident that vitamin D deficiency is common in various autoimmune diseases such as multiple sclerosis, Crohn's disease, type I diabetes mellitus, and systemic lupus erythematosus. However, further studies are required in order to confirm if vitamin D deficiency is a cause or the consequence of autoimmunity.

References

1. Sanlier N, Guney-Coskun M, Vitamin D. The immune system, and its relationship with diseases. Egyptian Pediatric Assoc Gazette. 2022;70(1):39.
2. Martens PJ, Gysemans C, Verstuyf A, Mathieu C. Vitamin d's effect on immune function. Nutrients. 2020;12(5):1248.
3. Carlberg C. Vitamin D and its target genes. Nutrients. 2022;14(7):1354.
4. Aranow C. Vitamin D and the immune system. J Investig Med. 2011;59(6):881–6.
5. Provvedini DM, Tsoukas CD, Deftos LJ, Manolagas SC. 1,25-Dihydroxyvitamin D3 receptors in human leukocytes. Science (80-). 1983;221(4616):1181–3.
6. Sassi F, Tamone C, D'Amelio P. Vitamin D: nutrient, hormone, and immunomodulator. Nutrients. 2018;10(11):1656.
7. Nguyen JA, Yates RM. Better together: current insights into phagosome-lysosome fusion. Front Immunol. 2021;25:12.
8. Møller SH, Wang L, Ho P-C. Metabolic programming in dendritic cells tailors immune responses and homeostasis. Cell Mol Immunol. 2022;19(3):370–83.
9. Josefowicz SZ, Lu L-F, Rudensky AY. Regulatory T cells: mechanisms of differentiation and function. Annu Rev Immunol. 2012;30(1):531–64.
10. Small AG, Harvey S, Kaur J, Putty T, Quach A, Munawara U, et al. Vitamin D upregulates the macrophage complement receptor immunoglobulin in innate immunity to microbial pathogens. Commun Biol. 2021;4(1):401.
11. Wang Q, He Y, Shen Y, Zhang Q, Chen D, Zuo C, et al. Vitamin D inhibits COX-2 expression and inflammatory response by targeting Thioesterase superfamily member 4. J Biol Chem. 2014;289(17):11681–94.
12. Wilson HM. SOCS proteins in macrophage polarization and function. Front Immunol. 2014;28:5.

13. Marsman C, Verstegen NJ, Streutker M, Jorritsma T, Boon L, ten Brinke A, et al. Termination of CD40L co-stimulation promotes human B cell differentiation into antibody-secreting cells. Eur J Immunol. 2022;52(10):1662–75.
14. Chen S, Sims GP, Chen XX, Gu YY, Chen S, Lipsky PE. Modulatory effects of 1,25-Dihydroxyvitamin D3 on human B cell differentiation. J Immunol. 2007;179(3):1634–47.
15. Mayne CG, Spanier JA, Relland LM, Williams CB, Hayes CE. 1,25-Dihydroxyvitamin D 3 acts directly on the T lymphocyte vitamin D receptor to inhibit experimental autoimmune encephalomyelitis. Eur J Immunol. 2011;41(3):822–32.
16. Roffe-Vazquez DN, Huerta-Delgado AS, Castillo EC, Villarreal-Calderón JR, Gonzalez-Gil AM, Enriquez C, et al. Correlation of Vitamin D with inflammatory cytokines, atherosclerotic parameters, and lifestyle factors in the setting of heart failure: a 12-month follow-up study. Int J Mol Sci. 2019;20(22):5811.
17. Grishkan IV, Fairchild AN, Calabresi PA, Gocke AR. 1,25-Dihydroxyvitamin D 3 selectively and reversibly impairs T helper-cell CNS localization. Proc Natl Acad Sci. 2013;110(52):21101–6.
18. Nath N, Prasad R, Giri S, Singh AK, Singh I. T-bet is essential for the progression of experimental autoimmune encephalomyelitis. Immunology. 2006;118(3):384–91.
19. Staeva-Vieira TP, Freedman LP. 1,25-Dihydroxyvitamin D3 inhibits IFN-γ and IL-4 levels during in vitro polarization of primary murine CD4+ T cells. J Immunol. 2002;168(3):1181–9.
20. van Hamburg JP, Mus A, de Bruijn MJW, de Vogel L, Boon L, Cornelissen F, et al. GATA-3 protects against severe joint inflammation and bone erosion and reduces differentiation of Th17 cells during experimental arthritis. Arthritis Rheum. 2009;60(3):750–9.
21. Sloka S, Silva C, Wang J, Yong VW. Predominance of Th2 polarization by Vitamin D through a STAT6-dependent mechanism. J Neuroinflammation. 2011;8(1):56.
22. Colin EM, Asmawidjaja PS, van Hamburg JP, Mus AMC, van Driel M, Hazes JMW, et al. 1,25-dihydroxyvitamin D 3 modulates Th17 polarization and interleukin-22 expression by memory T cells from patients with early rheumatoid arthritis. Arthritis Rheum. 2010;62(1):132–42.
23. Chen J, Bruce D, Cantorna MT. Vitamin D receptor expression controls proliferation of naïve CD8+ T cells and development of CD8 mediated gastrointestinal inflammation. BMC Immunol. 2014;15(1):6.
24. Sotirchos ES, Bhargava P, Eckstein C, Van Haren K, Baynes M, Ntranos A, et al. Safety and immunologic effects of high- vs low-dose cholecalciferol in multiple sclerosis. Neurology. 2016;86(4):382–90.
25. Bennett CL, Christie J, Ramsdell F, Brunkow ME, Ferguson PJ, Whitesell L, et al. The immune dysregulation, polyendocrinopathy, enteropathy, X-linked syndrome (IPEX) is caused by mutations of FOXP3. Nat Genet. 2001;27(1):20–1.
26. Van Belle TL, Vanherwegen A-S, Feyaerts D, De Clercq P, Verstuyf A, Korf H, et al. 1,25-Dihydroxyvitamin D3 and its analog TX527 promote a stable regulatory T cell phenotype in T cells from type 1 diabetes patients. PLoS One. 2014;9(10):e109194.
27. Zhang H, Wu H, Liu L, Li H, Shih DQ, Zhang X. 1,25-dihydroxyvitamin D 3 regulates the development of chronic colitis by modulating both T helper (Th)1 and Th17 activation. APMIS. 2015;123(6):490–501.
28. Veldman CM, Cantorna MT, DeLuca HF. Expression of 1,25-Dihydroxyvitamin D3 receptor in the immune system. Arch Biochem Biophys. 2000;374(2):334–8.
29. Lima GL, Paupitz J, Aikawa NE, Takayama L, Bonfa E, Pereira RMR. Vitamin D supplementation in adolescents and young adults with juvenile systemic lupus erythematosus for improvement in disease activity and fatigue scores: a randomized, double-blind. Placebo-Controlled Trial Arthritis Care Res (Hoboken). 2016;68(1):91–8.
30. Lemire JM, Archer DC. 1,25-dihydroxyvitamin D3 prevents the in vivo induction of murine experimental autoimmune encephalomyelitis. J Clin Invest. 1991;87(3):1103–7.
31. Mathieu C, Waer M, Laureys J, Rutgeerts O, Bouillon R. Prevention of autoimmune diabetes in NOD mice by 1,25 dihydroxyvitamin D3. Diabetologia. 1994;37(6):552–8.

32. McFarland HF, Martin R. Multiple sclerosis: a complicated picture of autoimmunity. Nat Immunol. 2007;8(9):913–9.
33. Wootla B, Eriguchi M, Rodriguez M. Is multiple sclerosis an autoimmune disease? Autoimmune diseases. 2012;16:2012.
34. Alharbi FM. Update in vitamin D and multiple sclerosis. Neurosciences J. 2015;20(4):329–35.
35. Kawasaki E. Type 1 diabetes and autoimmunity. Clin Pediatr Endocrinol. 2014;23(4):99–105.
36. Bendix M, Dige A, Jørgensen SP, Dahlerup JF, Bibby BM, Deleuran B, Agnholt J. Decrease in mucosal IL17A, IFNγ and IL10 expressions in active Crohn's disease patients treated with high-dose vitamin D alone or combined with infliximab. Nutrients. 2020;12(12):3699.
37. Fletcher J, Cooper SC, Ghosh S, Hewison M. The role of vitamin D in inflammatory bowel disease: mechanism to management. Nutrients. 2019;11(5):1019.
38. Salesi M, Farajzadegan Z. Efficacy of Vitamin D in patients with active rheumatoid arthritis receiving methotrexate therapy. Rheumatol Int. 2012;32(7):2129–33.
39. Dehghan A, Rahimpour S, Soleymani-Salehabadi H, Owlia MB. Role of vitamin D in flare ups of rheumatoid arthritis. Z Rheumatol. 2014;73(5):461–4.
40. Abou-Raya A, Abou-Raya S, Helmii M. The effect of Vitamin D supplementation on inflammatory and hemostatic markers and disease activity in patients with systemic lupus erythematosus: a randomized placebo-controlled trial. J Rheumatol. 2013;40(3):265–72.

Chapter 3
Assessment and Evaluation of Vitamin D

Piyush Pathak and Sant Ram

Abstract 1,25-dihydroxyvitamin D [1,25$(OH)_2$D], which is the major biologically active form of vitamin D, is a hormone that helps regulate the body's calcium and phosphate metabolism. Cholecalciferol (vitamin D_3) and ergocalciferols (vitamin D_2) are two different categories or families of vitamin D.

To determine the nutritional status of vitamin D in the blood, estimation of 25(OH)D in the blood is considered as an ideal way, while the measurement of 1,25(OH)2D is used in the diagnosis of vitamin D-dependent rickets.

Measurement of 25(OH)D and 1,25(OH)2D involves three steps, extraction, purification, and quantification. Techniques available to measure 25(OH)D include competitive protein binding assay, competitive and noncompetitive immunoassay, ultraviolet absorption, and mass spectrometry. For the estimation of 1,25(OH)2D Radioimmunoassay or RIA is commonly used. 25(OH)D in serum is considered normal between 10 and 65 ng/mL and 1,25(OH)2D is considered normal between 15 and 60 pg/mL. These are the most commonly used reference ranges, but the levels depend on the method used, ethnicity, sun exposure and other factors like pregnancy.

Screening for vitamin D deficiency is suggested in high-risk individuals which include breastfed infants, strict vegetarians, darker-skinned individuals, and the elderly. Supplementation with 50,000 IU/week for 12 weeks or 50,000 IU 3 times/week for 6 weeks is recommended in patients with 25(OH)D levels less than 20 ng/mL.

Keywords 25-hydroxyvitamin D · 1,25-dihydroxyvitamin D · Cholecalciferol · Ergocalciferol · Laboratory assays · Vitamin D deficiency

P. Pathak · S. Ram (✉)
Department of Biochemistry, PGIMER Chandigarh, Chandigarh, India

V. Kumar, S. Patel (eds.), *Vitamin D: A Comprehensive Guide for Medical Professionals*, https://doi.org/10.1007/978-981-96-6491-7_3

3.1 Introduction

Endogenously produced vitamin D or vitamin D absorbed from the food is changed into '25-hydroxyvitamin D' ['25(OH)D'], which is the circulating form, which later gets converted into '1,25-dihydroxyvitamin D' ['1,25(OH)$_2$D'], which is the major biologically active form. '1,25-dihydroxyvitamin D' is a hormone, and it helps in regulating calcium and phosphate metabolism in the body. 'Cholecalciferol' (vitamin D_3) and 'ergocalciferols' (vitamin D_2) are two different categories or families of vitamin D and both have hydroxy and dihydroxy forms. Vitamin D_3 synthesized in the skin is the parent compound of naturally occurring vitamin D, while vitamin D_2 is produced by yeast and it is the parent compound of another family [1].

Laboratory test to assess the vitamin D status of the patient includes measurement of 'vitamin D', '25(OH)D', and '1,25(OH)$_2$D', with the last two being the main metabolites of vitamin D. Sensitive and specific assays are available to measure both forms, which are hydroxyvitamin D and dihydroxyvitamin D. Ideally assays should measure metabolites of both D_2 and D_3 as both have biologically active forms in the human body.

A high-affinity transport protein which is specific for vitamin D and its metabolites and a group-specific component of serum or Gc-globulin, also known as 'vitamin D binding protein' (DBP) binds vitamin D, hydroxyvitamin D, and dihydroxyvitamin D in circulation [2].

3.2 Laboratory Tests

Measuring 25(OH)D in blood is the best way to determine the nutritional status of vitamin D in the blood due to its longer half-life of 2–3 weeks, hence shows less day-to-day variation, and 'hydroxyvitamin D' is the main circulating form in the blood. It is being emphasized that vitamin D nutritional status is assessed by measurement of '25(OH)D' and not '1,25(OH)$_2$D' [3, 4].

Measurement of '1,25(OH)$_2$D' is used in the diagnosis of 'vitamin D-dependent rickets' in both type 1 and 2 disease and diseases associated with the overproduction of '1,25(OH)$_2$D'.

Since biologically active '1,25(OH)$_2$D' is produced by both D_2 and D_3, an assay should measure metabolites of both D_2 and D_3 with equimolar reactivity. Food is supplemented with both D_2 and D_3, so there is no use in measuring D_2 and D_3 metabolites separately for the assessment of vitamin D nutritional status [4, 5].

Assay to measure '25(OH)D' and '1,25(OH)$_2$D' requires three steps, which are extraction, purification, and quantification [6].

Extraction In human circulation 'vitamin D binding protein' (DBP) binds both metabolites of vitamin D. The first step is extraction also known as deproteinization, which frees both metabolites of vitamin D from DBP. This step is usually done with acetonitrile in most of the commercially available instruments. Acetonitrile deproteinizes the specimen and denatures DBP. There are some other methods available for $1,25(OH)_2D$ which use immunoextraction. Previously organic solvents and solvent mixtures were used for deproteinization by two-phase, liquid–liquid partitions.

Purification This step separates different forms of vitamin D, lipids, and other interfering substances and is most commonly done by column chromatography.

The difference between 'vitamin D', '25(OH)D', and '$1,25(OH)_2D$' is the number of hydroxyl groups which give them different polarities, which is used to separate them. Dihydroxyvitamin D is the most polar and vitamin D is the least polar out of the three.

Reversed-phase C_{18} silica minicolumn and normal-phase silica minicolumn are used most commonly. Both are used in a single method to separate different metabolites of vitamin D, a modified method by using phase switching with a single nonend capped $C_{18}OH$ cartridge to eliminate the silica cartridge.

Quantification The method for quantification of hydroxyvitamin D and dihydroxyvitamin D are different and it is discussed below separately [7].

3.3 Measurement of '25-Hydroxyvitamin D'

Over the past few years, the most reliable indicator of vitamin D status is hydroxyvitamin D. Techniques available to measure '25(OH)D' are (i) competitive protein binding assay (CPBA), (ii) competitive and noncompetitive immunoassay (RIA, EIA, ICMA), (iii) ultraviolet absorption (after separation by HPLC), and (iv) mass spectrometry (after separation by chromatography).

CPBA or competitive immunoassays are the methods that are used most commonly in clinical laboratories for the measurement of '25(OH)D', and it is the most familiar method as well. These are easy to perform, use less volume of plasma, and reagents are widely available which make it the best choice.

RIA or radioimmunoassay was developed after the development of antiserum against bovine serum albumin (BSA) which is a conjugate of a vitamin D analog without a side chain. As a tracer, a radioiodinated vitamin D analog was used. Samples are deproteinized and analyzed directly without chromatography. RIA is less sensitive to interfering substances when compared with CPBA. Based on chemiluminescence (CLIA), an automated immunoassay method has been developed. The only concern with this assay is an underestimation of 25(OH)D in the presence of $25(OH)D_2$.

HPLC methods have used either normal phase chromatography, reverse phase chromatography, or a combination of both followed by UV absorption for quantification at 254 or 265 nm. This method required chromatography for extraction before HPLC. The sensitivity of UV detection is limited and it requires 0.5 to 1 mL of specimen. The HPLC method allows separation of '$25(OH)D_2$' and '$25(OH)D_3$'. HPLC is the method for the validation of the accuracy of immunoassays and CPBAs.

Another method for measurement of hydroxyvitamin D is LC-MS/MS assays which couple 'liquid chromatography' with 'tandem mass spectrometry' for the measurement of '25(OH)D' [8–10]. It is also one of the popular methods in clinical laboratories due to its high throughput. The '25(OH)D_2' and '25(OH)D_3' are quantified separately by this method, but it must be reported as the sum of two concentrations for the clinical setting use [11, 12].

3.4 Measurement of 1,25-Dihydoxyvitamin D

'1,25$(OH)_2$D' levels in circulation is significantly lower compared to 25(OH)D. The compound '1,25$(OH)_2$D' has high hydrophobicity and instability in circulation which makes its determination even more complicated. The most commonly used methods require extraction, chromatography, and measurement by radioreceptor assay (RRA) or RIA.

Radioreceptor assays are a less used method in clinical laboratories, and it is performed using receptors for vitamin D (VDR) from the calf thymus or chick intestine.

Radioimmunoassay or RIA is a commonly used and more convenient method for the estimation of 1,25$(OH)_2$D. The need for liquid scintillation is eliminated by the use of a radioiodinated tracer in this method. The need to prepare VDR from the calf thymus is eliminated in RIA by the use of an antiserum. Before analysis, the specimen is deproteinized, oxidized, and purified using silica cartridges. This method is commercially available with a C_{18}-OH extra clean cartridge. Another commercially available method uses selective immunoextraction of 1,25$(OH)_2$D [13, 14].

3.5 Specimen for Vitamin D Measurement

Serum is usually used for vitamin D metabolites measurement. Plasma can be used in assays with extraction and chromatography. The serum is stable for some time at room temperature or 4 °C; however, if analysis is delayed, the sample should be frozen. The sample does not require any special handling for the estimation of vitamin D and its metabolites.

3.6 Normal Values

Hydroxyvitamin D in serum is considered normal between 10 and 65 ng/mL (25–162 nmol/L), and 1,25$(OH)_2$D is considered normal between 15 and 60 pg/mL (36–144 pg/mL). These are the most commonly used reference ranges, but the actual reference ranges are dependent on the method.

The concentration of hydroxyvitamin D varies in different populations, with lower levels in African Americans when compared with non-Hispanic whites, but the clinical significance of the difference is uncertain. Circulating concentration of 25(OH)D also varies with sun exposure and shows seasonal variation. The concentration of 'vitamin D' metabolites also varies with age and is increased in some physiological conditions like pregnancy [15–17].

3.7 Assessment and Laboratory-Based Diagnosis

Groups with a high risk of vitamin D deficiency include breastfed infants, strict vegetarians, darker-skinned individuals, and the elderly. Screening for vitamin D deficiency is suggested in individuals with malabsorption syndromes, chronic kidney disease, hepatic failure, pregnant and lactating women, elderly patients with a history of falls, patients with signs and symptoms of rickets, osteomalacia, osteoporosis, and patients on antiseizure medications and granuloma forming disorders.

Other than vitamin D nutritional status, determination of hydroxyvitamin D is also useful for the differential diagnosis of hypocalcemia, hypercalcemia, and hypercalciuria [18].

Circulating 25(OH)D may decrease due to reduced availability of vitamin D from the diet, less conversion of 'vitamin D' to 'hydroxyvitamin D', accelerated metabolism, and urinary loss of hydroxyvitamin D.

Reduced availability occurs due to dietary deficiency, malabsorption due to different reasons, and inadequate sunlight exposure. Inadequate conversion to 25(OH)D is caused mainly by hepatocellular disease. Increased urinary loss of hydroxyvitamin D and DBP may occur in renal diseases including nephrotic syndrome. Certain drugs may induce drug metabolizing enzymes and increase the metabolism of vitamin D; active form examples include phenytoin and phenobarbital [19, 20].

Increased '1,25(OH)$_2$D' is associated with overproduction of '1,25(OH)$_2$D' in lymphoma (Activated macrophages are responsible for the conversion of hydroxyvitamin D to dihydroxyvitamin D), tuberculosis, sarcoidosis, Hodgkin's disease, fungal infection, and Wegener's granulomatosis. '1,25(OH)$_2$D' is also increased in 'vitamin D-dependent rickets type 2' due to nonfunctioning VDR (vitamin D receptor), and it may also be elevated in primary hyperparathyroidism [21–23].

Reduced '1,25(OH)$_2$D' indicates 'vitamin D-dependent rickets type 1', renal failure, hypercalcemia due to malignancies, hypoparathyroidism, pseudohypoparathyroidism, hyperphosphatemia, hypomagnesemia, and severe hepatocellular disease.

For the diagnosis of vitamin D deficiency, 'dihydroxyvitamin D' level is not an adequate tool as the circulating level of '1,25(OH)$_2$D' is often normal in vitamin D deficiency due to the compensation by hyperparathyroidism. Similarly, in the case of 'vitamin D' or '25(OH)D' intoxication, the circulating level of '1,25(OH)$_2$D' may be normal, increased, or decreased [24, 25].

3.8 Vitamin D Supplementation

Patients with '25(OH)D' levels less than 20 ng/mL are recommended to have 50,000 IU/week for 12 weeks or 50,000 IU 3 times/week for 6 weeks. Other than vitamin D deficiency, it is also recommended to give cholecalciferol 800–2000 IU per day in patients with a history of frequent falls and for prevention of maintenance of bone health in elderly patients. There are some shreds of evidence that vitamin D supplementation is also helpful in the treatment and prevention of hypertension, heart failure, diabetes mellitus, multiple sclerosis, depression, asthma, COPD, tuberculosis, and some cancers. The tolerable upper limit of vitamin D in adults is 4000 IU/day, which is less for children. Excess vitamin D can cause toxicity which manifests as hypercalcemia, renal failure, and ataxia [26, 27].

3.9 Conclusion

'25(OH)D' is the main circulating form of vitamin D, and '1,25$(OH)_2$D' is the biologically active form of vitamin D. Cholecalciferol (vitamin D_3) and ergocalciferols (vitamin D_2) are two families of vitamin D and both are present in hydroxy and dihydroxy forms in the body. Measuring '25(OH)D' in the blood is the best way to determine vitamin D's nutritional status. Measurement of 1,25$(OH)_2$D is used in the diagnosis of 'vitamin D-dependent rickets type 1' and 'vitamin D-dependent rickets type 2' and diseases associated with the overproduction of 1,25$(OH)_2$D. Assays to measure '25(OH)D' and '1,25$(OH)_2$D' require extraction, purification, and quantification. The purification step separates different forms of vitamin D, lipids, and other interfering substances and is most commonly done by chromatography. Hydroxyvitamin D can be measured by competitive protein binding assay, Competitive and noncompetitive immunoassay including RIA, EIA, ICMA, Ultraviolet absorption after separation by HPLC, and Mass spectrometry after separation by chromatography. 1,25$(OH)_2$D levels are measured by radioreceptor assay (RRA) or RIA. Serum is usually used for the measurement of vitamin D metabolites. Screening for vitamin D deficiency is suggested for high-risk groups including pregnant and lactating women, the elderly, and patients with signs and symptoms of rickets and osteomalacia. Circulating '25(OH)D' may decrease due to reduced availability of vitamin D from the diet, less conversion of vitamin D to hydroxyvitamin D, accelerated metabolism, and urinary loss of hydroxyvitamin D. Measurement of '1,25$(OH)_2$D' is useful in the diagnosis of 'vitamin D-dependent rickets'.

References

1. Bikle DD. Vitamin D metabolism, mechanism of action, and clinical applications. Chem Biol. 2014;21:319–29. https://doi.org/10.1016/J.CHEMBIOL.2013.12.016.
2. Pike JW, Christakos S. Biology and mechanisms of action of the vitamin D hormone. Endocrinol Metab Clin N Am. 2017;46:815–43. https://doi.org/10.1016/J.ECL.2017.07.001.
3. Berger MM, Shenkin A, Schweinlin A, Amrein K, Augsburger M, Biesalski HK, et al. ESPEN micronutrient guideline. Clin Nutr. 2022;41:1357–424. https://doi.org/10.1016/J.CLNU.2022.02.015.
4. DeLuca HF. Overview of general physiologic features and functions of vitamin D. Am J Clin Nutr. 2004;80:80. https://doi.org/10.1093/AJCN/80.6.1689S.
5. Jones G, Kaufmann M. Diagnostic aspects of vitamin D: clinical utility of vitamin D metabolite profiling. JBMR Plus. 2021;5:5. https://doi.org/10.1002/JBM4.10581.
6. Heaney RP. Assessing vitamin D status. Curr Opin Clin Nutr Metab Care. 2011;14:440–4. https://doi.org/10.1097/MCO.0B013E328348ED85.
7. Hollis BW. Measuring 25-hydroxyvitamin D in a clinical environment: challenges and needs. Am J Clin Nutr. 2008;88:88. https://doi.org/10.1093/AJCN/88.2.507S.
8. Glendenning P, Fraser WD. 25-OH-vitamin D assays. J Clin Endocrinol Metab. 2005;90:3129. https://doi.org/10.1210/JC.2004-2399.
9. Holick MF, Glendenning P, Fraser WD. 25-OH-vitamin D assays [2]. J Clin Endocrinol Metab. 2005;90:3128–9. https://doi.org/10.1210/jc.2005-0162.
10. Maunsell Z, Wright DJ, Rainbow SJ. Routine isotope-dilution liquid chromatography-tandem mass spectrometry assay for simultaneous measurement of the 25-hydroxy metabolites of vitamins D2 and D3. Clin Chem. 2005;51:1683–90. https://doi.org/10.1373/CLINCHEM.2005.052936.
11. Stepman HCM, Vanderroost A, Van Uytfanghe K, Thienpont LM. Candidate reference measurement procedures for serum 25-hydroxyvitamin D3 and 25-hydroxyvitamin D2 by using isotope-dilution liquid chromatography-tandem mass spectrometry. Clin Chem. 2011;57:441–8. https://doi.org/10.1373/CLINCHEM.2010.152553.
12. Vogeser M, Kyriatsoulis A, Huber E, Kobold U. Candidate reference method for the quantification of circulating 25-hydroxyvitamin D3 by liquid chromatography-tandem mass spectrometry. Clin Chem. 2004;50:1415–7. https://doi.org/10.1373/CLINCHEM.2004.031831.
13. Nikooyeh B, Samiee SM, Farzami MR, Alavimajd H, Zahedirad M, Kalayi A, et al. Harmonization of serum 25-hydroxycalciferol assay results from high-performance liquid chromatography, enzyme immunoassay, radioimmunoassay, and immunochemiluminescence systems: a multicenter study. J Clin Lab Anal. 2017;31 https://doi.org/10.1002/jcla.22117.
14. Snellman G, Melhus H, Gedeborg R, Byberg L, Berglund L, Wernroth L, et al. Determining vitamin D status: a comparison between commercially available assays. PLoS One. 2010;5 https://doi.org/10.1371/JOURNAL.PONE.0011555.
15. Lips P. Relative value of 25(OH)D and 1,25(OH)2D measurements. J Bone Miner Res. 2007;22:1668–71. https://doi.org/10.1359/JBMR.070716.
16. Lips P. Which circulating level of 25-hydroxyvitamin D is appropriate? J Steroid Biochem Mol Biol. 2004;89–90:611–4. https://doi.org/10.1016/j.jsbmb.2004.03.040.
17. Dawson-Hughes B, Heaney RP, Holick MF, Lips P, Meunier PJ, Vieth R. Estimates of optimal vitamin D status. Osteoporos Int. 2005;16:713–6. https://doi.org/10.1007/S00198-005-1867-7.
18. Binkley N, Carter GD. Toward clarity in clinical vitamin D status assessment: 25(OH)D assay standardization. Endocrinol Metab Clin N Am. 2017;46:885–99. https://doi.org/10.1016/j.ecl.2017.07.012.
19. Hollis BW. Assessment and interpretation of circulating 25-hydroxyvitamin D and 1,25-dihydroxyvitamin D in the clinical environment. Endocrinol Metab Clin N Am. 2010;39:271–86. https://doi.org/10.1016/J.ECL.2010.02.012.

20. Hollis BW. Assessment of circulating 25(OH)D and 1,25(OH)2D: emergence as clinically important diagnostic tools. Nutr Rev. 2007;65 https://doi.org/10.1111/J.1753-4887.2007.TB00348.X.
21. Holick MF. Vitamin D status: measurement, interpretation, and clinical application. Ann Epidemiol. 2009;19:73–8. https://doi.org/10.1016/J.ANNEPIDEM.2007.12.001.
22. Pilz S, Zittermann A, Trummer C, Theiler-Schwetz V, Lerchbaum E, Keppel MH, et al. Vitamin D testing and treatment: a narrative review of current evidence. Endocr Connect. 2019;8:R27–43. https://doi.org/10.1530/EC-18-0432.
23. Pludowski P, Takacs I, Boyanov M, Belaya Z, Diaconu CC, Mokhort T, et al. Clinical practice in the prevention, diagnosis and treatment of vitamin D deficiency: a central and eastern European expert consensus statement. Nutrients. 2022;14:14. https://doi.org/10.3390/NU14071483.
24. Pramyothin P, Holick MF. Vitamin D supplementation: guidelines and evidence for subclinical deficiency. Curr Opin Gastroenterol. 2012;28:139–50. https://doi.org/10.1097/MOG.0B013E32835004DC.
25. Ebeling PR, Adler RA, Jones G, Liberman UA, Mazziotti G, Minisola S, et al. Management of endocrine disease: therapeutics of vitamin D. Eur J Endocrinol. 2018;179:R239–59. https://doi.org/10.1530/EJE-18-0151.
26. Bouillon R, Manousaki D, Rosen C, Trajanoska K, Rivadeneira F, Richards JB. The health effects of vitamin D supplementation: evidence from human studies. Nat Rev Endocrinol. 2022;18:96–110. https://doi.org/10.1038/S41574-021-00593-Z.
27. Haines ST, Park SK. Vitamin D supplementation: what's known, what to do, and what's needed. Pharmacotherapy. 2012;32:354–82. https://doi.org/10.1002/PHAR.1037.

Chapter 4
Physiology and Pharmacology of Vitamin D

Amol N. Patil

Abstract Vitamin D is a crucial fat-soluble vitamin synthesized in the skin upon exposure to ultraviolet B (UVB) radiation. It exists in two main forms: D2 (ergocalciferol) from plant sources and D3 (cholecalciferol) from animal sources and sunlight. These forms undergo hydroxylation in the liver to 25-hydroxyvitamin D [25(OH)D], the major circulating form, and further hydroxylation in the kidneys to the active form, 1,25-dihydroxyvitamin D [1,25(OH)2D]. Vitamin D enhances the intestinal absorption of these minerals, ensuring proper bone mineralization. Parathyroid hormone (PTH) is pivotal in this regulation, as low calcium levels stimulate PTH secretion, which in turn promotes the conversion of 25(OH)D to 1,25(OH)2D, increasing calcium and phosphorus levels in the blood. Vitamin D has immunomodulatory functions, influencing the innate and adaptive immune systems. It enhances the pathogen-fighting effects of monocytes and macrophages and decreases the proliferation of pro-inflammatory cytokines. Deficiency in vitamin D can lead to rickets in children, characterized by bone deformities and growth retardation, and osteomalacia in adults, causing bone pain and muscle weakness. Chronic deficiency may contribute to osteoporosis. Symptoms include fatigue, bone pain, muscle weakness, and mood changes. Conversely, vitamin D toxicity, though rare, can lead to hypercalcemia, causing nausea, vomiting, weakness, and serious complications like kidney damage. Optimal levels of vitamin D for health are generally considered to be a serum 25(OH)D concentration of 30–50 ng/mL. Maintaining these levels through appropriate sun exposure, diet, and supplementation is crucial for bone health and overall well-being.

A. N. Patil (✉)
Department of Pharmacology, Post Graduate Institute of Medical Education and Research, Chandigarh, India
e-mail: patil.amol@pgimer.edu.in

V. Kumar, S. Patel (eds.), *Vitamin D: A Comprehensive Guide for Medical Professionals*, https://doi.org/10.1007/978-981-96-6491-7_4

Keywords Vitamin D · Synthesis · Preparations · Sources · Metabolic pathways · Parathyroid hormone (PTH) · Calcium and phosphorus levels · Immunomodulatory functions · Deficiency · Toxicity · Optimal levels

Learning Goals

- Explain how vitamin D is synthesized, differentiate between different types of vitamin D_2 and D_3, and identify sources of the same.
- Outline the metabolic pathways and describe the roles of each form in calcium and phosphorus regulation.
- Understand the role of parathyroid hormone (PTH), calcium, and phosphorus levels in regulating vitamin D activation.
- Describe the role of vitamin D in enhancing the effect on the absorption of calcium and phosphorus in the gut, vitamin D in bone health, and the immunomodulatory functions of vitamin D.
- Deficiency, the symptoms, and consequences of vitamin D deficiency in children and adults.
- Rickets and osteomalacia.
- Describe the vitamin D toxicity.

Evaluate current scientific evidence regarding the optimal levels of vitamin D for health.

4.1 Physiology and Biochemistry of Vitamin D

Vitamin D is a **fat-soluble** vitamin that plays a crucial role in various physiological processes within the human body. There are two main forms of vitamin D, vitamin D_2 (ergocalciferol; molecular weight 396.6 g/mol; C28H44O) and vitamin D_3 (cholecalciferol; molecular weight 384.6 g/mol; C27H44O), differing with respect to their side chain (Fig. 4.1). Here are some key aspects of the physiology of vitamin D:

Source: Vitamin D_2 is found in a few plant sources (e.g., mushrooms) but is mostly produced on a commercial scale by the irradiation of yeast. The type of vitamin D_2 formed through this method is used to fortify foods and produce supplements. Vitamin D_3 has several dietary sources acting on the parent compound or ingested in the form of deep-sea fatty fish, egg yolks, liver, or supplements in addition to **ultraviolet (UV)** exposure happening naturally.

Fig. 4.1 Two main forms of vitamin D

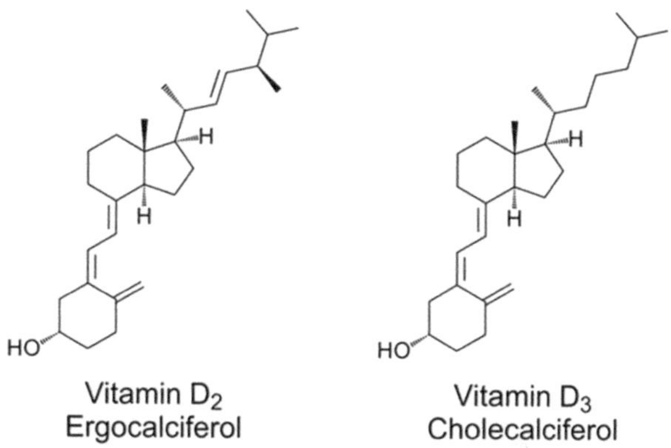

4.1.1 Synthesis

Sun Exposure The parent compound is formed in the Malpighian layer of skin during a relatively minor route of cholesterol synthesis. Ultraviolet radiation with wavelengths between 290 and 315 nm causes the bond between the ninth and tenth positions of the steroid ring to open, forming a compound called secosterol. UV radiation mediates this conversion. Vitamin D is a derivative of 7-dehydrocholesterol, also called ergosterol, the parent compound. This further undergoes cis-to-trans isomerization by the formation of a trans bond between the fifth and sixth carbon atoms, leading to the formation of vitamin D_3, or cholecalciferol. The involvement of UV radiation in the process has led to vitamin D being nicknamed the **sunshine vitamin**.

Dietary Sources Vitamin D can also be obtained from certain foods, such as fatty fish, egg yolks, and fortified dairy products. The ingested vitamin D can be either D_2 or D_3 in these cases. Vitamin D can be absorbed in a passive manner in the small intestine, particularly in the jejunum. Once absorbed, vitamin D is incorporated into chylomicrons, the lipoprotein particles. These chylomicrons are then transported via the lymphatic system to the bloodstream.

4.1.2 Activation

Both vitamin D_2 and D_3 undergo hydroxylation in the liver to form 25-hydroxyvitamin D [25(OH)D], also known as calcidiol. A mitochondrial hydroxylase enzyme introduces a hydroxyl group at the 25 positions. This reaction requires energy in the form of NADPH and oxygen. The 25-hydroxy cholecalciferol product is the **inactive storage form** of cholecalciferol and is stored in the liver. The active form of vitamin D, known as calcitriol, is produced through a second hydroxylation in the kidneys, converting 25(OH)D to 1,25-dihydroxyvitamin D [1,25(OH)2D]. This conversion is tightly regulated by the body's parathyroid hormone (**PTH**), calcium, and phosphorus levels. The 1,25(OH)2D has a structure similar to other steroid hormones produced in the body [1].

4.2 Pharmacodynamics/Mechanism of Action of Vitamin D

The mechanism of action of vitamin D involves its active form, 1,25-dihydroxyvitamin D [1,25(OH)2D], also known as calcitriol, interacting with specific receptors in target tissues. The primary target organs for the actions of calcitriol are the intestines, bones, kidneys, and parathyroid glands. Here is an overview of the mechanism of action of vitamin D:

Action on the Gut

Calcitriol binds to the vitamin D receptors (VDR) in the **nucleus** of the intestinal cells.

This binding activates the transcription of genes involved in the synthesis of proteins that facilitate the absorption of calcium and phosphorus from the intestine into the bloodstream. Calcium absorption primarily occurs in the duodenum (the first part of the small intestine). There, calcium is absorbed through both **active and passive** transport mechanisms. TRPV6 (Transient Receptor Potential Cation Channel, Subfamily V, Member 6) is a calcium-selective channel that facilitates the active transport of calcium across the apical membrane of enterocytes in the duodenum. Vitamin D promotes calcium absorption by enhancing the synthesis of calcium-binding proteins and the expression of TRPV6. Calcitriol also stimulates calcium absorption through the intestines' transcellular pathway. NaPi-IIb (Sodium-Phosphate Cotransporter Type IIb) is the primary transporter responsible for the active absorption of phosphate in the small intestine, particularly in the duodenum and jejunum. Similar to its role in calcium absorption, calcitriol stimulates the synthesis of proteins involved in phosphate transport, including NaPi-IIb. This enhances the absorption of phosphate in the intestines. Besides important transporters, the fibroblast growth factor 23 (FGF23) plays a significant role in the regulation of phosphate and vitamin D metabolism. FGF23 plays a crucial role in phosphate homeostasis by reducing phosphate reabsorption in the kidneys. This is achieved by downregulating the expression of sodium–phosphate cotransporters (NaPi-IIa and NaPi-IIc) in the renal tubules, leading to increased phosphate excretion [2].

Action on the Bone

Calcitriol plays a crucial role in bone health by promoting the deposition of calcium and phosphorus in bones. Vitamin D receptors (VDRs) are present in bone tissue in addition to several other organs of the body. When activated by the active form of vitamin D, calcitriol, VDRs act as transcription factors. They bind to specific regions of DNA, influencing the expression of genes related to calcium and phosphate metabolism, bone remodeling, and mineralization. In osteoblasts, VDR activation stimulates the synthesis of bone matrix formation/mineralization proteins, including **osteocalcin**. VDRs may also influence osteoclasts, cells responsible for bone resorption, and help maintain the balance between bone formation and resorption. Besides VDRs, CaSRs also mediate vitamin D action on the bone. CaSR is expressed in both osteoblasts and osteoclasts. It influences bone turnover by modulating the activity of these cells in response to changes in extracellular calcium concentration.

Additionally, calcitriol helps regulate osteoclasts' activity (cells responsible for bone resorption), maintaining a balance between bone formation and breakdown. FGF23 is a hormone primarily produced by osteocytes in bone and has important effects on the kidneys and vitamin D metabolism. Its main actions include promoting phosphate excretion by the kidneys and inhibiting vitamin D activation [3].

Vitamin D is essential for bone formation and maintenance. It helps prevent conditions like rickets (in children) and osteomalacia (in adults). Both conditions are related to inadequate mineralization of bone tissue, primarily due to a deficiency of vitamin D, calcium, or phosphate. However, they differ in terms of the affected populations and the manifestation of symptoms. Rickets is primarily a childhood disease that occurs during the growing years when the bones are still developing. It can also result from genetic factors or disorders affecting vitamin D metabolism. Pediatric patients present with delayed growth and development, softening and weakening of the bones, leading to deformities such as bowed legs or knock knees, pain in the spine, pelvis, and legs, or muscle weakness. On the other hand, osteomalacia is more common in adults, especially in the elderly population. It may also be associated with certain gastrointestinal disorders that affect the absorption of these nutrients. Adult patients' presentation includes bone pain, often in the lower back, pelvis, and legs, muscle weakness, fractures that may occur due to weakened bones, difficulty in walking, a waddling gait, etc. In both conditions, vitamin D deficiency is a key factor. When there is insufficient vitamin D, calcium and phosphate levels in the blood drop, leading to inadequate mineralization of the bone matrix [4].

Action of Kidney

Calcitriol promotes the reabsorption of calcium in the kidneys, reducing urinary calcium loss. In the renal tubules, calcitriol influences the expression of proteins involved in calcium reabsorption. One crucial protein affected by vitamin D is the TRPV5 (transient receptor potential vanilloid type 5) calcium channel. TRPV5 is responsible for the active transcellular reabsorption of calcium in the distal convoluted tubules. On the other hand, FGF23 inhibits the activity of 1α-hydroxylase, an enzyme responsible for converting inactive vitamin D (25-hydroxyvitamin D) to its active form (1,25-dihydroxyvitamin D or calcitriol) in the kidneys. By inhibiting this enzyme, FGF23 reduces the production of active vitamin D. FGF23 also stimulates the activity of 24-hydroxylase, an enzyme that catabolizes both 25-hydroxyvitamin D and 1,25-dihydroxyvitamin D. This further contributes to the overall reduction in active vitamin D levels. FGF23 exerts its actions by binding to fibroblast growth factor receptors (FGFRs), particularly FGFR1 and FGFR3. This binding activates downstream signaling pathways that ultimately influence cellular responses. FGF23 does not bind directly to FGFRs but requires the presence of a **co-receptor** known as Klotho. Klotho, a co-receptor, is expressed in various tissues, including the kidneys and parathyroid glands. The FGF23-Klotho interaction is crucial for FGF23 signaling [5].

Parathyroid Gland Action
Calcitriol regulates parathyroid hormone (PTH) secretion through negative feedback mechanisms. CaSR is expressed on the surface of parathyroid cells, particularly in the chief cells of the parathyroid glands. CaSR plays a crucial role in sensing extracellular calcium levels. When blood calcium levels are low, CaSR is not activated. Low CaSR activity leads to increased synthesis and release of parathyroid hormone (PTH) from the parathyroid glands. When blood calcium levels are elevated, CaSR is activated. Activated CaSR inhibits the release of PTH from the parathyroid glands. Low blood calcium levels or low dietary calcium intake stimulate the release of PTH from the parathyroid glands. PTH, in turn, stimulates the synthesis of active vitamin D (calcitriol) in the kidneys. Calcitriol can modulate the sensitivity of CaSR in parathyroid cells. Higher levels of calcitriol may decrease the responsiveness of CaSR to changes in extracellular calcium, contributing to the inhibition of PTH release when blood calcium levels are elevated. The interplay between CaSR, PTH, and vitamin D forms a feedback loop that helps maintain calcium homeostasis [6].

4.2.1 Action on the Immune System

The presence of VDRs in immune cells highlights the immunomodulatory role of vitamin D. Immune cells, including T lymphocytes, B lymphocytes, monocytes, and dendritic cells, express vitamin D receptors. VDRs are found in both the nucleus and cytoplasm of these immune cells. Calcitriol has immunomodulatory effects, influencing both innate and adaptive immune responses. It affects the function of various immune cells, such as T lymphocytes and macrophages, and plays a role in the regulation of inflammatory processes. Vitamin D can contribute to the induction of immune tolerance, helping to prevent excessive immune responses. Vitamin D can modulate the differentiation and function of T cells. It may inhibit the development of pro-inflammatory Th1 cells and promote the generation of regulatory T cells (Tregs), which play a role in immune tolerance. In B cells, Vitamin D can influence the differentiation and antibody production of B cells. It can affect B cell proliferation and class switching. Calcitriol is involved in the regulation of cell growth and differentiation in various tissues. It influences the expression of genes related to cellular proliferation and differentiation. However, the exact mechanisms and the role of vitamin D in these conditions are complex and still an active area of research.

Vitamin D is involved in the modulation of the immune system, and its deficiency has been associated with an increased risk of autoimmune diseases and infections. Some autoimmune diseases, such as multiple sclerosis, rheumatoid arthritis, and type 1 diabetes, have been associated with lower levels of vitamin D. However, correlation does not necessarily imply causation, and the exact nature of this relationship is still under investigation. Vitamin D has been shown to influence the function of various immune cells, including T cells and B cells. Deficiency may contribute to an increased risk of autoimmune conditions. Vitamin D has been shown to possess antimicrobial properties and plays a role in the body's defense against infections. It can enhance the production of **antimicrobial peptides**, part of the innate immune system's response to infections. There is evidence suggesting that vitamin D deficiency may be associated with an increased susceptibility to respiratory infections, including viral infections like influenza and respiratory tract infections. Some studies have explored the potential protective effects of vitamin D against acute respiratory infections [7].

4.3 Cell Differentiation and Apoptosis

Vitamin D is thought to have a role in cell differentiation and apoptosis (programmed cell death). The active form of vitamin D, calcitriol, interacts with vitamin D receptors (VDRs) in various cells throughout the body. This interaction influences gene expression and signaling pathways, contributing to cell growth, differentiation, and apoptosis regulation. It promotes the differentiation of **osteoblasts**, which are cells responsible for bone formation. It can inhibit the differentiation of pro-inflammatory T cells while promoting the differentiation of regulatory T cells, contributing to immune system balance. Vitamin D can affect the cell cycle by influencing gene expression in cell proliferation and differentiation. Proper cell cycle regulation is essential for normal cellular function and tissue development. Calcitriol is thought to induce apoptosis in cancer cells, contributing to the inhibition of tumor growth. Vitamin D receptors are observed in the brain, and vitamin D has also been implicated in neuroprotection. It may influence apoptosis in neural cells, potentially playing a role in neurological health and the prevention of neurodegenerative diseases. Some studies suggest that vitamin D may have protective effects on the cardiovascular system, and apoptosis regulation in vascular cells is one aspect of this. The effects of vitamin D on cell differentiation and apoptosis can vary among different cell types and tissues. The presence of vitamin D receptors in various tissues contributes to tissue-specific responses [8].

Overall, the mechanism of action of vitamin D involves binding its active form, calcitriol, to vitamin D receptors in target tissues. This interaction leads to the transcriptional activation of genes involved in calcium and phosphorus homeostasis, bone health, kidney function, parathyroid hormone regulation, immune system modulation, and cellular growth and differentiation. The overall effect is maintaining the balance of essential minerals in the body and supporting various physiological processes.

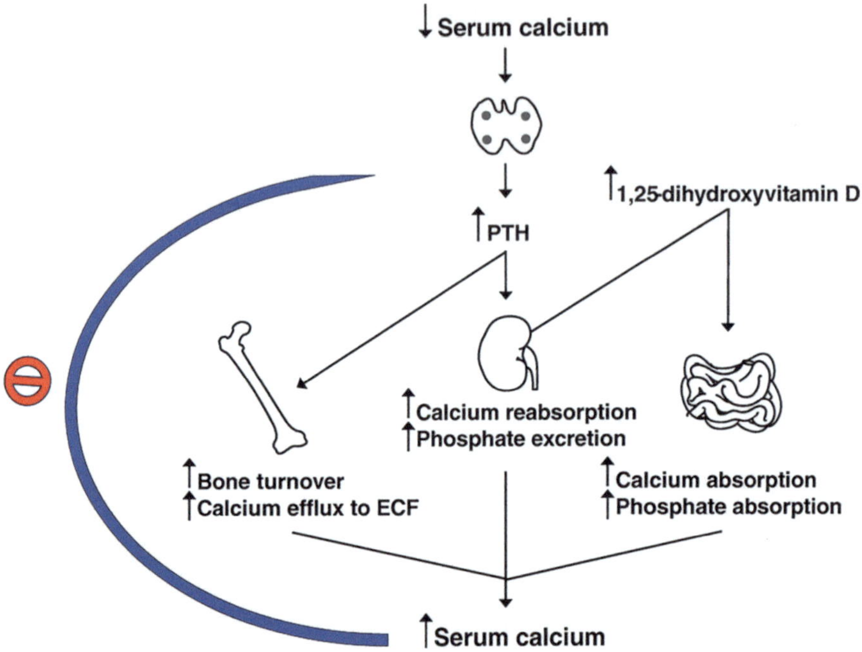

Fig. 4.2 Tight interplay between PTH (mastermind), vitamin D, calcium, and phosphate

4.3.1 Regulation

Feedback mechanisms regulate vitamin D synthesis. Parathyroid hormone (PTH) stimulates the conversion of 25(OH)D to 1,25(OH)2D, while high levels of calcium and phosphorus inhibit this conversion (Fig. 4.2) [9].

4.3.2 Deficiency

Vitamin D deficiency can lead to various health issues, including bone diseases like rickets and osteomalacia. It has also been linked to an increased risk of certain chronic diseases, autoimmune disorders, and infections described above. Vitamin D status can be assessed by measuring the prohormone 25(OH) D(the storage form), which is an indicator of supply rather than function. The most stable and plentiful metabolite of vitamin D in human serum, 25(OH)D, has a half-life of about 3 weeks, making it the most suitable indicator of vitamin D status. A guide for the assessment is presented in Table 4.1 [10].

Adult reference ranges for the active form 1,25(OH)$_2$ D are different for different genders. For males, it is 18–64 pg/mL, while for females, it ranges a bit longer, i.e., 18–78 pg/mL. Here, as per molecular weight mentioned at the beginning of the

Table 4.1 Usability of vitamin 25 hydroxy vitamin D level in clinical diagnosis

25 Hydroxy vitamin D level	Label used
$\leq$ 10 ng/ml	Deficient
11–20 ng/ml	Insufficient
$\geq$ 20 ng/ml	Optimal

To convert into nmol/L from ng/ml, we are required to multiply the ng/ml value by 2.496

chapter, for easy unit conversion 1 pmol/L = 0.42 pg/mL; conversely, 1 pg/mL = 2.4 pmol/L. Decreased levels of 1,25(OH)2 D can result from chronic kidney disease, heritable disorders, tumor-induced osteomalacia, HIV protease inhibitors, or severe vitamin D deficiency. Low 1,25(OH)2 D levels have been shown to present even in the early stages of kidney failure [11].

It is important to note that while vitamin D is essential for health, excessive intake can lead to toxicity, causing hypercalcemia and other adverse effects. Therefore, it is recommended to maintain adequate but not excessive levels of vitamin D through a balanced diet, sunlight exposure, and, if necessary, supplementation under the guidance of a healthcare professional [12].

4.4 Various Routes of Administration and Respective Preparations of Vitamin D

Vitamin D is available in various preparations, including different forms of the vitamin and various dosage forms. The two main forms of vitamin D are vitamin D_2 (ergocalciferol) and vitamin D_3 (cholecalciferol). Vitamin D_3 supplements are available over the counter.

Oral Drug Formulations Oral administration is the most common and convenient way to take vitamin D supplements. Vitamin D is available in various oral forms, including tablets, capsules, chewable tablets, and liquid formulations. Vitamin D is commonly available in oral supplements, including **tablets, capsules, and liquid formulations (syrup)**. These supplements may contain vitamin D_2, D_3, or a combination.

Fortified Foods Many foods, particularly dairy products, orange juice, and cereals, are fortified with vitamin D to enhance dietary intake. Fortified foods often contain vitamin D_3 form.

Injectable Vitamin D Injectable forms of vitamin D, usually vitamin D_2 or D_3, can be administered via **intramuscular** injection. This route may be preferred in situations where oral administration is **not** feasible or when a **rapid** increase in vitamin D levels is needed. **Intravenous** administration of vitamin D is **less** common and is typically reserved for specific medical situations, such as severe deficiency or impaired absorption. Intravenous administration allows for direct delivery of vitamin D into the bloodstream. Injectable forms of vitamin D, usually vitamin D_2 or D_3, may be prescribed for individuals with **severe** deficiencies who may have difficulty absorbing oral supplements.

Topical Preparations Some vitamin D analogs are formulated into topical creams or ointments for the treatment of certain skin conditions (topical creams or ointments in psoriasis). These preparations are typically used under medical supervision (prescription-only drug). In India, they may carry a red line on the container, designating it to be prescribed with prescription only.

Combination Preparations Vitamin D is sometimes included in multivitamin supplements or combined with other minerals like calcium in formulations designed to support bone health (e.g., multivitamin syrup for children) [13].

Note In cases of severe vitamin D deficiency or specific medical conditions, healthcare providers may prescribe **higher-strength** vitamin D supplements or analogs.

It is important for individuals to follow healthcare provider recommendations regarding the type and dosage of vitamin D supplements, as excessive intake can lead to **toxicity**. Additionally, adequate dietary intake and sensible sun exposure are essential components of maintaining optimal vitamin D levels in the body. Always consult with a healthcare professional before starting or altering vitamin D supplementation.

Vitamin D Dosing The Recommended Dietary Allowance (RDA) varies by age, and the values provided here are from the US Recommended Dietary Allowances (RDAs):

- Infants (up to 12 months): 400 IU (10 mcg) per day.
- Children (1–18 years): 600 IU (15 mcg) per day.
- Adults (19–70 years): 600 IU (15 mcg) per day.
- Adults (71 years and older): 800 IU (20 mcg) per day.
- Pregnant and breastfeeding women: 600 IU (15 mcg) per day.

In some cases, healthcare providers may recommend higher doses for individuals with specific health conditions or risk factors for vitamin D deficiency. This may include doses of 1000 IU to 2000 IU or more per day, often under medical supervision [14].

Many over-the-counter vitamin D supplements in India provide doses ranging from 400 IU to 2000 IU per tablet or capsule. Depending on the specific product, these supplements are often taken once daily or as prescribed by a healthcare professional. Vitamin D drops/oral liquid formulations are another common oral supplementation for infants and young children.

Common Concentration Drops are often available in concentrations like 400 IU per drop or 600 IU per drop. Drops can be directly administered or mixed with food or liquid.

It is important to note that individual vitamin D needs can vary, and testing of vitamin D levels (25-hydroxyvitamin D) may be done to determine if supplementation is necessary. Healthcare providers may adjust recommendations based on individual health status, underlying medical conditions, and other factors [13].

4.5 Drug Interactions of Vitamin D

Vitamin D can interact with certain medications, affecting their absorption, metabolism, or efficacy. It is important to be aware of potential drug interactions to ensure proper management of health conditions. Here are some common drug interactions involving vitamin D:

Anticonvulsants: Anticonvulsant drugs, such as phenytoin, carbamazepine, and phenobarbital, induce hepatic cytochrome P450 enzymes. These enzymes are responsible for the metabolism of vitamin D. Increased activity of these enzymes can accelerate the breakdown of both vitamin D_2 and D_3, leading to lower circulating levels of active vitamin D metabolites. Anticonvulsants may also interfere with the absorption of vitamin D in the intestines. For example, phenytoin has been associated with a decrease in vitamin D absorption, which can contribute to lower levels of circulating vitamin D. Some anticonvulsant drugs can directly interfere with the hydroxylation of vitamin D in the liver and kidneys. This process is essential for the conversion of vitamin D into its active form (calcitriol). Reduced hydroxylation can lead to lower levels of biologically active vitamin D. [15]

Glucocorticoids: The impact of glucocorticoids on vitamin D metabolism can vary among individuals, and factors such as the specific glucocorticoid, dosage, duration of use, and individual patient characteristics may influence the extent of the interaction. Glucocorticoids can reduce the absorption of calcium in the intestines, which is a process facilitated by active vitamin D. Glucocorticoids can suppress the activity of the renal enzyme responsible for converting inactive vitamin D (calcidiol) into its active form (calcitriol). Calcitriol is crucial for regulating calcium and phosphorus metabolism. Glucocorticoids may inhibit the activity of osteoblasts, which are bone-forming cells. Osteoblasts play a role in the production of a hormone called osteocalcin, which helps in vitamin D metabolism. Inhibition of osteoblast function can indirectly affect vitamin D-related processes. Glucocorticoids may increase renal excretion of calcium, leading to a loss of calcium from the body. This effect can contribute to disturbances in calcium balance and may be associated with bone loss. Glucocorticoids can influence the secretion of parathyroid hormone (PTH), which is involved in the regulation of calcium and phosphorus levels. The relationship between glucocorticoids and PTH is complex; alterations in PTH levels can indirectly impact vitamin D metabolism [16].

Weight Loss Medications:

- Orlistat works by inhibiting the activity of pancreatic lipase, an enzyme responsible for breaking down dietary fats into absorbable forms. This leads to a reduction in the absorption of dietary fats, and this effect can extend to fat-soluble vitamins, including vitamin D. Supplementation may be needed.

Cholestyramine and Colestipol:

- Bile acid sequestrants like cholestyramine and colestipol, used to lower cholesterol levels, can interfere with the absorption of fat-soluble vitamins, including vitamin D [17].

Calcium Channel Blockers:

- Some calcium channel blockers, such as verapamil, may increase vitamin D levels in the blood by reducing its breakdown. The mechanism is not clearly understood [18].

Thiazide Diuretics:

- Thiazide diuretics, commonly used to treat hypertension, can increase calcium levels in the blood and enhance the effects of vitamin D [19].

Mineral Oil:

- When taken as a laxative, mineral oil can reduce the absorption of fat-soluble vitamins, including vitamin D [20].

Certain Weight Loss Surgeries (considered equivalent to a maintained drug effect):

- Procedures such as gastric bypass surgery can impact the absorption of fat-soluble vitamins, necessitating careful monitoring and potential supplementation [21].

Certain Anti-HIV Medications:

- Some antiretroviral medications used in HIV treatment, such as certain protease inhibitors, may affect vitamin D metabolism [22].

It is crucial for individuals taking any of these medications or undergoing certain medical procedures to discuss vitamin D supplementation with their healthcare provider. Monitoring vitamin D levels and adjusting supplementation accordingly can help prevent deficiencies or excesses, which can have adverse health effects. Always consult with a healthcare professional for personalized advice based on your specific health conditions and medications.

4.6 Population Pharmacokinetics of Vitamin D

Population pharmacokinetics refers to the study of the variability in drug (or substance) disposition among individuals in a population. When applied to vitamin D, population pharmacokinetics helps understand how factors such as age, gender, genetics, health status, and other demographic variables can influence the absorption, distribution, metabolism, and elimination of vitamin D within a diverse group of people.

Here are some considerations (model covariates for precise vitamin D dosing) in the context of population pharmacokinetics of vitamin D:

Age and Life Stage: Vitamin D metabolism can vary with age. For example, infants and the elderly may have different absorption rates and requirements. Growth, bone development, and hormonal changes throughout life can influence vitamin D pharmacokinetics.

Genetic Variability: Genetic factors can impact the activity of enzymes involved in the metabolism of vitamin D. Polymorphisms in genes encoding these enzymes may result in interindividual differences in vitamin D metabolism. Variations in vitamin D receptor (VDR) genes may also influence the responsiveness of target tissues to vitamin D.

Body Composition: Vitamin D is a fat-soluble vitamin, and its distribution in the body can be affected by factors such as body fat percentage. Individuals with higher body fat may have a different distribution of vitamin D, potentially affecting its bioavailability.

Health Status: Health conditions, especially those affecting the liver and kidneys, can impact vitamin D metabolism. Chronic diseases, malabsorption disorders, and kidney dysfunction can alter vitamin D levels and requirements.

Geographic Location: Sunlight exposure, a major source of vitamin D synthesis in the skin, varies with geographic location. People living at higher latitudes with less sunlight may have a higher risk of vitamin D deficiency.

Ethnicity and Skin Pigmentation: Individuals with darker skin pigmentation may produce less vitamin D in the skin in response to sunlight exposure compared to those with lighter skin. Ethnicity and skin color can influence baseline vitamin D levels.

Dietary Habits: Dietary intake of vitamin D varies among populations based on cultural and dietary habits. Differences in dietary vitamin D intake can contribute to variations in vitamin D levels.

Medications and Interactions: Medications that affect liver and kidney function, or interact with vitamin D metabolism, can influence its pharmacokinetics. Understanding drug interactions is crucial in assessing vitamin D status in populations [23].

Excretion
The end products of vitamin D metabolism are excreted primarily in the bile. The excretion of vitamin D and its metabolites is a slow process, contributing to the **long half-life** of vitamin D in the body.

Half-Life The half-life of vitamin D in the blood is relatively long, estimated to be several **weeks** for 25(OH)D and several hours for calcitriol [24].

Research in population pharmacokinetics of vitamin D aims to identify patterns and sources of variability to provide better guidance for vitamin D supplementation and public health recommendations. Individualized approaches considering these factors may be essential for optimizing vitamin D status across diverse populations.

4.7 Adverse Effects and Toxicity of Vitamin D

While vitamin D is essential for various physiological functions, excessive intake can lead to vitamin D toxicity, also known as hypervitaminosis D. Adverse effects associated with vitamin D therapy or excessive supplementation include:

1. Hypercalcemia: The most significant risk of vitamin D toxicity is an elevated level of calcium in the blood (hypercalcemia). Excess vitamin D enhances the absorption of calcium from the intestines and may lead to the deposition of calcium in soft tissues, causing a range of symptoms such as nausea, vomiting, weakness, frequent urination, and in severe cases, kidney damage.
2. Kidney Damage: Prolonged hypercalcemia can contribute to the development of kidney stones and impair kidney function.
3. Gastrointestinal Issues: High doses of vitamin D may cause gastrointestinal symptoms, including nausea, vomiting, and diarrhea.
4. Dehydration: Excessive calcium excretion in the urine due to hypercalcemia may contribute to dehydration.
5. Cardiovascular Effects: Some studies have suggested a potential link between excessive vitamin D intake and an increased risk of cardiovascular events. However, more research is needed to establish a clear connection.

It is important to note that vitamin D toxicity is rare and is usually the result of extremely high doses of vitamin D supplements rather than dietary sources or sunlight exposure. The Recommended Dietary Allowance (RDA) for vitamin D varies by age, sex, and life stage, and it is important to follow these guidelines to avoid excessive intake.

Individuals with certain medical conditions, such as hyperparathyroidism, sarcoidosis, or certain granulomatous diseases, may be more susceptible to vitamin D toxicity. Additionally, those taking medications that affect vitamin D metabolism should be cautious about vitamin D supplementation and consult with their healthcare provider.

If you suspect vitamin D toxicity or experience symptoms related to excessive vitamin D intake, it is important to seek medical attention promptly. Monitoring vitamin D levels through blood tests and following healthcare provider recommendations for supplementation can help prevent adverse effects. Always consult with a healthcare professional before starting or making significant changes to any supplementation regimen [12].

Vitamin D During Pregnancy
Vitamin D deficiency has been linked to an increased risk of preterm birth. Adequate levels of vitamin D may contribute to a lower risk of preterm labor and delivery. Vitamin D is important for the overall health of the mother, contributing to her bone health and potentially reducing the risk of complications such as osteoporosis later in life. Vitamin D is also important for lactating mothers, as it affects the vitamin D content in breast milk. Breastfed infants may rely on maternal vitamin D status, so maintaining adequate levels is crucial for both the mother and the nursing infant. Emerging research suggests a potential role of vitamin D in fetal brain development. Adequate vitamin D levels during pregnancy may contribute to optimal cognitive development in the child.

It is important for pregnant women to discuss their vitamin D status with healthcare providers and, if necessary, take supplements to ensure they meet recommended levels. However, individual requirements may vary, and healthcare professionals can provide personalized advice based on factors such as diet, sunlight exposure, and overall health. It is crucial for pregnant women to receive prenatal care and follow the guidance of healthcare providers to support a healthy pregnancy and the well-being of both mother and baby.

In summary, a thorough understanding of the physiology of vitamin D is essential for maintaining overall health, preventing various diseases, and optimizing the functioning of multiple systems within the body.

References

1. Calcium I of M (US) C to RDRI for VD, Ross AC, Taylor CL, Yaktine AL, Valle HBD. Overview of Vitamin D. In: Dietary reference intakes for Calcium and Vitamin D [Internet]. National Academies Press (US); 2011 [cited 2024 Mar 1]. Available from: https://www.ncbi.nlm.nih.gov/books/NBK56061/
2. DeLuca HF. Vitamin D-dependent calcium transport. Soc Gen Physiol Ser. 1985;39:159–76.
3. Bhattarai HK, Shrestha S, Rokka K, Shakya R. Vitamin D, calcium, parathyroid hormone, and sex steroids in bone health and effects of aging. J Osteoporos. 2020;17(2020):9324505.
4. DeLuca HF. The metabolism and functions of vitamin D. Adv Exp Med Biol. 1986;196:361–75.
5. Cochran M, Coates PTH, Morris HA. The effect of calcitriol on fasting urine calcium loss and renal tubular reabsorption of calcium in patients with mild renal failure–actions of a permissive hormone. Clin Nephrol. 2005;64(2):98–102.
6. Carrillo-López N, Fernández-Martín JL, Cannata-Andía JB. The role of calcium, calcitriol and their receptors in parathyroid regulation. Nefrol Publicacion Soc Espanola Nefrol. 2009;29(2):103–8.

7. Aranow C. Vitamin D and the immune system. J Investig Med Off Publ Am Fed Clin Res. 2011;59(6):881–6.
8. Samuel S, Sitrin MD. Vitamin d's role in cell proliferation and differentiation. Nutr Rev. 2008 Oct;66(10 Suppl 2):S116–24.
9. Goltzman D, Mannstadt M, Marcocci C. Physiology of the calcium-parathyroid hormone-Vitamin D Axis. Front Horm Res. 2018;50:1–13.
10. Sizar O, Khare S, Goyal A, Givler A. Vitamin D deficiency. In: StatPearls [Internet]. Treasure Island (FL): StatPearls Publishing; 2024. [cited 2024 Mar 1]. Available from: http://www.ncbi.nlm.nih.gov/books/NBK532266/.
11. Kim MH, Lee J, Ha J, Jo K, Lim DJ, Lee JM, et al. Gender specific association of parathyroid hormone and vitamin D with metabolic syndrome in population with preserved renal function. Sci Rep. 2018;18(8):1149.
12. Asif A, Farooq N. Vitamin D toxicity. In: StatPearls [Internet]. Treasure Island (FL): StatPearls Publishing; 2024. [cited 2024 Mar 1]. Available from: http://www.ncbi.nlm.nih.gov/books/NBK557876/.
13. Lhamo Y, Chugh PK, Tripathi CD. Vitamin D supplements in the Indian market. Indian J Pharm Sci. 2016;78(1):41–7.
14. Ramasamy I, Vitamin D. Metabolism and guidelines for Vitamin D supplementation. Clin Biochem Rev. 2020;41(3):103–26.
15. Tomlinson B, Young RP, Ng MC, Anderson PJ, Kay R, Critchley JA. Selective liver enzyme induction by carbamazepine and phenytoin in Chinese epileptics. Eur J Clin Pharmacol. 1996;50(5):411–5.
16. Favus MJ, Kimberg DV, Millar GN, Gershon E. Effects of cortisone administration on the metabolism and localization of 25-hydroxycholecalciferol in the rat. J Clin Invest. 1973 Jun;52(6):1328–35.
17. Bansal AB, Al Khalili Y. Orlistat. In: StatPearls [Internet]. Treasure Island (FL): StatPearls Publishing; 2024. [cited 2024 Mar 1]. Available from: http://www.ncbi.nlm.nih.gov/books/NBK542202/.
18. Fox J, Della-Santina CP. Oral verapamil and calcium and vitamin D metabolism in rats: effect of dietary calcium. Am J Phys. 1989 Nov;257(5 Pt 1):E632–8.
19. Griebeler ML, Kearns AE, Ryu E, Thapa P, Hathcock MA, Melton LJ, et al. Thiazide-associated hypercalcemia: incidence and association with primary hyperparathyroidism over two decades. J Clin Endocrinol Metab. 2016;101(3):1166–73.
20. Gal-Ezer S, Shaoul R. The safety of mineral oil in the treatment of constipation–a lesson from prolonged overdose. Clin Pediatr (Phila). 2006;45(9):856–8.
21. Toninello P, Montanari A, Bassetto F, Vindigni V, Paoli A. Nutritional support for bariatric surgery patients: the skin beyond the fat. Nutrients. 2021;13(5):1565.
22. Cozzolino M, Vidal M, Arcidiacono MV, Tebas P, Yarasheski KE, Dusso AS. HIV-protease inhibitors impair vitamin D bioactivation to 1,25-dihydroxyvitamin D. AIDS Lond Engl. 2003;17(4):513–20.
23. Whiting DR, Guariguata L, Weil C, Shaw J. IDF diabetes atlas: global estimates of the prevalence of diabetes for 2011 and 2030. Diabetes Res Clin Pract. 2011 Dec;94(3):311–21.
24. Bikle DD. Vitamin D: production, metabolism and mechanisms of action. In: Feingold KR, Anawalt B, Blackman MR, Boyce A, Chrousos G, Corpas E, et al., editors. Endotext [Internet]. South Dartmouth (MA): MDText.com, Inc.; 2000. [cited 2024 Mar 1]. Available from: http://www.ncbi.nlm.nih.gov/books/NBK278935/.

Chapter 5
Vitamin D in Women's Health

Swati Swati and Jyotsna Rani

Abstract Although vitamin D is related historically to skeletal functions, recently its role in multiple functions of body have been discovered. It has a positive role in follicular development and selection for ovulation. Its immune-mediated actions have a role in implantation and endometriosis. Low levels of vitamin D have been linked to pathogenesis of polycystic ovarian syndrome (PCOS). Deficiency during pregnancy leads to poorer fetomaternal outcomes, e.g. preeclampsia, prematurity, gestational diabetes, anaemia, etc. Not just reproductive aged women have been linked to deficiency of vitamin D; in adolescent girls it has been correlated to premenstrual syndrome, while in postmenopausal women its supplementation has positive effects on the skeletal system, urogenital system, and neurological system as well. Right from birth to death vitamin D plays several roles in female body systems.

Keywords Vitamin D · Women · Pregnancy · Effects · Supplementation

Learning Objectives
- To understand the role of vitamin D in various diseases specific to female.
- To raise awareness about established and proposed benefits of vitamin D supplementation in girls and women.
- To be aware of harm(s) due to excessive levels of vitamin D in women.

S. Swati (✉)
Department of Obstetrics & Gynecology, AIIMS, Deoghar, Jharkhand, India

J. Rani
Department of Obstetrics & Gynecology, Dr. B R Ambedkar State Institute of Medical Sciences, Mohali, Punjab, India

V. Kumar, S. Patel (eds.), *Vitamin D: A Comprehensive Guide for Medical Professionals*, https://doi.org/10.1007/978-981-96-6491-7_5

5.1 Introduction

There are 13 vitamins: A, C, D, E, K, and the B series, which are B_1 (thiamine), B_2 (riboflavin), B_3 (niacin), B_5 (pantothenic acid), B_6 (pyridoxine), B_7 (biotin), B_9 (folic acid), and B_{12} (cobalamin). These are considered essential for the human body [1]. Cholecalciferol or Vitamin D_3 is the natural form of vitamin D, and it is produced in the skin from 7-dehydrocholesterol. It is synthesized in the skin, which is the most important source of vitamin D, and this is dependent upon the intensity of the ultraviolet irradiation which varies with season and latitude. The other source is a diet rich in dairy product, fish oil, or fortified food products.

Vitamin D_3 is not biologically active itself, and it is transported to the liver via blood by vitamin D binding protein (DBP; which binds vitamin D and its metabolites in serum). Here it is hydroxylated at C-25 to generate 25-hydroxyvitamin D_3 [25(OH)D3] which is the main circulating form of vitamin D [2]. This hydroxylation is done by several cytochrome P-450 enzymes (CYPs) including CYP2R1, CYP27A1, and CYP2D25. Serum concentration of this one is the most reliable biomarker of vitamin D status in the body [3–5]. After that 25(OH)D3 is taken to the proximal renal tubule where it is hydroxylated again at the position of carbon 1 of the A ring by renal 25(OH)D 1α hydroxylase (mitochondrial CYP27B1) resulting in the formation of 1,25(OH)2D3 which is the main functional and hormonally active form of vitamin D. It is responsible for most of the biologic actions of vitamin D [6, 7]. Human and animal studies have shown that with aging there is decreased capacity of the kidney to convert 25(OH)D3 to 1,25(OH)2D3 and also an increase in the catabolism of 1,25(OH)2D3 by CYP24A [8, 9].

Although optimum level of vitamin D has been a topic of controversy, 25(OH)D levels <20 ng/mL are considered as Vitamin D deficiency by several societies and groups of experts. Insufficiency is accepted as levels between 25(OH)D 20–29 ng/mL and $\geq$ 30 ng/mL is usually taken as sufficient levels and for certain other situations some authors propose reaching >40 ng/mL levels [10–12] (Fig. 5.1).

Vitamin D is related historically to skeletal functions. However, in the last couple of years there has been colossal recognition of vitamin D as its role in pathway or management of almost every disease system. This chapter will review the role of vitamin D in fertility, pregnancy, lactation, and overall women's health in relation to its deficiency or excess in body.

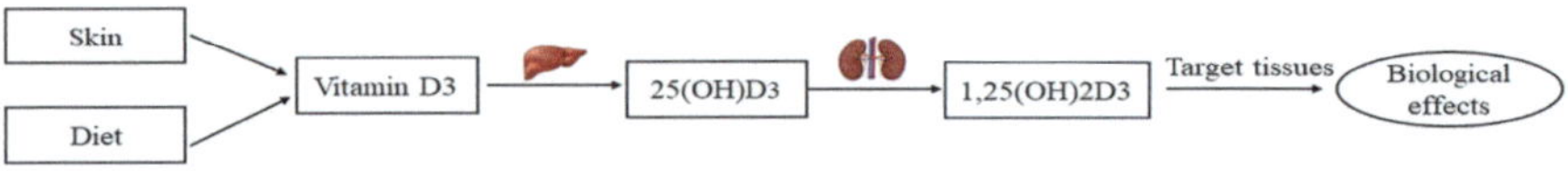

Fig. 5.1 Metabolism and effects of vitamin D

5.2 Role of Vitamin D in Fertility

Reproductive systems of both sexes have high distribution of vitamin D receptors (VDRs) and the enzymes involved in its activation and metabolism [13–15]. Aromatase enzymes are stimulated and activated by vitamin D in ovary and thus it influences estrogen and progesterone levels. It has a role in transcription of dehydroepiandrosterone sulfotransferase [13, 16]. With beginning of each menstrual cycle, a cohort of oocytes are recruited and one matures to be ovulated. Anti-Mullerian Hormone (AMH) has a role in this process, thereby controlling ovarian reserve. Transition from primordial follicles to primary follicles is inhibited by AMH. The regressive effect on the granulose cells is mediated by the AMH receptor (AMHR-II), so higher levels of AMH suppresses the ovarian maturation. Vitamin D has been shown to decrease the mRNA AMH expression, inhibit the AMHR-II expression, and increase the FSH receptor gene expression in vitro [17, 18]. A strong negative correlation has been seen between 25OHD levels (particularly with 25(OH)D > 30 ng/mL levels) in blood and follicular fluid and the AMH levels in a study on women with infertility [19]. Thus, vitamin D has a positive role in follicular development and selection for ovulation.

In endometrial cells, vitamin D favours implantation by regulating the expression of the Homeobox-A10 protein (HOXA10), that is relevant for the development of the uterus and endometrium [13, 16]. Vitamin D is known to regulate immune response. It decreases inflammatory response induced by IL-1β and TNF-α and production of TGF-β and IL-8 is increased. These immune mediated actions play a role in implantation and endometriosis [12, 20].

An inverse relationship was found between levels of 25(OH)D and severity of endometriosis [20].

Success rate for IVF and ICSI have been found to be less with insufficient or deficient levels of Vitamin D which was better for those having sufficient levels of the same. These were shown in studies using both self- and donor oocytes [21, 22].

Low levels of vitamin D have been linked to pathogenesis of polycystic ovarian syndrome (PCOS) including insulin resistance, metabolic syndrome, hyperandrogenemia, metabolic and endocrine disorders. Lower levels are related to oxidative stress and pro-inflammatory milieu in women with PCOS [23]. Theories also proposed are vitamin D receptor variants, and vitamin D binding protein gene variants which have not been proven by studies [24, 25].

5.3 Pregnancy and Vitamin D

Evidence has shown lower levels of vitamin D to be associated with poor pregnancy outcomes [26]. Preeclampsia and prematurity are the leading contributors to maternal and neonatal mortalities, respectively. Do they have a relation with lower maternal serum vitamin D levels? The answer is yes, as deficient state of this essential vitamin has also been correlated with several other complications in pregnancies as discussed below (Fig. 5.2).

Preterm Birth: Studies have correlated maternal vitamin D deficiency and insufficiency to increase the likelihood of preterm births by the tune of 1.5–2 times. [27–29]

Preeclampsia: Low levels of vitamin D have been implicated as an independent risk factor for preeclampsia because it acts as an endocrine modulator of renin angiotensin aldosterone system (RAAS) by having a suppressant effect [30]. In addition it suppresses smooth muscle proliferation, modulates endothelial dysfunction, and thus delays impairing of vascular health. [31]

Gestational Diabetes Mellitus (GDM): It has also been shown to be more prevalent in vitamin D deficient pregnant women. Systematic review and dose-response meta-analysis have shown that risk of GDM decreased by 2% with each 4 ng/ml increase in serum 25(OH)D levels [32]. Mechanism proposed are vitamin D actions in beta cell and insulin resistance [33].

Small for Gestation (SGA): A meta-analysis of observational studies has shown increased risk of SGA babies in vitamin D deficient women by 1.4 times [34]. 2000 IU per day supplementation of vitamin D have shown significant weight gain in foetuses of women having pregnancies with type 1 (early) foetal growth restriction before 32 weeks of gestation [35].

Labour and Postpartum: Muscle cells (smooth/skeletal) contain vitamin D receptors, hence in the state of its deficiency/insufficiency pelvic floor dysfunction affects smooth progress of labour and increases chances of caesarean section and also postpartum incontinence [36, 37].

Anaemia in Pregnancy: Anaemia and vitamin D deficiency often coexist during pregnancy and lactation. Hepcidin is a hormone which regulates systemic iron homeostasis, and storage and distribution of iron in tissues. Vitamin D has been

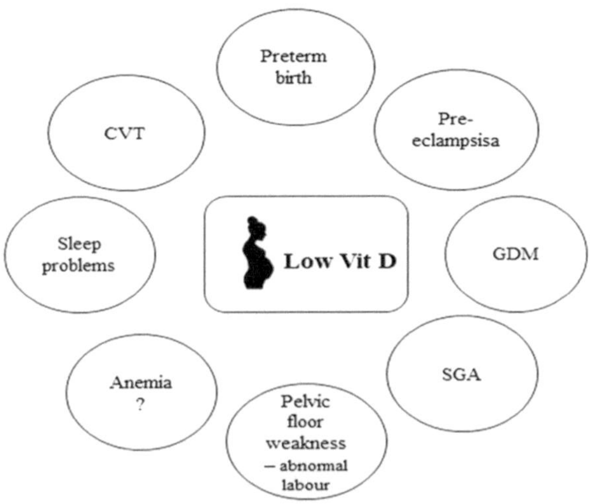

Fig. 5.2 Effects of vitamin D deficiency during pregnancy

found to regulate hepcidin production and hepcidin–ferroportin axis in macrophages [38]. Vitamin D deficiency might be considered as a risk factor for anaemia but adequate evidences are still awaited.

Sleep in Pregnancy: Pregnant women need adequate sleep. Several women in antenatal OPD complain of having inadequate sleep in terms of quantity and quality. A study in African American and Hispanic pregnant women between 24- and 32-week gestation having vitamin D deficiency were assessed by the Pittsburgh Sleep Quality Index. Vitamin D deficiency was found to have an association with poor sleep quality, and hence supplementation with vitamin D may be a simple solution to this [39].

Cerebral Venous Sinus Thrombosis (CVT): Pregnancy is a hypercoagulable state, yet all women do not develop CVT. Apart from inherited thrombophilia, we have little knowledge about markers for predicting this deadly complication in pregnant women. A prospective study found low levels of vitamin D and high C-reactive protein/albumin ratio (CAR) in pregnant women with CVT, so they have proposed vitamin D deficiency and high CAR as markers for predicting CVT in pregnancy [40].

5.4 Lactation

Mothers with inadequate vitamin D levels in blood will secrete breast milk having low levels of the same, thereby increasing the risk of vitamin D deficiency in newborn and infants. Extra supplementation, i.e. 2000 IU per day of the same, is needed during lactation to maintain the health of both mother and baby [41].

We have seen the effects of vitamin D deficiency/insufficiency in pregnancy and lactation. A systematic review and meta-analysis estimated the pooled prevalence of vitamin D deficiency in 4088 healthy pregnant women in India, and the combined estimate of deficiency was 32.35% [95% CI, (12.58–117.48)]. Although the study had high heterogeneity (tau2 = 0.39, I2 = 100%) and high risk of bias, this study emphasized upon the need of screening in antenatal women based on vitamin D deficiency in almost one-third healthy pregnant women [42].

5.5 Vitamin D in Postmenopausal Women

Both ageing and withdrawal of estrogen lead to decreased muscle and bone mass after menopause. Studies have shown the beneficial role of coadministration of vitamin D (800–2000 IU/day) and calcium (1000–1200 mg/day) for 3 to 5 years to prevent fracture in postmenopausal women after repleting vitamin D status with higher weekly or daily doses. This dose of both has no bad effect on cardiovascular health and nephrolithiasis [43].

In addition, vitamin D deficiency has been shown to be associated with diabetes, infections, and cardiovascular and neurodegenerative diseases in the general population [44]. While prescribing vitamin D to menopausal women, they should be counselled about these added advantages and also about benefits in genitourinary symptoms of menopause [45].

5.6 Prevention of Breast Cancer

1,25-dihydroxyvitamin D3 also known as calcitriol which is the hormonally active metabolite of vitamin D has been found to inhibit growth and induce differentiation of several malignant cells including breast cancer cells. This action is by stimulation of apoptosis and induction of cell cycle arrest. It also reduces biological activity of prostaglandins and decreases expression of aromatase in breast cancer cells and breast adipose tissue. Additionally, calcitriol downregulates the expression of alpha receptors of estrogen in breast cancer cells. These findings do suggest a preventive role of calcitriol for breast malignancy [46].

5.7 Genitourinary and Sexual Health

A narrative review has emphasized the role of vitamin D in urogenital health as vitamin D receptors are present and expressed in the superficial layers of urogenital organs. The review found evidence in clinical studies showing positive effects in genitourinary syndrome of menopause by having effects on vaginal pH, dryness, sexual function, libido, and reduction of urinary tract infections [45].

Vitamin D is an immune modulator, and some authors propose its role in the defence mechanism of vaginal mucosa, thereby influencing acquisition and progression of several reproductive tract infections including HIV. Those having HIV infection whose disease progressed fast to AIDS had lower serum levels of vitamin D [47].

5.8 Premenstrual Syndrome (PMS)-Related Mood Disorders in Adolescents

PMS is a common disorder impacting lifestyle and performances of adolescent girls and young women. Apart from several modalities of treatment, vitamin D supplementation is also an effective measure to treat PMS. Adolescent girls having severe mood disorder with PMS who were found deficient in vitamin D were found to have much improvement in mood score when given vitamin D (200,000 IU stat followed by 25,000 IU fortnightly) for 4 months versus placebo in a randomized controlled trial [48].

5.9 Hypervitaminosis D

While Vitamin D deficiency/insufficiency is a public health concern, physicians, gynaecologists, and obstetricians must be aware of effects from excess levels of vitamin D in blood. Hypervitaminosis D leads to cell damage, so overtreatment with vitamin D, especially in combination of excess calcium, have been implicated to cause nephrolithiasis, renal malfunction, dehydration, cancers, and possibly cardiovascular diseases [49].

Optimum bone mass is achieved with optimum levels to vitamin D, meaning both its deficiency and excess leads to bone loss.

To treat hypervitaminosis D cotreatment with calcitonin 4 U/Kg intramuscularly (repeated 12 hourly for 48 hours) and intravenous bisphosphonates (zolendronate 4 mg over 15 minutes) should be used [50].

5.10 Conclusion

Vitamin D plays a potential role in women's health right from adolescence to postmenopause through the reproductive years. Several subspecialities of Gynaecology and Obstetrics should focus upon the preventive and curative role of this wonder vitamin as it has been found in pathogenesis of most of the diseases specific to females. At the same time one should stay alert so as to not overtreat and cause hypervitaminosis D.

References

1. Duckworth S, Mistry HD, Chappell LC. Vitamin supplementation in pregnancy. Obstetrician Gynaecolt. 2012;14:175–8.
2. Christakos S, Dhawan P, Verstuyf A, Verlinden L, Carmeliet G. Vitamin D: metabolism, molecular mechanism of action, and pleiotropic effects. Physiol Rev. 2016;96(1):365–408. https://doi.org/10.1152/physrev.00014.2015. PMID: 26681795; PMCID: PMC4839493
3. Bikle D, Adams J, Christakos S. Vitamin D: production, metabolism and clinical requirements. In: Rosen C, editor. Primer Metab Bone Dis. Hoboken, NJ: Wiley; 2013. p. 235–45.
4. Heaney RP, Horst RL, Cullen DM, Armas LA. Vitamin D3 distribution and status in the body. J Am Coll Nutr. 2009;28:252–6.

5. Holick MF. Vitamin D deficiency. N Engl J Med. 2007;357:266–81.
6. Jones G, Prosser DE, Kaufmann M. Cytochrome P450-mediated metabolism of vitamin D. J Lipid Res. 2014;55:13–31.
7. Jones G, Prosser DG. The activating enzymes of vitamin D. Vitamin. 2011;D:23–42.
8. Armbrecht HJ, Zenser TV, Davis BB. Effect of age on the conversion of 25-hydroxyvitamin D3 to 1,25-dihydroxyvitamin D3 by kidney of rat. J Clin Invest. 1980;66:1118–23.
9. Matkovits T, Christakos S. Variable in vivo regulation of rat vitamin D-dependent genes (osteopontin, Ca,Mg-adenosine triphosphatase, and 25-hydroxyvitamin D3 24-hydroxylase): implications for differing mechanisms of regulation and involvement of multiple factors. Endocrinology. 1995;136:3971–82.
10. Rosen CJ, Abrams SA, Aloia JF, Brannon PM, Clinton SK, Durazo-Arvizu RA, Gallagher JC, Gallo RL, Jones G, Kovacs CS, et al. IOM committee members respond to Endocrine Society vitamin D guideline. J Clin Endocrinol Metab. 2012;97:1146–52. https://doi.org/10.1210/jc.2011-2218.
11. Holick MF, Binkley NC, Bischoff-Ferrari HA, Gordon CM, Hanley DA, Heaney RP, Murad MH, Weaver CM. Endocrine Society. Evaluation, treatment, and prevention of vitamin D deficiency: an Endocrine Society clinical practice guideline. J Clin Endocrinol Metab. 2011;96:1911–30. https://doi.org/10.1210/jc.2011-0385.
12. Mansur JL, Oliveri B, Giacoia E, Fusaro D, Costanzo PR. Vitamin D: before, during and after pregnancy: effect on neonates and children. Nutrients. 2022;14(9):1900. https://doi.org/10.3390/nu14091900. PMID: 35565867; PMCID: PMC9105305
13. Lerchbaum E, Obermayer-Pietsch B. Vitamin D and fertility: a systematic review. Eur J Endocrinol. 2012;166:765–78. https://doi.org/10.1530/EJE-11-0984.
14. Blomberg JM, Nielsen JE, Jogensen A, Rajperts-De ME, Krinstensen DM, Jorgensen N, Skakkebaek NE, Juul A, Leffers H. Vitamin D receptor and vitamin D metabolizing enzymes are expressed in the human male reproductive tract. Hum Reprod. 2010;25:1303–11. https://doi.org/10.1093/humrep/deq024.
15. Aquila S, Guido C, Middea E, Perrotta I, Bruno S, Pellegrino M, Andó S. Human male gamete endocrinology: 1 alpha, 25-dihydroxyvitamin D3 (1,25(OH)2D3) regulates different aspects of human sperm biology and metabolism. Reprod Biol Endocrinol. 2009;7:140–52. https://doi.org/10.1186/1477-7827-7-140.
16. Voulgaris N, Papanastasiou L, Piaditis G, Angelousi A, Kaltsas G, Mastorakos G, Kassi E. Vitamin D and aspects of female fertility. Hormones. 2017;16:5–21.
17. Arslan S, Akdevelioglu Y. The relationship between female reproductive functions and vitamin D. J Am Coll Nutr. 2018;37:546–51. https://doi.org/10.1080/07315724.2018.1431160.
18. Merhi Z, Doswell A, Krebs K, Cipolla M. Vitamin D alters genes involved in follicular development and steroidogenesis in human cumulus granulosa cells. J Clin Endocrinol Metab. 2014;99:1137–45. https://doi.org/10.1210/jc.2013-416.
19. Bednarska-Czerwinska A, Olszak-Wasik K, Olejek A, Czerwinski M, Tukiendorf AA. Vitamin D and anti-Mullerian hormone levels in infertility treatment: the change-point problem. Nutrients. 2019;11:1053. https://doi.org/10.3390/nu11051053.
20. Harris HR, Chavarro JE, Malspeis S, Willett WC, Missmer SA. Dairy-food, calcium, magnesium, and vitamin D intake and endometriosis: a prospective cohort study. Am J Epidemiol. 2013;177:420–30. https://doi.org/10.1093/aje/kws247.
21. Paffoni A, Ferrari S, Vigano P, Pagliardini L, Papaleo E, Candiani M, Tirelli A, Fedele L, Somigliana E. Vitamin D deficiency and infertility: insights from in vitro fertilization cycles. J Clin Endocrinol Metab. 2014;99:2372–6. https://doi.org/10.1210/jc.2014-1802.
22. Rudick BJ, Ingles SA, Chung K, Stanczyk FZ, Paulson RJ, Bendikson KA. Influence of vitamin D levels on in vitro fertilization outcomes in donor-recipient cycles. Fertil Steril. 2014;101:447–52. https://doi.org/10.1016/j.fertnstert.2013.10.008.
23. Sparic R, Andjic M, Vergara D, Morciano A, D'Oria O, Baldini GM, Malvasi A, Tinelli A. PCOS and vitamin D: a clinical appraisal. Arch Gynecol Obstet. 2023; https://doi.org/10.1007/s00404-023-07227-x. Epub ahead of print

24. Santos BR, Lecke SB, Spritzer PM. Genetic variant in vitamin D-binding protein is associated with metabolic syndrome and lower 25-hydroxyvitamin D levels in polycystic ovary syndrome: a cross-sectional study. PLoS One. 2017;12(3):e0173695. https://doi.org/10.1371/journal.pone.0173695. PMID: 28278285; PMCID: PMC5344505
25. Haldar D, Agrawal N, Patel S, Kambale PR, Arora K, Sharma A, Tripathi M, Batra A, Kabi BC. Association of VDBP and CYP2R1 gene polymorphisms with vitamin D status in women with polycystic ovarian syndrome: a north Indian study. Eur J Nutr. 2018;57(2):703–11. https://doi.org/10.1007/s00394-016-1357-z. Epub 2016 Dec 23
26. Wei SQ. Vitamin D and pregnancy outcomes. Curr Opin Obstet Gynecol. 2014;26(6):438–47. https://doi.org/10.1097/GCO.0000000000000117.
27. Eggel-Hort B, Maisonneuve E, Gonzalez RE, Baud D. Vitamine D et grossesse [Vitamin D and pregnancy]. Rev Med Suisse. 2021;17:1774–8.
28. Miliku K, Vinkhuyzen A, Blanken LM, McGrath JJ, Eyles DW, Burne TH, Hofman A, Tiemeier H, Steegers EA, Gaillard R, et al. Maternal vitamin D concentrations during pregnancy, fetal growth patterns, and risks of adverse birth outcomes. Am J Clin Nutr. 2016;103:1514–22. https://doi.org/10.3945/ajcn.115.123752.
29. Gernand AD, Simhan HN, Baca KM, Caritis S, Bodnar LM. Vitamin D, pre-eclampsia, and preterm birth among pregnancies at high risk for pre-eclampsia: an analysis of data from a low-dose aspirin trial. BJOG. 2017;124:1874–82. https://doi.org/10.1111/1471-0528.14372.
30. Li YC, Kong J, Wei M, Chen ZF, Liu SQ, Cao LP. 1,25-Dihydroxyvitamin D3 is a negative endocrine regulator of the reninangiotensin system. J Clin Invest. 2002;110:229–38. https://doi.org/10.1172/JCI0215219.
31. Cardús A, Parisi E, Gallego C, Aldea M, Fernández E, Valdivielso JM. 1,25-Dihydroxyvitamin D3 stimulates vascular smooth muscle cell proliferation through a VEGF-mediated pathway. Kidney Int. 2006;69:1377–84. https://doi.org/10.1038/sj.ki.5000304.
32. Sadeghian M, Asadi M, Rahmani S, Akhavan ZM, Sadeghi O, Hosseini SA, Zare JA. Circulating vitamin D and the risk of gestational diabetes: a systematic review and dose-response meta-analysis. Endocrine. 2020;70:36–47. https://doi.org/10.1007/s12020-020-02360-y.
33. Alvarez JA, Ashraf A. Role of vitamin D in insulin secretion and insulin sensitivity for glucose homeostasis. Int J Endocrinol. 2010;2010:351385. https://doi.org/10.1155/2010/351385.
34. Hu Z, Tang L, Xu HL. Maternal vitamin D deficiency and the risk of small for gestational age: a meta-analysis. Iran J Public Health. 2018;47(12):1785–95. PMID: 30788292; PMCID: PMC6379614
35. Jakubiec-Wisniewska K, Huras H, Kolak M. Effect of vitamin D supplementation on the fetal growth rate in pregnancy complicated by fetal growth restriction. Children (Basel). 2022;9(4):549. https://doi.org/10.3390/children9040549. PMID: 35455593; PMCID: PMC9025187
36. Hay-Smith J, Mørkved S, Fairbrother KA, Herbison GP. Pelvic floor muscle training for prevention and treatment of urinary and faecal incontinence in antenatal and postnatal women. Cochrane Database Syst Rev. 2008;8:CD007471.
37. Salvesen KA, Mørkved S. Randomised controlled trial of pelvic floor muscle training during pregnancy. BMJ. 2004;329:378–80. https://doi.org/10.1136/bmj.38163.724306.3A.
38. Adams JS, Hewison M. Unexpected actions of vitamin D: new perspectives on the regulation of innate and adaptive immunity. Nat Clin Pract Endocrinol Metab. 2008;4:80–90.
39. Woo J, Penckofer S, Giurgescu C, Yeatts PE. Vitamin D deficiency and sleep quality in minority pregnant women. MCN Am J Matern Child Nurs. 2020;45(3):155–60. https://doi.org/10.1097/NMC.0000000000000610.
40. Yevgi R, Bilge N, Simsek F, Eren A, Cimilli Senocak GN. Vitamin D levels and C-reactive protein/albumin ratio in pregnant women with cerebral venous sinus thrombosis. J Thromb Thrombolysis. 2022;53(2):532–9. https://doi.org/10.1007/s11239-021-02541-0. Epub 2021 Aug 3

41. Thandrayen K, Pettifor JM. Maternal vitamin D status: implications for the development of infantile nutritional rickets. Endocrinol Metab Clin North Am. 2010;39(2):303–20, table of contents. https://doi.org/10.1016/j.ecl.2010.02.006.
42. Jeyakumar A, Shinde V, Ravindran R. Pooled estimate of vitamin D deficiency among pregnant women in India: a systematic review and meta-analysis. J Health Popul Nutr. 2021;40(1):28. https://doi.org/10.1186/s41043-021-00253-y. PMID: 34187594; PMCID: PMC8243731
43. Anagnostis P, Livadas S, Goulis DG, Bretz S, Ceausu I, Durmusoglu F, Erkkola R, Fistonic I, Gambacciani M, Geukes M, Hamoda H, Hartley C, Hirschberg AL, Meczekalski B, Mendoza N, Mueck A, Smetnik A, Stute P, van Trotsenburg M, Rees M, Lambrinoudaki I. EMAS position statement: vitamin D and menopausal health. Maturitas. 2023;169:2–9. https://doi.org/10.1016/j.maturitas.2022.12.006. Epub 2022 Dec 21. Anagnostis P
44. Athyros VG, Adamidou F, Florentin M, Karagiannis A. Vitamin D and cardiovascular disease: a novel agent for reducing cardiovascular risk? Curr Vasc Pharmacol. 2010;8:720–30.
45. Hassanein MM, Huri HZ, Abduelkarem AR, Baig K. Therapeutic effects of vitamin D on vaginal, sexual, and urological functions in postmenopausal women. Nutrients. 2023;15(17):3804. https://doi.org/10.3390/nu15173804. PMID: 37686835; PMCID: PMC10490181
46. Krishnan AV, Swami S, Feldman D. Vitamin D and breast cancer: inhibition of estrogen synthesis and signaling. J Steroid Biochem Mol Biol. 2010;121(1–2):343–8. https://doi.org/10.1016/j.jsbmb.2010.02.009. Epub 2010 Feb 13
47. Adeyemi OM, Agniel D, French AL, Tien PC, Weber K, Glesby MJ, Villacres MC, Sharma A, Merenstein D, Golub ET, et al. Vitamin D deficiency in HIV-infected and HIV-uninfected women in the United States. J Acquir Immune Defic Syndr. 2011;57:197–204. https://doi.org/10.1097/QAI.0b013e31821ae418.
48. Tartagni M, Cicinelli MV, Tartagni MV, Alrasheed H, Matteo M, Baldini D, De Salvia M, Loverro G, Montagnani M. Vitamin D supplementation for premenstrual syndrome-related mood disorders in adolescents with severe Hypovitaminosis D. J Pediatr Adolesc Gynecol. 2016;29(4):357–61. https://doi.org/10.1016/j.jpag.2015.12.006. Epub 2015 Dec 24
49. Brouwer-Brolsma EM, Bischoff-Ferrari HA, Bouillon R, Feskens EJ, Gallagher CJ, Hypponen E, et al. Vitamin D: do we get enough? Osteoporos Int. 2013;24:1567–77.
50. Asif A, Farooq N. Vitamin D toxicity. In: StatPearls [Internet]. Treasure Island (FL): StatPearls Publishing; 2023. PMID: 32491799.

Chapter 6
Vitamin D and Child Health-Role in Normal Growth and Development

Manoj Kumar

Abstract Vitamin D is critical for the healthy growth and development of children, influencing several key physiological processes. It enhances the absorption of calcium and phosphorus, which are essential for bone mineralization and skeletal development, thereby preventing conditions like rickets. Adequate vitamin D levels support proper bone density and reduce the risk of fractures. Additionally, vitamin D plays a role in modulating the immune system, helping to protect against infections and potentially lowering the risk of autoimmune diseases. It also contributes to muscle function, supporting physical development and motor skills. Children typically acquire vitamin D through sunlight exposure, diet, and supplements. Ensuring optimal vitamin D levels is crucial for promoting overall health, preventing deficiencies, and supporting long-term well-being.

Keywords Vitamin D · Child health · Vitamin D deficiency · Growth and development

Learning Objectives

- To comprehend the significance of vitamin D in child growth and development.
- To understand the significance of vitamin D in various acute and chronic pediatric diseases.
- To raise knowledge about the established and suggested benefits of vitamin D supplementation in children.
- To be aware of the risks associated with an overabundance of vitamin D in a child.

M. Kumar (✉)
Department of Pediatrics, All India Institute of Medical Sciences, Patna, Bihar, India

V. Kumar, S. Patel (eds.), *Vitamin D: A Comprehensive Guide for Medical Professionals*, https://doi.org/10.1007/978-981-96-6491-7_6

6.1 Introduction

It is truly quoted by Buckminster Fuller that "Every child is born a genius." So, it is the responsibility of the physician as well as the parents to take care of the budding future of the nation. This can truly be done by proper monitoring of growth, development, and behavior attainment. The timely monitoring of child progress, identification of developmental delay or anomalies, and provision of needful assistance play an important role. The terms "growth" and "development" represent two distinct aspects of change dynamics. The growth represents the quantitative domain, whereas development represents the qualitative domain.

Growth: It refers to the increase in intracellular material with net increase in the size or mass of tissues with hypertrophy or cell size increase contributes to a lesser amount in this process.

Development: The acquisition of a wide range of day-to-day abilities necessary for an individual's optimal functioning and execution of various skills is an important aspect of development. So this functional maturity of skills is referred to as development which signifies the nervous system maturation and myelination process [1].

Only lately has it been recognized that the prenatal dietary environment influences developmental plasticity, which may contribute to the later development of chronic disorders such as obesity, diabetes, allergic condition, neurodevelopmental and intellectual disabilities, and other conditions.

The growth and development is required for all aspects of human existence, including physical, emotional, cognitive, and sexual development, from conception to birth, neonatal adaption, infancy, childhood, puberty, and young adulthood.

The idea is that a shortage of nutrients during critical stages of development increases the risk of chronic disorders in adults as per epidemiologic observations of cohorts. Nutrition in the form of minerals and vitamins is critical for growth in early life. A proper and diversified diet after weaning is required for healthy growth of newborns as macronutrients and micronutrients. Micronutrients are introduced in little amounts, but they are necessary for optimal growth in the pediatric stage, particularly zinc, iron, vitamin D, folic acid, and other minerals [2].

Globally, almost 1 billion people are vitamin D deficient [3]. The Ministry of Health and Family Welfare, Government of India, performed the Comprehensive National Nutrition Survey in 2016–18, comprising about 35,000 children aged 1 to 19 years from all over India. In this survey, vitamin D deficiency was found to be 14% among children aged 1–4 years, 18% among school-age children (5–9 years), and 24% among teenagers (10–19 years), which was lower than the proportions reported in hospital and community-based studies from India. The gender gap was most pronounced in adolescents, with 34% of girls deficient compared to 14% of boys [4].

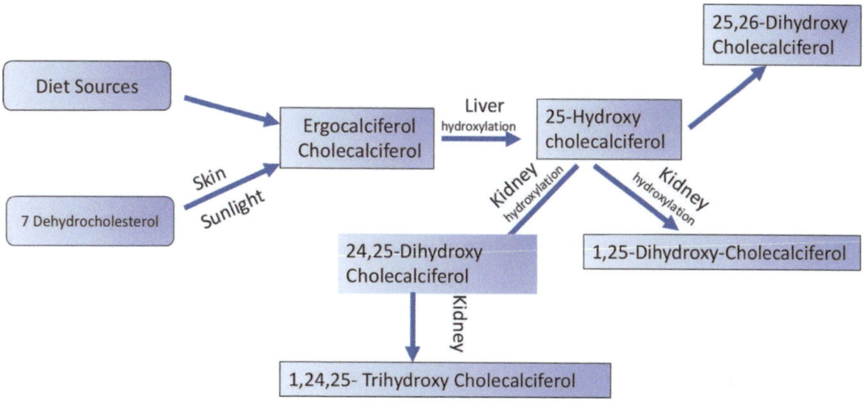

Fig. 6.1 Flow chart illustrating the metabolic pathway of vitamin D synthesis and conversion. It begins with "Diet Sources" and "7-Dehydrocholesterol" leading to "Ergocalciferol" and "Cholecalciferol" through skin and sunlight exposure. These compounds undergo liver hydroxylation to form "25-Hydroxycholecalciferol." Further kidney hydroxylation results in "1,25-Dihydroxycholecalciferol" and "24,25-Dihydroxycholecalciferol." "25,26-Dihydroxycholecalciferol" is also formed. Arrows indicate the direction of conversion, with labels for each hydroxylation step

6.2 Metabolism of Vitamin D

Vitamin D_3, is synthesized in the skin dermis from 7-dehydrocholesterol upon exposure to sunlight (UV B light of wavelength 290-315 nm) [5]. Most vitamin D absorption occurs in the distal small intestine, with a slightly more rapid uptake occurring in the duodenum. About 80% of dietary vitamin D is absorbed after incorporation into micelles; it then undergoes passive (facilitated) diffusion into the enterocytes with fat and bile salts. Once in the enterocyte, vitamin D is incorporated into chylomicrons and moves to the lymphatic system from where it is transported to the blood and finally to the liver. Some of the vitamin D in chylomicrons is transferred to a vitamin D-binding protein (DBP) for transport to extrahepatic (muscle and adipose) tissue. Since vitamin D is a fat-soluble vitamin, its absorption is impaired in conditions of fat malabsorption. Vitamin D undergoes two hydroxylation reactions. The first hydroxylation reaction occurs in the endoplasmic reticulum of the hepatocyte when vitamin D is delivered to the liver and converted by the enzyme 25-hydroxylase to 25-hydroxyvitamin D, (25OHD also known as calcidiol). 25OHD, because of its long half-life (3 weeks), is considered as the biochemical marker of vitamin D status in the body. The second and final hydroxylation reaction occurs in the kidneys, catalyzed by 25-hydroxy-D3–1-a-hydroxylase, to form 1,25-dihydroxyvitamin D, [{1,25$(OH)_2D_3$}, also known as calcitriol/cholecalcitriol], the physiologically active form of vitamin D. This active form of vitamin affects the calcium homeostasis through its action on the intestine, kidney, and bones [6], (Fig. 6.1).The various functions of vitamin D are enlisted in Fig. 6.2.

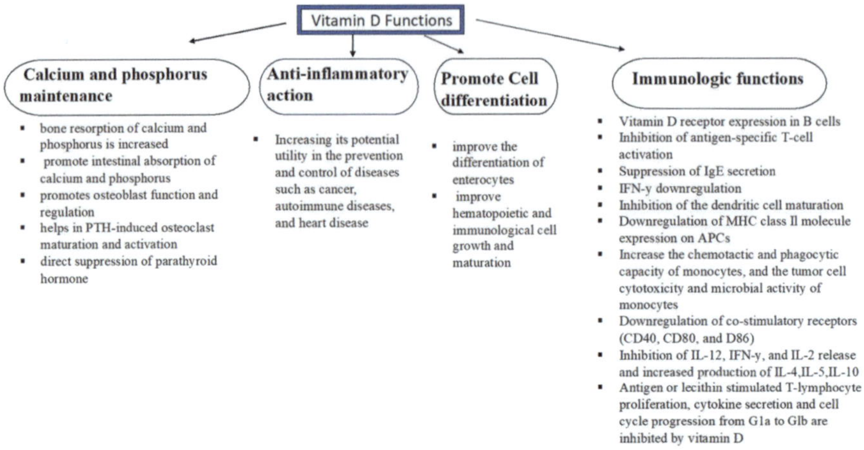

Fig. 6.2 Flow chart illustrating the functions of Vitamin D. The central box labeled "Vitamin D Functions" branches into four categories: "Calcium and phosphorus maintenance," "Anti-inflammatory action," "Promote Cell differentiation," and "Immunologic functions." Each category lists specific roles. For example, "Calcium and phosphorus maintenance" includes promoting intestinal absorption and osteoblast function. "Anti-inflammatory action" highlights disease prevention. "Promote Cell differentiation" focuses on cell growth. "Immunologic functions" details receptor expression and immune response modulation

6.3 Vitamin D Deficiency

Vitamin D deficiency, insufficiency, and sufficiency are defined by the Global Consensus Guidelines and previous Indian Academy of Pediatrics Guidelines as blood 25-hydroxy vitamin D values of 12 ng/mL, 12–20 ng/mL, and > 20 ng/mL, respectively [7]. Individual vitamin D status is determined by a variety of genetic, behavioral, and demographic factors such as age, gender, skin pigmentation, sunshine exposure, latitude, sunscreen use, dietary habits, and supplement intake [8].

6.4 Vitamin D Toxicity

Serum 25(OH)D levels >100 ng/mL with hypercalcemia and/or hypercalciuria can be used to diagnose vitamin D toxicity, where levels between 50 and100 ng/mL should be treated with care. Due to the increasing implications of vitamin D, the chances of landing in vitamin D toxicity is to be taken care of properly. Toxicity can cause lethargy, hypotonia, confusion, disorientation, depression, psychosis, hallucinations, and coma as well as interferes with renal concentrating mechanisms, resulting in polyuria, dehydration, and hypernatremia. Hypercalcemia can also cause acute renal failure, nephrolithiasis, and nephron-calcinosis, all of which can lead to chronic renal failure. Arrhythmias or dehydration are common causes of death.

6.5 Role of Vitamin D in Growth and Development

6.5.1 Newborns and Vitamin D

Breast milk is a poor source of vitamin D (approximately 12–60 IU/L), and complementary feeding choices after 6 months of life are rarely enriched or rich in vitamin D. Hence vitamin D supplementation plays a significant role. The influence of vitamin D in a child's growth and development is a complex issue, as evidenced by the many conflicting findings. Research indicates that poor vitamin D status in early childhood may not be an important limiting factor for linear growth [9]. However, vitamin D supplementation of low birthweight infants can result in better vitamin D status and benefit weight and length growth at 6 months. Moreover, the growth promotion effect of vitamin D may vary according to the stage of development, as studies suggest that prenatal vitamin D exposure may have a greater effect on muscle strength than on muscle mass in the development of offspring [9, 10]. Low maternal vitamin D concentrations were associated with lower birthweight, and optimum maternal serum 25(OH)D concentrations during pregnancy were associated with lower fat mass and higher bone mass in infants [10].

Other studies on vitamin D supplementation on various neonatal outcomes stated that maternal vitamin D deficiency (20 ng/mL) was found to have a significant connection with low birth weight, small for gestational age, and preterm birth, but no link between NICU admission, head circumference, or newborn deaths/stillbirth was established [11].

In another study, it was seen that at 6 months, infants receiving 400 IU/day had higher mean seated, prone, and composite Alberta Infant Motor Scale scores than infants getting 800 IU/day and 1200 IU/day, which imply that larger vitamin D supplementation doses in infancy are not necessary to optimize motor skill acquisition [12].

The Oxford Centre for Evidence-Based Medicine (OCEBM) classification graded from 1 to 5 for Level of evidence (LOE) of each recommendation [13].

Recommendation: Preterm newborns can tolerate 400 IU per day. Doses of 800 IU/day can attain desired biochemical levels more quickly, but there is insufficient evidence and safety data to create a separate dose guideline for preterm newborns [LOE 3].

400 IU/day vitamin D is recommended as supplementation for infants (0–1 years of age). Doses more than 400 IU provide no further skeletal benefit during childhood. Above doses, such as 1600 IU daily or above, can cause toxicity (LOE2).

6.5.2 Anthropometric and Vitamin D

Stunting, wasting, underweight, overweight, and obesity anthropometric indices provide specific information on growth and body composition for measuring nutritional status. Vitamin D plays a critical role in maintaining calcium homeostasis and bone health, and its deficiency during pregnancy or infancy has been associated with adverse growth in children, including rickets characterized by abnormalities of the growth plate and delayed growth. Vitamin D supplementation was associated with a mild improvement in linear growth in children under the age of five. 240 participants in 3 trials found no significant effect on length/height Z-scores. [14] Larger, long-term studies are needed to clarify whether oral vitamin D supplementation affects linear growth in children under the age of five, in both healthy and high-risk children.

In another study with 10,450 participants, the results showed that every 10 nmol/L increase in serum 25(OH)D concentration was linked with a 0.15 cm/year increase in height growth velocity (P 0.001) and a 7% reduction in the probability of low bone mineral density [9]. Vitamin D supplementation during pregnancy or infancy is associated with lower BMI and BMI z-score in offspring at 3–6 years of age [10].

Recommendation: A dose of 400 IU daily was found beneficial to achieve serum 25(OH)D levels >20 ng/mL and for prevention of rickets in infants. Rickets and vitamin D deficiency should be treated for 12 weeks with oral cholecalciferol (2000 IU for children under 1 year of age and 3000 IU for older children). Intermittent regimens may be considered if children above the age of 6 months are unable to adhere to regular dosing.

6.5.3 Neurodevelopmental Disorders and Vitamin D

The neuroprotective impact of vitamin D is thought to be secondary to its effects on neuronal calcium homeostasis, neuromediator release, neurotrophin synthesis, and oxidative damage prevention. The American Academy of Developmental Medicine and Dentistry Vitamin D Task Force examined the evidence of 25(OH)D concentrations that promote the health of people with developmental disorders. Maintaining optimal serum 25(OH)D concentrations aid skeletal development in newborns, children, and adolescents, as well as musculoskeletal health and neuromuscular coordination in adult patients [15]. Inadequate vitamin D levels in autistic children are prevalent due to genetic and/or environmental factors and may be a risk factor for developing autistic spectral disorders (ASDs). Vitamin D supplementation may be beneficial in reducing symptoms in attention deficit hyperactivity disorders (ADHDs) and children with autistic spectral disorder with low vitamin levels. Furthermore, vitamin D supplementation during pregnancy may be appropriate as a preventative measure for both ADHD and ASD or a measure to minimize the prevalence of ASD. Furthermore, randomized clinical trials are required to confirm the usefulness of vitamin D and the autistic spectrum of disease [16]. In these children, adequate vitamin D status is related with improved symptomatology and should be maintained with supplementation. Because the American Academy of Paediatrics supports vitamin D supplementation during infancy and youth, pediatricians and family practitioners should assess the available information on autism and vitamin D and take appropriate action [17].

Recommendation: The literature does not support vitamin D's direct therapeutic effect in ADHD and ASD. Maintaining optimum vitamin D status is recommended in these illnesses. (LOE 2).

The Endocrine Society and the Institute of Medicine recommend an average vitamin D intake of 600–1000 IU/d between 1 and 18 years of age. Vitamin D supplementation is only indicated in cases of poor sun exposure or specific conditions [3].

Further research is needed to fully understand the relationship between vitamin D and growth, and to assess the effectiveness of vitamin D supplementation in early life on children's bone health, lean mass, and adiposity [18, 10].

In one study, 1489 participants were recruited in an original Delhi Infant Vitamin D Supplementation Randomized Controlled Trial (referred to as DIVIDS-1) during their infancy, with a follow-up to 6 months, which came with the inference that vitamin D supplementation at one recommended daily allowance in the first 6 months of life increased growth and reduced stunting in low-birth weight term infants [19]. In a subsequent study, i.e., DIVIDS-2 study, the 912 DIVIDS children were followed up till aged 3–6 years, to determine long-term effects. However, the advantage of early supplementation on length was not sustained at age 3–6 years. Vitamin D supplementation in infancy reduced BMI at age 3–6 years; however, no long-term effects of early vitamin D supplementation were seen on body composition, bone health, or motor development. The trial though had some limitations as nearly half of the participants were present in the DIVIDS-2 trial which could have resulted in the observation that no long-term effect was seen. Further research on this account may be needed [19].

Beyond its usual role in calcium homeostasis, vitamin D exhibits pleiotropic effects. Hundreds of genes containing vitamin D receptor response elements regulate cell cycling and proliferation, differentiation, and death, either directly or indirectly. Nongenomic effects of vitamin D molecules on cell activity are also observed. Vitamin D's noncalcemic effects affect normal and pathological cell proliferation, carcinogenesis, immunological function, and cardiovascular physiology. This chapter looks at several of the diverse processes by which vitamin D affects cellular development and differentiation, as well as cell-specific variables that influence vitamin D responsiveness [20].

6.6 Infectious Diseases and Vitamin D

6.6.1 Diarrhea and Vitamin D

Diarrhea has a significant impact on growth and development of child health, but its association with vitamin D level is inconclusive. The majority of observational and interventional research have found a link between vitamin D level and the frequency of episodes and severity of diarrhea. However, no correlation with vitamin deficiency or therapeutic efficacy of vitamin D in acute diarrhea in children of normal or underweight weight could be found. A recent meta-analysis including data from two studies found no advantage to vitamin D supplementation in children with diarrhea [21].

Recommendation: In the case of acute childhood diarrhea, vitamin D supplementation is not beneficial. [Level of Evidence (LOE-1)].

6.6.2 Acute Lower Respiratory Tract Infection and Vitamin D

There is emerging evidence that vitamin D has immune-modulating properties and aids in the enhancement of innate immunity [22]. So, if a child is suffering from repeated infection like acute respiratory tract infection (pneumonia), tuberculosis, diarrhea, HIV, etc., then this suffering is going to impair the growth and development of the child and hence may contribute to morbidity and mortality of the pediatric population [23]. Although several observational studies and systematic reviews have revealed a link between vitamin D deficiency and acute lower respiratory tract infection (ALRTI), the same biological effect of vitamin D has not been fully demonstrated by experimental research and RCT systematic reviews [24].

Recommendation: Vitamin D levels in children with pneumonia are lower, and a lower vitamin D level may be associated with a higher frequency and severity of pneumonia. (Level Of Evidence, LOE-1). Routine vitamin D supplementation has no benefit in the treatment of pediatric pneumonia. (LOE 1).

6.6.3 Mycobacterium Infection and Vitamin D

There have been several observational studies on the link between vitamin D deficiency and tuberculosis in adults, but there has been less research in children. A meta-analysis of five clinical trials, one of which included children, found no evidence of a therapeutic benefit of vitamin D in the treatment of tuberculosis. A meta-analysis of eight clinical studies including 1787 individuals with active pulmonary tuberculosis found that vitamin D improves sputum smear and culture conversion. But the results were inconclusive. The precise mechanism of action of vitamin D as an immunological modulator is unknown [25, 26].

Recommendation: Vitamin D levels in children with tuberculosis are much decreased. (LOE 1). There is no benefit to routine vitamin D supplementation in children with TB. (LOE 1).

6.6.4 Human Immunodeficiency Virus Infection and Vitamin D

In human immunodeficiency virus infection, vitamin D deficiency reduces innate and adaptive immunological responses, increases inflammation, and increases infection susceptibility due to changes in monocyte and T cell activity. Vitamin D plays an important role in host defense against infections by activating genes and pathways that boost innate and adaptive immunity [27]. By attaching to the vitamin D receptor (VDR) and activating and regulating several cellular pathways, vitamin D mediates its biological effects. A systematic review that included 30 studies, 2 of which were conducted in children, indicated that there was high heterogeneity in included studies and that more controlled trials were needed to demonstrate an association between the two. [28, 29].

Recommendation: The therapeutic impact of vitamin D supplementation in pediatric human immunodeficiency virus infection is not established. (LOE 2).

6.7 Noninfectious Diseases and Vitamin D

6.7.1 Obesity and Vitamin D

Several explanations have been offered to explain the probable relationship between vitamin D levels and obesity and metabolic syndrome. Most studies show a minor and inconsequential effect, if any, on a few indicators such as central obesity(abdominal) and insulin release [30, 31].

Recommendation: Obesity and blood sugar control factors are linked to vitamin D status (LOE 2). For weight and metabolic consequences, routine vitamin D administration is not indicated in obese children and adolescents. (LOE 1).

6.7.2 Atopic Dermatitis and Vitamin D

Vitamin D has impacts on keratinocyte proliferation and differentiation, reduction of inflammatory responses, expression of antimicrobial peptides, permeability barrier integrity, cathelicidin synthesis, and suppression of IgE-mediated cutaneous reactions [32]. The significance of vitamin D in the etiology of atopic dermatitis is not completely understood. An observational study of 138 Norwegian atopic dermatitis patients found no link between low dietary vitamin D and clinical disease severity scores [33].

Recommendation: Routine vitamin D supplementation in children with atopic dermatitis is not recommended because supplementation has not been shown to help atopic dermatitis. (LOE 2).

6.7.3 Allergy and Asthma and Vitamin D

A meta-analysis of 27 studies, 18 of which included children, revealed a beneficial relationship between vitamin D and lung function in both adults and children with asthma. The evidence that vitamin D deficiency increases the risk of asthma and allergies is growing. However, the optimal dosage of vitamin D that reduces both the chance of developing and the severity of these illnesses is yet unknown. Before clear recommendations can be made, the results of ongoing clinical trials of vitamin D supplementation will be required [34].

Recommendation: Vitamin D levels are lower in children with asthma, which may be linked to decreased lung function and asthma exacerbation (LOE 1). Nonetheless, vitamin D supplementation in children with asthma has not been shown to be effective (LOE 1).

6.7.4 Renal Disease and Vitamin D

The use of high-dose corticosteroids on a regular basis may jeopardize metabolic bone health even further, so vitamin D supplementation is suggested by international and national guidelines largely for the restoration of bone health in nephrotic patients, which has been documented in many studies as improvement in biochemical and radiographic indicators of bone health [35, 36, 37].

Recommendation: In order to keep serum 25(OH)D levels over 20 ng/mL, children with nephrotic syndrome should take vitamin D and calcium supplements on a regular basis (LOE 3).

6.7.5 Hepatic Disease and Vitamin D

In patients with cholestasis, vitamin D deficiency is connected with intraluminal bile acid deficit, which affects vitamin D absorption. Furthermore, hepatic conversion of vitamin D to 25-hydroxyvitamin D (25-OHD) is hindered in some patients with severe hepatic impairment. The North American Society for Paediatric Gastroenterology, Hepatology, and Nutrition advocate vitamin D supplementation for pediatric patients with cholestasis, with dose adjustment based on serum 25-OHD levels [38]. Several earlier studies found that the severity of vitamin D

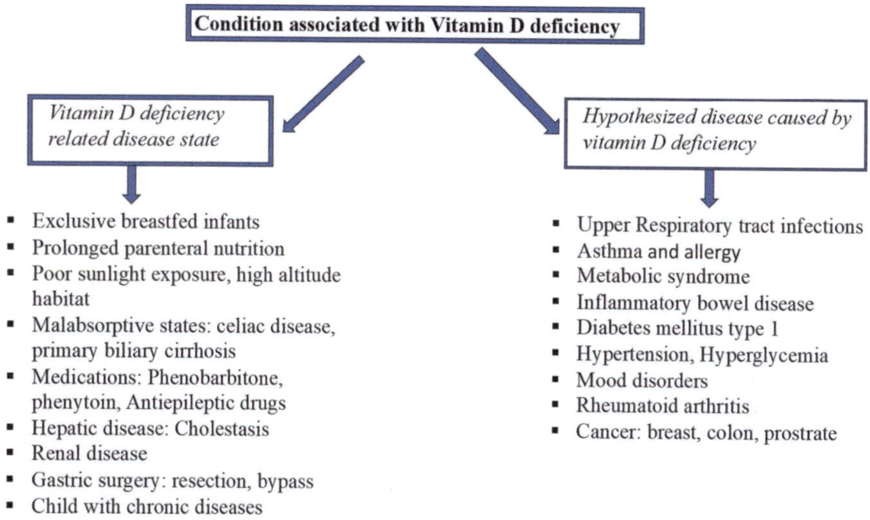

Fig. 6.3 Flow chart illustrating conditions associated with Vitamin D deficiency. The chart is divided into two main categories: "Vitamin D deficiency related disease state" and "Hypothesized disease caused by vitamin D deficiency." The first category includes factors like exclusive breastfeeding, prolonged parenteral nutrition, poor sunlight exposure, malabsorptive states, certain medications, hepatic and renal diseases, gastric surgery, and chronic diseases in children. The second category lists potential diseases such as upper respiratory tract infections, asthma, metabolic syndrome, inflammatory bowel disease, diabetes mellitus type 1, hypertension, mood disorders, rheumatoid arthritis, and certain cancers

deficiency in cholestasis patients was associated with poor clinical outcomes such as an insufficient response to ursodeoxycholic acid, the development of cirrhosis, liver-related death, and the necessity for liver transplantation. The various condition associated with vitamin D deficiency along with related and hypothesized disease states is given in Fig. 6.3.

6.8 Conclusion

In children with underlying high-risk illnesses, routine vitamin D supplementation may be given. It is not suggested to examine seemingly healthy youngsters for vitamin D insufficiency on a regular basis. Asymptomatic children should be checked only if they are at risk for vitamin D deficiency (for example, children on long-term anticonvulsants or glucocorticoids; chronic kidney illness, malabsorption states, children with impairments, chronic inflammatory diseases, and so on).

Many new aspects of vitamin D have emerged, including novel roles in reproduction, parathyroid function, skin proliferation and differentiation, myeloid cell differentiation, and as anticancer agents. Without a doubt, the vitamin D field will continue to grow in the coming decade as novel functions of the vitamin D system are being discovered.

References

1. Balasundaram P, Avulakunta ID. Human growth and development. In: StatPearls [Internet]. Treasure Island (FL): StatPearls Publishing; 2023 [cited 2023 Dec 22]. Available from: http://www.ncbi.nlm.nih.gov/books/NBK567767/
2. Savarino G, Corsello A, Corsello G. Macronutrient balance and micronutrient amounts through growth and development. Ital J Pediatr. 2021;47(1):109.
3. Holick MF, Vitamin D. Deficiency. N Engl J Med. 2007;357(3):266–81.
4. Comprehensive National Nutrition Survey (2016–18) reports :: National Health Mission [Internet]. [cited 2023 Dec 31]. Available from: https://nhm.gov.in/index1.php?lang=1&level=2&sublinkid=1332&lid=713
5. Fischer PR, Thacher TD, Pettifor JM. Pediatric vitamin D and calcium nutrition in developing countries. Rev Endocr Metab Disord. 2008;9(3):181–92.
6. Kulie T, Groff A, Redmer J, Hounshell J, Schrager S. Vitamin D: an evidence-based review. J Am Board Fam Med. 2009;22(6):698–706.
7. Gupta P, Dabas A, Seth A, Bhatia VL, Khadgawat R, Kumar P, et al. Indian academy of pediatrics revised (2021) guidelines on prevention and treatment of Vitamin D deficiency and rickets. Indian Pediatr. 2022;59(2):142–58.
8. Mithal A, Wahl DA, Bonjour JP, Burckhardt P, Dawson-Hughes B, Eisman JA, et al. Global vitamin D status and determinants of hypovitaminosis D. Osteoporos Int J Establ Result Coop Eur Found Osteoporos Natl Osteoporos Found USA. 2009;20(11):1807–20.
9. Xiao P, Cheng H, Wang L, Hou D, Li H, Zhao X, et al. Relationships for vitamin D with childhood height growth velocity and low bone mineral density risk. Front Nutr [Internet] 2023 [cited 2023 Dec 18];10. Available from: https://www.frontiersin.org/articles/10.3389/fnut.2023.1081896
10. Ma K, Wei SQ, Bi WG, Weiler HA, Wen SW. Effect of Vitamin D supplementation in early life on children's growth and body composition: a systematic review and meta-analysis of randomized controlled trials. Nutrients. 2021;13(2):524.
11. van der Pligt P, Willcox J, Szymlek-Gay EA, Murray E, Worsley A, Daly RM. Associations of maternal Vitamin D deficiency with pregnancy and neonatal complications in developing countries: a systematic review. Nutrients. 2018;10(5):640.
12. Wicklow B, Gallo S, Majnemer A, Vanstone C, Comeau K, Jones G, et al. Impact of Vitamin D supplementation on gross motor development of healthy term infants: a randomized dose-response trial. Phys Occup Ther Pediatr. 2016;36(3):330–42.
13. OCEBM Levels of Evidence [Internet]. [cited 2024 Jan 6]. Available from: https://www.cebm.ox.ac.uk/resources/levels-of-evidence/ocebm-levels-of-evidence
14. Huey SL, Acharya N, Silver A, Sheni R, Yu EA, Peña-Rosas JP, et al. Effects of oral vitamin D supplementation on linear growth and other health outcomes among children under five years of age. Cochrane Database Syst Rev. 2020;12(12):CD012875.
15. Grant WB, Wimalawansa SJ, Holick MF, Cannell JJ, Pludowski P, Lappe JM, et al. Emphasizing the health benefits of Vitamin D for those with neurodevelopmental disorders and intellectual disabilities. Nutrients. 2015;7(3):1538–64.
16. Jia F, Shan L, Wang B, Li H, Miao C, Xu Z, et al. Bench to bedside review: possible role of vitamin D in autism spectrum disorder. Psychiatry Res. 2018;260:360–5.
17. Cannell JJ. Vitamin D and autism, what's new? Rev Endocr Metab Disord. 2017;18(2):183–93.
18. Esposito S, Leonardi A, Lanciotti L, Cofini M, Muzi G, Penta L. Vitamin D and growth hormone in children: a review of the current scientific knowledge. J Transl Med. 2019;17(1):87.
19. Trilok-Kumar G, Kaur M, Rehman AM, Arora H, Rajput MM, Chugh R, et al. Effects of vitamin D supplementation in infancy on growth, bone parameters, body composition and gross motor development at age 3-6 years: follow-up of a randomized controlled trial. Int J Epidemiol. 2015;44(3):894–905.
20. Samuel S, Sitrin MD. Vitamin D's role in cell proliferation and differentiation: Nutrition Reviews©. Nutr Rev. 2008;66(s2):S116–24.

21. Yakoob MY, Salam RA, Khan FR, Bhutta ZA. Vitamin D supplementation for preventing infections in children under five years of age. Cochrane Database Syst Rev. 2016;2016(11):CD008824.
22. White JH, Vitamin D. Signaling, infectious diseases, and regulation of innate immunity. Infect Immun. 2008;76(9):3837–43.
23. Somnath SH, Biswal N, Chandrasekaran V, Jagadisan B, Bobby Z. Therapeutic effect of vitamin D in acute lower respiratory infection: a randomized controlled trial. Clin Nutr ESPEN. 2017;20:24–8.
24. Jolliffe DA, Camargo CA, Sluyter JD, Aglipay M, Aloia JF, Ganmaa D, et al. Vitamin D supplementation to prevent acute respiratory infections: a systematic review and meta-analysis of aggregate data from randomised controlled trials. Lancet Diabetes Endocrinol. 2021;9(5):276–92.
25. Xia J, Shi L, Zhao L, Xu F. Impact of vitamin D supplementation on the outcome of tuberculosis treatment: a systematic review and meta-analysis of randomized controlled trials. Chin Med J. 2014;127(17):3127–34.
26. Wu HX, Xiong XF, Zhu M, Wei J, Zhuo KQ, Cheng DY. Effects of vitamin D supplementation on the outcomes of patients with pulmonary tuberculosis: a systematic review and meta-analysis. BMC Pulm Med. 2018;18(1):108.
27. Jiménez-Sousa MÁ, Martínez I, Medrano LM, Fernández-Rodríguez A, Resino S. Vitamin D in human immunodeficiency virus infection: influence on immunity and disease. Front Immunol. 2018;12(9):458.
28. Eckard AR, O'Riordan MA, Rosebush JC, Lee ST, Habib JG, Ruff JH, et al. Vitamin D supplementation decreases immune activation and exhaustion in HIV-1-infected youth. Antivir Ther. 2018;23(4):315–24.
29. Giusti A, Penco P. Vitamin D deficiency in HIV-infected patients: a systematic review. Nutr Diet Suppl. 2011;101
30. Bhatt SP, Misra A, Sharma M, Guleria R, Pandey RM, Luthra K, et al. Vitamin D insufficiency is associated with abdominal obesity in urban Asian Indians without diabetes in North India. Diabetes Technol Ther. 2014;16(6):392–7.
31. Duan L, Han L, Liu Q, Zhao Y, Wang L, Wang Y. Effects of Vitamin D supplementation on general and central obesity: results from 20 randomized controlled trials involving apparently healthy populations. Ann Nutr Metab. 2020;76(3):153–64.
32. Wadhwa B, Relhan V, Goel K, Kochhar AM, Garg VK. Vitamin D and skin diseases: a review. Indian J Dermatol Venereol Leprol. 2015;81:344.
33. Solvoll K, Soyland E, Sandstad B, Drevon CA. Dietary habits among patients with atopic dermatitis. Eur J Clin Nutr. 2000;54(2):93–7.
34. Litonjua AA. Vitamin D deficiency as a risk factor for childhood allergic disease and asthma. Curr Opin Allergy Clin Immunol. 2012;12(2):179–85.
35. Selewski DT, Chen A, Shatat IF, Pais P, Greenbaum LA, Geier P, et al. Vitamin D in incident nephrotic syndrome: a Midwest pediatric nephrology consortium study. Pediatr Nephrol Berl Ger. 2016;31(3):465–72.
36. Trautmann A, Vivarelli M, Samuel S, Gipson D, Sinha A, Schaefer F, et al. IPNA clinical practice recommendations for the diagnosis and management of children with steroid-resistant nephrotic syndrome. Pediatr Nephrol Berl Ger. 2020;35(8):1529–61.
37. Vasudevan A, Thergaonkar R, Mantan M, Sharma J, Khandelwal P, Hari P, et al. Consensus guidelines on Management of Steroid-Resistant Nephrotic Syndrome. Indian Pediatr. 2021;58(7):650–66.
38. Mouzaki M, Bronsky J, Gupte G, Hojsak I, Jahnel J, Pai N, et al. Nutrition support of children with chronic liver diseases: a joint position paper of the north American Society for Pediatric Gastroenterology, Hepatology, and nutrition and the European Society for Pediatric Gastroenterology, Hepatology, and nutrition. J Pediatr Gastroenterol Nutr. 2019;69(4):498–511.

Chapter 7
Rickets and Other Childhood Disorders Related to Vitamin D

Shubham Charaya, Renu Suthar, and Suresh Kumar Angurana

Abstract Vitamin D plays an important role in calcium homeostasis and bone health, modulation and regulation of inflammation, immune function, cell growth, neuromuscular function, and glucose metabolism. Due to several modifiable and nonmodifiable factors, vitamin D deficiency (VDD) is common in children in both developed and developing countries. Rickets is a common manifestation in children with VDD. Diagnosis of rickets and VDD relies on dietary intake, sunlight exposure, vitamin D supplementation, clinical features of rickets, and measurement of serum 25-hydroxy vitamin D [25(OH)D] levels. The treatment and prevention of VDD include vitamin D supplementation, food fortification, and adequate exposure to sunlight.

Keywords Vitamin D · Rickets · 1,25(OH)D · Bone health · Malnutrition

7.1 Introduction

Vitamin D is a fat-soluble vitamin which is essential for maintaining calcium homeostasis and bone health. Other important roles of vitamin D include modulation and regulation of inflammation, immune function, cell growth, neuromuscular function, and glucose metabolism. Vitamin D is derived from the diet or produced endogenously by dermal synthesis. On exposure to sunlight, vitamin D (inactive form) is produced in the skin which is then converted into active form in liver and

S. Charaya · R. Suthar · S. K. Angurana (✉)
Department of Pediatrics, Advanced Pediatrics Centre (APC), Postgraduate Institute of Medical Education and Research (PGIMER), Chandigarh, India

V. Kumar, S. Patel (eds.), *Vitamin D: A Comprehensive Guide for Medical Professionals*, https://doi.org/10.1007/978-981-96-6491-7_7

kidneys. Vitamin D deficiency (VDD) is common throughout the world including both developed and developing nations [1–6]. VDD causes several clinical conditions, rickets being most common in children, which causes thinning, brittleness, and deformity of bones. In adults, VDD leads to osteomalacia [7]. The measurement of serum 25-hydroxy vitamin D [25(OH)D] levels is used to quantify vitamin D, and this reflects both endogenous and dietary levels. Prevention and treatment of VDD includes supplementation with vitamin D, food fortification, and adequate exposure to sunlight [8, 9].

7.2 Prevalence

VDD in healthy children has been estimated to be ranging from 9 to 24% in high-income countries [2, 3] and 36–90% in developing and low-middle income countries (LMICs) [1, 10]. Prevalence of VDD is very high in children with underlying diseases like malabsorption syndromes, renal failure, congenital heart diseases, and among critically ill children [11–14].

7.3 Vitamin D Chemistry

Calciferol, another name for vitamin D, is a general term that describes a class of fat-soluble substances that contain cholesterol with four rings. 25(OH)D is the most widely distributed form of vitamin D, and 1,25-dihydroxyvitamin D [1, 25(OH)D] is the active form of vitamin D. 1,25(OH)D uses intracellular receptors to control the transcription of genes. Its functional receptor, vitamin D receptors (VDRs), are present in all nucleated cells. Intestinal absorption of calcium and enterocyte differentiation are its major biological functions. It also decreases absorption of phosphate from the intestine, suppress PTH secretion, modulate functions of osteoblasts, and promotes PTH-induced activation of osteoclasts and resorption of the bone [15].

7.4 Vitamin D Metabolism

As a prohormone, vitamin D is produced by the skin in response to ultraviolet B (UV-B) radiation in sunlight and can also be absorbed through food or supplementation. It is then activated in the liver [to 25(OH)D] and kidneys [to 1,25(OH)D]. ***Vitamin D_2*** (ergocalciferol) is present in plant dietary sources. Exposure of ergosterol to irradiation in plants lead to formation of ergocalciferol. ***Vitamin D_3*** (cholecalciferol) is derived mainly from animals sources and also through supplements. When 7-dehydrocholestrol is exposed to UV-B radiation (wavelength 290–310 nm),

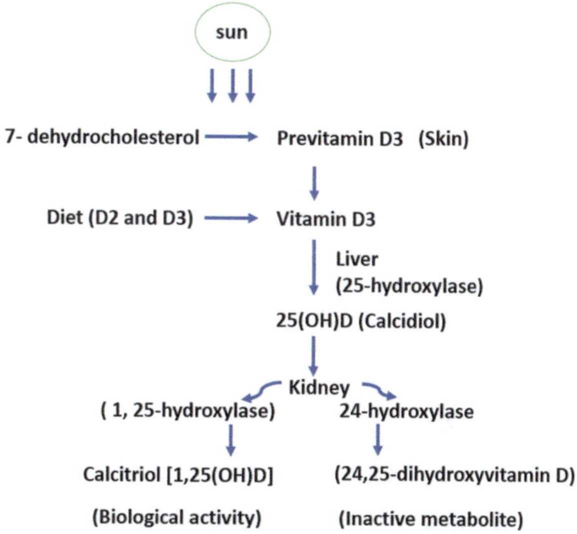

Fig. 7.1 Vitamin D synthesis and metabolism

previtamin D is formed which after isomerization converts into vitamin D_3. ***Calcidiol*** is the storage form of vitamin D. Vitamin D_3 after binding to VDBP (vitamin D binding protein) is transported to the liver where it undergoes 25-hydroxylation and forms 25(OH)D (calcidiol). The measurement of serum 25(OH)D is used to determine the level of vitamin D. ***Calcitriol*** [1,25(OH)D] is the most active and effective form of vitamin D which is formed in the kidney with the help of 1,25-hydroxylase which acts on 25(OH)D to form 1,25(OH)D. There are multiple other sites for 1-alpha-hydroxylation apart from the kidney, such as the lymphatic system, large intestine (colon), skin, osteoblasts, and macrophages, indicating an autocrine–paracrine role. Calcitriol helps the body to absorb calcium through the digestive tract, controls bone resorption, and decreases the amount of calcium and phosphate excreted in the urine. Parathyroid hormone (PTH), hypophosphatemia, and growth hormone drive this process. The synthesis of vitamin D is shown in Fig. 7.1.

7.5 Vitamin D Sources

Vitamin D is found in limited dietary products including some organ meats, egg yolk, and fish rich in oil such as mackerel, salmon etc. Vitamin D found in diet is absorbed through micelle by enterocytes and then packaged into chylomicrons. A major natural source of vitamin D is its dermal synthesis on exposure to UV-B rays in sunlight. Duration of sunlight exposure required for adequate production of vitamin D is difficult to predict as it varies with type of skin, application of screen, season, time of day, etc. [16].

7.6 Recommended Daily Vitamin D Intake

In infants, vitamin D supplementation in doses of 400 IU/day is recommended [8, 9]. After infancy, daily requirements of 400–600 IU/day should be met through diet and sunlight exposure to prevent VDD. It will maintain the 25(OH)D concentrations >20 ng/ml (>50 nmol/L) [8, 9].

7.7 Pathogenesis of VDD

The thickness of growth plate depends on two processes, i.e., proliferation and hypertrophy of chondrocyte; and growth plate cartilage calcification with subsequent invasion of blood supply and formation of primary bone spongiosa. Deficits in calcium or phosphorus can cause the growth plate cartilage to become less mineralized, which can impede or delay vascular penetration [7, 17]. In these situations, growth plate cartilage accumulates, and its expansion occur without calcification. Furthermore, the growth plate's chondrocytes deorganize and lose their columnar orientation leading to hypertrophic zone enlargement and poor mineralization. The mineralization defect causes osteoid build-up in the bone tissue beneath the growth plate. These changes alter the skeletal site geometry and lead to increase in diameter as an attempt to compensate for decreased bone strength. If left untreated for a longer duration, structural deformities occur [15].

7.8 Risk Factors for VDD

Vitamin D Insufficiency in Mothers Low levels of vitamin D in mothers are associated with low levels of vitamin D in the infants. Vitamin D is transferred from the mother to the fetus through the placenta [8].

Prematurity Since premature newborns don't have enough time to absorb vitamin D from the mother's placenta, they have particularly low vitamin D concentrations. Since the majority of the fetal skeleton calcification occurs in the last trimester, shortage of vitamin D in the mother during this time can cause fetal vitamin D deficiency, which in extreme cases can result in fetal rickets [15].

Exclusive Breastfeeding Breast milk from mothers who don't get enough vitamin D has low levels of vitamin D (15–50 IU/L or 0.4–1.2 μg/L). Only 10–40 IU/day (0.25–1 μg/day) of vitamin D is consumed by exclusively breastfed neonates who drink 750 mL of breast milk on an average per day when they are not exposed to the sun or receive vitamin D supplements. For the majority of neonates, sunlight exposure alone rarely results in adequate vitamin D levels. Fortification of infant formula provides enough vitamin D for infants on formula milk.

Low Dietary Intake The main dietary sources of vitamin D like oily fishes, liver oil, organ meats, and yolk found in egg are not consumed by children enough to maintain required 25(OH)D levels in the absence of other sources.

Skin Pigmentation and Low Sun Exposure The amount of time and quality of sunshine (specifically UV-B rays) determines the synthesis of vitamin D. Skin that is more pigmented produces less vitamin D because melanin functions as a natural sunblock. For those with fair skin, 10–15 min of sun exposure (to the hands, arms, and face, or the arms and legs) in the spring, summer, and fall between 1000–1500 h will be enough for adequate cutaneous vitamin D synthesis. To achieve the same level of 25(OH)D concentrations, persons with very dark skin pigmentation require six to ten times as much sun exposure as those with light skin pigmentation. Most people with medium skin pigmentation only need three times as much sun exposure as those with light skin pigmentation. Season and latitude of habitation are important variables that also impact the skin's ability to produce vitamin D. Because of the diffused sunshine in the winter, VDD is more common in winters than it is in the summer. Altitude, cloud cover, and sunscreen use all have an impact on UV-B exposure. The synthetic capacity of vitamin D can be reduced by up to 95% when using sunscreen with an SPF of 30 [18, 19].

Obesity Because vitamin D is sequestered in fat, there is an inverse relationship between obesity and 25(OH)D concentrations. Therefore, compared to adolescents of normal weight, adolescents who are obese have increased vitamin D requirements.

Medications Certain antiepileptics and antiretroviral drugs can cause VDD by increasing metabolism of 25(OH)D and 1,25(OH)D. Patients taking glucocorticoids have increased vitamin D requirements because these drugs decrease absorption of calcium from the intestine which depends on vitamin D. Certain antifungals like ketoconazole also increase vitamin D requirements by blocking 1-alpha hydroxylation.

Malabsorption and Other Medical Disorders Fat malabsorption can cause decreased levels of vitamin D because its absorption is dependent on chylomicrons. Thus, patients having inflammatory bowel disease, celiac disease, cholestasis, insufficiency of exocrine pancreas (like in cystic fibrosis), and after resection of gut are prone to develop VDD [8, 12, 13]. VDD may result from deficiencies in 25-hydroxylation and 1-hydroxylation, which are linked to liver and kidney diseases, respectively.

Genetic Disorders Inherited disorders causing deficiency or resistance to vitamin D are as follows:

- *1-alpha-hydroxylase deficiency:* It is due to CYP27B1 gene mutation (previously known as vitamin D-dependent rickets type 1A). It is an autosomal recessive disorder which manifests as early-onset rickets, enamel hypoplasia in some cases along with hypocalcemia, hypophosphatemia, and secondary hyperparathyroidism [20].

- *25-hydroxylase deficiency.* It was previously known as vitamin D-dependent rickets type 1B). It is an uncommon cause of VDD caused by CYP2R1 gene mutation leading to decreased levels of 25-hydroxylase. They have low levels of 25(OH)D and normal level of 1,25(OH)D. Homozygous mutation manifests in more severe forms and shows very less sensitivity to vitamin D treatment even in higher doses [21].
- *Hereditary Resistance to Vitamin D* (HRVD): A mutation in the VDR (vitamin D receptor gene) results in hereditary resistance to vitamin D, also referred to as vitamin D-dependent rickets type 2. Serum calcium and phosphate levels are decreased, although levels of 25(OH)D and 1,25(OH)D are within a normal to high range. Affected children develop rickets in their first two years of life, and roughly two-thirds of cases develop alopecia and ectodermal abnormalities as a result of low vitamin D receptor activation inside keratinocytes and is a marker of disease severity [20].

Conditions with Low 25(OH)D but without VDD Low levels of vitamin D-binding protein (VDBP) can lead to low serum concentrations of 25(OH)D in certain situations, such as nephrotic syndrome. As serum free 25(OH)D concentration (a physiologically significant indicator of bioavailable 25OHD) may be normal, this finding does not always imply a VDD.

7.9 Clinical Features of VDD

7.9.1 Skeletal Manifestations

When bone tissue fails to mineralize properly, it leads to softening of bones and skeletal abnormalities. This condition appears in children as rickets, a manifestation of VDD. Rickets initially manifests in areas of rapid growth such as distal forearm, knee, and costochondral junctions. Severe rickets can result in tetanic spasms, hypocalcemic seizures, developmental delay, malformed teeth, and failure to thrive in addition to bone deformities. Deformity also depends on the age and weight bearing pattern. Forearms (wrist widening) and the posterior bowing of the distal tibia are common abnormalities in infants. Toddlers, who has begun to walk, typically exhibit an exaggeration of the normal physiologic bowing of the legs (genu varum) (Fig. 7.2). The elder kid may exhibit windswept deformity, which consists of one leg with valgus deformity and the other with varus deformity, or valgus abnormalities of the legs (knock knees) [15]. Clinical features of rickets are highlighted in Table 7.1.

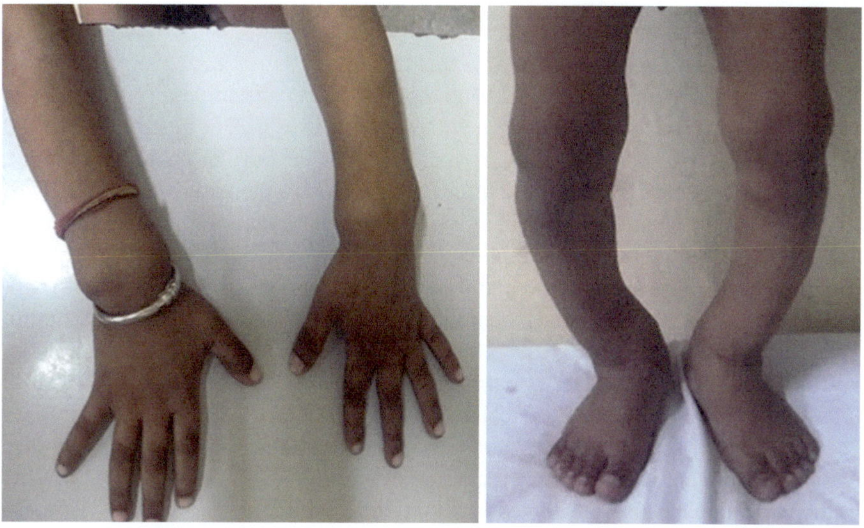

Fig. 7.2 Clinical signs of rickets including wrist widening, double malleolus sign, and genu varus

Table 7.1 Clinical signs and symptoms of rickets

Symptoms	Infants <6 months: Hypocalcemic seizure Older children: Delayed milestones and skeletal deformities Adolescent: Lower limb pains, carpopedal spasms, and knock knees
General signs	Failure to thrive, listlessness, protruding abdomen, muscle weakness and hypotonia, fractures, dilated cardiomyopathy, increased intracranial pressure
Skull	Craniotabes (early sign), frontal bossing, delayed closure of anterior Fontanelle
Teeth	Delayed eruption and enamel hypoplasia
Thorax	Rachitic rosary, pectus excavatum (pigeon chest), pectus carinatum, Harrison groove
Extremities	Arms: Wrist widening Leg: Genu varum, genu valgum, saber shin deformity, green stick fractures, and wind-swept deformity Ankle: Double malleolus
Ligaments and muscles	Laxity and decreased tone, myopathy
Spine	Kyphoscoliosis

7.10 Extraskeletal Manifestations

Vitamin D has a significant role in immunomodulation and immunity [22]. Lower vitamin D levels are found in children with pneumonia as well as tuberculosis. VDD may be associated with an increase in incidence and severity of pneumonia [9, 23, 24]. In patients with HIV, lower vitamin D levels lead to increased susceptibility of infection due to impaired T cell function, decreased immune response (both innate and adaptive), and more inflammation [25]. VDD has been associated with poor lung functions in asthma. In obese children and metabolic syndrome VDD is common. There is an association between vitamin D status and clinical presentation in atopic dermatitis [26]. There is also increased prevalence of VDD in children with autism spectrum disorder (ASD) and attention deficit hyperkinetic disorder (ADHD) [27].VDD is also common in critically ill children and is associated with increased susceptibility to sepsis and need for ventilator support and also in some studies associated with clinical outcomes including mortality [11, 23, 28, 29]. It is also one of the preventable causes of dilated cardiomyopathy [30].

7.11 Investigations

Methods used for measuring vitamin D levels are liquid chromatography coupled to ultraviolet (LC-UV) or tandem mass spectrometry (LC MS/MS) [31]. Vitamin D status is categorized as deficient <12 ng/mL (<30 nmol/L), insufficient 12–20 ng/mL (30–50 nmol/L), and sufficient >20 ng/ml (>50 mmol/L) based on serum 25(OH)D levels [9, 18]. Additional tests for rickets include serum levels of calcium, phosphate, parathyroid hormone, creatinine, alkaline phosphatase, and urine calcium. Biochemical findings of rickets and other related disorders are mentioned in Table 7.2.

Radiology The radiograph shows growth plate enlargement, loss of the growth plate-metaphysis boundary, and elimination of the provisional zone of calcification. These changes are more noticeable in areas with high demand of minerals. As a result, the distal ulna in the upper limb most clearly displays the initial indications

Table 7.2 Biochemical findings of rickets and other related disorders

Disorder	Ca	Pi	ALP	PTH	25-D	1,25-D	Urine Ca	Urine Pi
Vitamin D deficiency	N, ↓	↓	↑	↑	↓	↓, N, ↑	↓	↑
Dietary ca deficiency	N, ↓	↓	↑	↑	N, ↓	↑	↓	↑
VDDR, type 1	N, ↓	↓	↑	↑	N	↓	↓	↑
VDDR, type 2	N, ↓	↓	↑	↑	N	↑↑	↓	↑
Chronic renal failure	N, ↓	↑	↑	↑	N	↓	N, ↓	↓

Ca Calcium, *ALP* Alkaline phosphatase, *Pi* Phosphate, *PTH* Parathyroid hormone, *VDDR* Vitamin D-dependent rickets, *25-D* 25-hydroxyvitamin D, *1,25-D* 1,25-hydoxyvitamin D

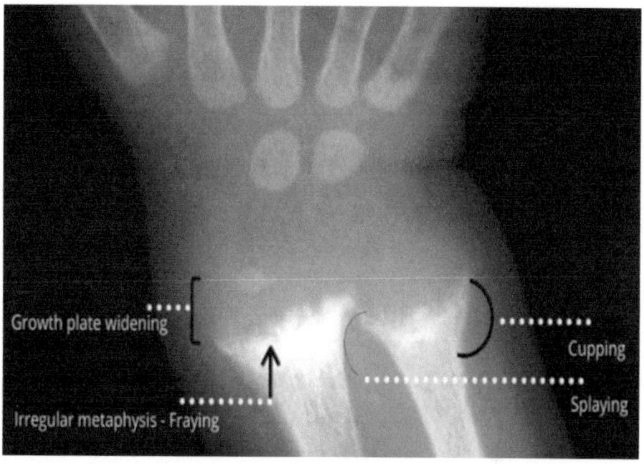

Fig. 7.3 Radiological features of rickets

of compromised mineralization. The metaphysis above and below the knees are the most useful sites in the lower extremities. The characteristic changes of rickets on radiographs include (Fig. 7.3):

- Fraying: The metaphysis edge becomes less defined.
- Cupping: The edge of the metaphysis becomes more concave as opposed to convex or flat.
- Splaying: Coarse trabeculation of the diaphysis.
- Generalized rarefaction.

7.12 Management of Nutritional Rickets

Regardless of their 25(OH)D levels, all children with nutritional rickets should receive vitamin D. All children on treatment for VDD should also receive appropriate doses of calcium and phosphorus. Even with adequate vitamin D intakes, bone mineralization may not improve in the absence of calcium and phosphorus [18, 32].

The vitamin D dosage regimen according to age is listed in Table 7.3 [33]. Schedules that are intermittent or daily can effectively treat nutritional rickets. Daily doses are preferred over bolus dosages since they are more physiological. In some cases, therapy with large cumulative doses spread out over several weeks or months may be more feasible if it is suspected that there may be issues with daily vitamin D therapy compliance. To reduce the danger of hypervitaminosis, bolus doses corresponding to the daily dose might be given at intervals of 2–4 weeks [34].

Radiologic healing starts to occur in most children with nutritional VDD at around four weeks. Radiographs should be obtained at 4 and 12 weeks after starting vitamin D therapy to look for evidence of recovery. Once full healing has occurred, a maintenance dose of vitamin D should be initiated. Blood calcium, phosphate, alkaline phosphatase, and serum 25 (OH)D levels should be measured at twelve weeks after beginning vitamin D therapy in order to assess toxicity and response [9].

Table 7.3 Doses of vitamin D and calcium according to the age for treatment of VDD and rickets

Age	Vitamin D dosage	Calcium dosage
<6 months	2000 IU/day for 12 weeks	50–75 mg/kg (max 500 mg/day)
6–12 months	2000 IU/day for 12 weeks or 30,000 every fortnight for 6 weeks or 60,000 monthly for 3 doses	50–75 mg/kg (max 500 mg/day)
>12 months	3000 IU/day for 12 weeks or 60,000 every fortnight for 5 doses	50–75 mg/kg (max 500 mg/day)

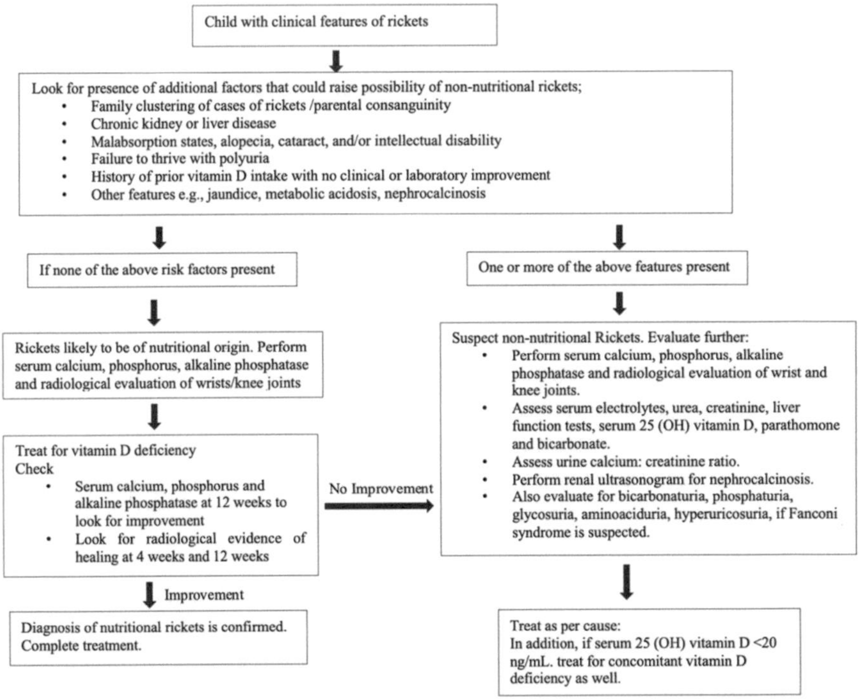

Fig. 7.4 Approach to management of rickets

If 12 weeks have passed without full radiological healing and biochemical parameter normalization (serum 25(OH)D > 20 ng/mL), therapeutic dosages of calcium and vitamin D should be maintained. Meanwhile children should be evaluated for nonnutritional causes of rickets as per Fig. 7.4.

Vitamin D Toxicity: It is defined as serum 25(OH)D levels >100 ng/mL along with hypercalcemia and/or hypercalciuria and can result from oversupplementation. Symptoms of acute intoxication are due to hypercalcemia and include confusion, polyuria, polydipsia, anorexia, vomiting, and muscle weakness. Chronic intoxication may cause bone demineralization. Nephrocalcinosis, which is detectable by renal ultrasonography, can result from hypercalciuria.

Children who suffer from severe diseases may have short height and lasting abnormalities. Orthopedic intervention may be beneficial for certain patients with leg abnormalities, although this is often saved for situations in which there is strong indication that the deformity is causing functional problems and the metabolic bone disease has healed.

7.13 Management of Nonnutritional Rickets

Malabsorption High doses of vitamin D are required in cases of malabsorption disorders. 25(OH)D is superior to vitamin D_3. (25–50 μg/day or 5–7 μg/kg/day) due to its better absorption. The dose is modified in response to observations of 25(OH)D serum levels. As an alternative, calcitriol may be utilized since it also has superior absorption in cases of fat malabsorption.

Vitamin D–Dependent Rickets Type 1A It responds to 1,25(OH)D (calcitriol) treatment over an extended period. Lower amounts are administered once the rickets has healed, with initial doses being 0.25–2 μg/day. Ensuring appropriate intake of calcium is crucial, particularly during the initial phases of therapy. The dosage of calcitriol is controlled to maintain low-normal serum calcium, normal serum phosphorus, and high-normal serum PTH levels in order to prevent hypercalciuria and nephrocalcinosis. Consequently, periodic evaluation of urine calcium excretion is an important part of patient monitoring, with a goal of less than 4 mg/kg/day [35].

Vitamin D–Dependent Rickets Type 1B It may respond to pharmacological doses of vitamin D (3,000 U/day) because of residual mutant protein activity or alternative enzymes with 25-hydroxylase activity.

Hereditary Resistance to Vitamin D When given exceptionally high dosages of calcitriol [1,25(OH)D], some patients, particularly those without alopecia, respond. In individuals suffering from vitamin D-dependent rickets, this response is brought on by a partially functioning vitamin D receptor. High-dose vitamin D and oral calcium should be tried for three to six months. Although some individuals may need doses as high as 50–60 μg/day, the recommended beginning dose of 1,25(OH)D is 2 μg/day. The daily dosage of calcium is 1000–3000 mg. If high-dose vitamin D is ineffective for a patient, long-term intravenous calcium therapy may be used before switching to very high-dose oral calcium supplements [20].

Chronic Kidney Disease *The* goal of therapy is to control secondary hyperparathyroidism and prevent bone disease (osteitis fibrosa cystica and mixed osteodystrophy). In CKD stage 2 to 4 with elevated PTH and 25(OH)D level < 30 ng/ml, ergocalciferol or cholecalciferol should be given, and if the levels are >30 ng/ml, calcitriol should be used. In CKD stage 5 calcitriol is given until PTH is reduced to a range between 2 and 3 times the upper limit of normal [36]. Just as crucial as taking activated vitamin D, controlling the serum phosphorus level is also important by using a combination of dietary phosphorus restriction and oral phosphate binders, since hyperphosphatemia is a stimulator for PTH release. Normal serum calcium levels should also be maintained in these patients [36].

7.14 Prevention

For every infant, a daily dosage of 400 IU of vitamin D supplementation is advised and after infancy daily consumption of 400–600 IU should come from food and exposure to sunlight to prevent VDD [8, 9]. Routine supplementation of at least 400 IU/day is recommended in certain high-risk conditions like chronic liver and kidney disease, malabsorptive states, disorders with extensive cutaneous involvement, nonambulatory conditions such as neuromuscular diseases or cerebral palsy, prolonged steroid therapy, and on antiepileptic drugs and ketoconazole [9]. It is advised that newborns and older children with body surface areas of 15–40% spend at least 15–30 min each day in the sun at midday (1000–1500 h) for preventing VDD across different regions and seasons [37]. Staple food fortification helps protect against rickets and raise children's and teenagers' vitamin D levels [18].

7.15 Conclusion

VDD is very common in children and is mostly nutritional in origin. It frequently presents with classical signs like craniotabes, rachitic rosary, genu varum, genu valgum, double malleoli sign, Harrison sulcus, etc. Initial evaluation include serum 25(OH)D levels along with calcium, phosphorus, and alkaline phosphatase. Rachitic findings are easily visible on radiographs like cupping, fraying, and widening of the growth plate. Treatment includes vitamin D and calcium supplementation. It is crucial to monitor response in order to determine the duration of the therapy and to perform further investigations for nonnutritional factors. Prevention is important, and routine supplementation is recommended during infancy and in certain high-risk conditions. Fortification with vitamin D should be considered in low-middle income countries due to high prevalence of VDD in a healthy population.

Acknowledgments None.

Funding None.

Conflict/Declaration of Interest None.

References

1. Ministry of Health and Family Welfare (MoHFW), Government of India, UNICEF and Population Council. 2019. Comprehensive National Nutrition Survey (CNNS) National Report. New Delhi [Internet]. Available from: https://nhm.gov.in/WriteReadData/l892s/1405796031571201348.pdf
2. Mansbach JM, Ginde AA, Camargo CA. Serum 25-hydroxyvitamin D levels among US children aged 1 to 11 years: do children need more vitamin D? Pediatrics. 2009;124(5):1404–10.
3. Braegger C, Campoy C, Colomb V, Decsi T, Domellof M, Fewtrell M, et al. Vitamin D in the healthy European paediatric population. J Pediatr Gastroenterol Nutr. 2013 Jun;56(6):692–701.
4. Angurana SK, Mahajan V. Vitamin D deficiency among healthy children in developing countries: an epidemic being recognized. Public Health Nutr. 2016;19(8):1526–7.
5. Angurana SK, Angurana RS, Mahajan G, Kumar N, Mahajan V. Prevalence of vitamin D deficiency in apparently healthy children in North India. J Pediatr Endocrinol Metab JPEM. 2014 Nov;27(11–12):1151–6.
6. Angurana S. Vitamin D deficiency in children: is there a need for routine supplementation? Indian J Endocrinol Metab. 2018;22(5):714.
7. Ross AC, editor. Modern nutrition in health and disease. 11th ed. Philadelphia: Wolters Kluwer Health/Lippincott Williams & Wilkins; 2014. p. 1616.
8. Munns CF, Shaw N, Kiely M, Specker BL, Thacher TD, Ozono K, et al. Global consensus recommendations on prevention and Management of Nutritional Rickets. J Clin Endocrinol Metab. 2016;101(2):394–415.
9. Gupta P, Dabas A, Seth A, Bhatia VL, Khadgawat R, Kumar P, et al. Indian academy of pediatrics revised (2021) guidelines on prevention and treatment of Vitamin D deficiency and rickets. Indian Pediatr. 2022;59(2):142–58.
10. Green RJ, Samy G, Miqdady MS, El-Hodhod M, Akinyinka OO, Saleh G, et al. Vitamin D deficiency and insufficiency in Africa and the Middle East, despite year-round sunny days. S Afr Med J. 2015;105(7):603.
11. McNally JD, Nama N, O'Hearn K, Sampson M, Amrein K, Iliriani K, et al. Vitamin D deficiency in critically ill children: a systematic review and meta-analysis. Crit Care. 2017;21(1):287.
12. Pappa HM, Bern E, Kamin D, Grand RJ. Vitamin D status in gastrointestinal and liver disease. Curr Opin Gastroenterol. 2008;24(2):176–83.
13. Yadav K, Singh M, Angurana SK, Attri SV, Sharma G, Tageja M, et al. Evaluation of micronutrient profile of north Indian children with cystic fibrosis: a case–control study. Pediatr Res. 2014;75(6):762–6.
14. Angurana SK. Serum Vitamin D status and outcome among critically ill children admitted to the pediatric intensive care unit in South India: correspondence. Indian J Pediatr. 2016;83(8):910–1.
15. Chang S-W, Lee H-C. Vitamin D and health – the missing vitamin in humans. Pediatr Neonatol. 2019;60(3):237–44.
16. Terushkin V, Bender A, Psaty EL, Engelsen O, Wang SQ, Halpern AC. Estimated equivalency of vitamin D production from natural sun exposure versus oral vitamin D supplementation across seasons at two US latitudes. J Am Acad Dermatol. 2010;62(6):929.e1–9.
17. Ross AC, Taylor CL, Yaktine AL, Del Valle HB. Institute of medicine (US) committee to review dietary reference intakes for vitamin D and calcium. Dietary reference intakes for calcium and vitamin D [Internet]. Washington (DC): National Academies Press (US); 2011 [cited 2024 Jan 4]. (The National Academies Collection: Reports funded by National Institutes of Health). Available from: http://www.ncbi.nlm.nih.gov/books/NBK56070/.
18. Roth DE, Abrams SA, Aloia J, Bergeron G, Bourassa MW, Brown KH, et al. Global prevalence and disease burden of vitamin D deficiency: a roadmap for action in low- and middle-income countries. Ann N Y Acad Sci. 2018;1430(1):44–79.

19. Roth DE, Martz P, Yeo R, Prosser C, Bell M, Jones AB. Are national vitamin D guidelines sufficient to maintain adequate blood levels in children? Can J Public Health Rev Can Sante Publique. 2005;96(6):443–9.
20. Malloy PJ, Feldman D. Genetic disorders and defects in vitamin d action. Endocrinol Metab Clin N Am. 2010;39(2):333–46, table of contents.
21. Ozden A, Doneray H, Turkyilmaz A. Two novel CYP2R1 mutations in a family with vitamin D-dependent rickets type 1b. Endocrine. 2021;72(3):852–64.
22. White JH, Vitamin D. Signaling, infectious diseases, and regulation of innate immunity. Infect Immun. 2008;76(9):3837–43.
23. Ponnarmeni S, Kumar Angurana S, Singhi S, Bansal A, Dayal D, Kaur R, et al. Vitamin D deficiency in critically ill children with sepsis. Paediatr Int Child Health. 2016;36(1):15–21.
24. Parameswaran P, Vaidya PC, Attri SV, Angurana SK, Vignesh P, Singh M. Vitamin D deficiency: prevalence and association with intrathoracic tuberculosis in Indian children. Indian J Pediatr. 2021;88(3):276.
25. Jiménez-Sousa MÁ, Martínez I, Medrano LM, Fernández-Rodríguez A, Resino S. Vitamin D in human immunodeficiency virus infection: influence on immunity and disease. Front Immunol. 2018;12(9):458.
26. Kim M, Kim S-N, Lee Y, Choe Y, Ahn K. Vitamin D status and efficacy of Vitamin D supplementation in atopic dermatitis: a systematic review and meta-analysis. Nutrients. 2016;8(12):789.
27. Grant W, Wimalawansa S, Holick M, Cannell J, Pludowski P, Lappe J, et al. Emphasizing the health benefits of Vitamin D for those with neurodevelopmental disorders and intellectual disabilities. Nutrients. 2015;7(3):1538–64.
28. Razavi Khorasani N, Moazzami B, Zahedi Tajrishi F, Mohammadpour Z, Rouhi F, Alizadeh-Navaei R, et al. The association between low levels of Vitamin D and clinical outcomes in critically-ill children: a systematic review and meta-analysis. Fetal Pediatr Pathol. 2020;39(6):503–17.
29. Vitamin D. Deficiency in critically ill. Indian J Crit Care Med. 2017;21(8):541–2.
30. Kumar S, Randhawa MS, Angurana SK, Nallasamy K, Bansal A, Kumar MR, Sachdeva N, Jayashree M. Clinical profile, intensive care needs and outcome of children with dilated cardiomyopathy associated with vitamin D deficiency: a 5-year PICU experience. Indian J Crit Care Med. 2023;27(7):510–4.
31. Tai SS-C, Bedner M, Phinney KW. Development of a candidate reference measurement procedure for the determination of 25-Hydroxyvitamin D_3 and 25-Hydroxyvitamin D_2 in human serum using isotope-dilution liquid chromatography−tandem mass spectrometry. Anal Chem. 2010;82(5):1942–8.
32. Aggarwal V, Seth A, Marwaha RK, Sharma B, Sonkar P, Singh S, et al. Management of Nutritional Rickets in Indian children: a randomized controlled trial. J Trop Pediatr. 2013;59(2):127–33.
33. https://iapindia.org/pdf/Ch-013-Rickets.pdf. In. Available from: https://iapindia.org/pdf/Ch-013-Rickets.pdf
34. Mondal K, Seth A, Marwaha RK, Dhanwal D, Aneja S, Singh R, et al. A randomized controlled trial on safety and efficacy of single intramuscular versus staggered Oral dose of 600 000IU Vitamin D in treatment of nutritional rickets. J Trop Pediatr. 2014;60(3):203–10.
35. Edouard T, Alos N, Chabot G, Roughley P, Glorieux FH, Rauch F. Short- and long-term outcome of patients with pseudo-vitamin D deficiency rickets treated with calcitriol. J Clin Endocrinol Metab. 2011;96(1):82–9.
36. Shroff R, Wan M, Nagler EV, Bakkaloglu S, Fischer D-C, Bishop N, et al. Clinical practice recommendations for native vitamin D therapy in children with chronic kidney disease stages 2-5 and on dialysis. Nephrol Dial Transplant Off Publ Eur Dial Transpl Assoc - Eur Ren Assoc. 2017;32(7):1098–113.
37. Harinarayan C. How to treat Vitamin D deficiency in sun-drenched India – guidelines. J Clin Sci Res. 2018;7(3):131.

Chapter 8
Vitamin D and the Geriatric Population

Sitanshu Barik and Amol Dubepuria

Abstract The geriatric population is characterized by wide ranges of age (60 years onwards), type of ageing (successful and unsuccessful), comorbidities, polypharmacy, and reduced mobility, but paradoxically, with the improvement in medical sciences, the lifespan of an average human being is still increasing. In this backdrop, it is absolutely essential to explore the role of vitamin D in the said population and lay down guidelines which shall help medical professionals to rationalize its supplementation in the said population. This chapter discusses the prevalence of vitamin D deficiency in the elderly, effect of ageing on vitamin D metabolism, effects of vitamin D deficiency with its various drug interactions, and its preventive and treatment strategies.

Keywords Vitamin D · Elderly · Geriatric

8.1 Introduction

Vitamin D has been actively researched in the recent years and its deficiency has been found to be endemic in all age groups irrespective of race and region. Its role in the immune and musculoskeletal system has been emphasized as well as its role in the pathogenesis of various autoimmune diseases. It also influences the morbidity and mortality in various diseases [1]. Numerous reports have presented conflicting reports on the role of vitamin D supplementation in the elderly. The geriatric population is characterized by wide ranges of age (60 years onwards), type of ageing (successful and unsuccessful), comorbidities, polypharmacy, and reduced mobility,

S. Barik (✉) · A. Dubepuria
AIIMS, Nagpur, India

V. Kumar, S. Patel (eds.), *Vitamin D: A Comprehensive Guide for Medical Professionals*, https://doi.org/10.1007/978-981-96-6491-7_8

but paradoxically, with the improvement in medical sciences, the lifespan of an average human being is still increasing. In this backdrop, it is absolutely essential to explore the role of vitamin D in the said population and lay down guidelines which shall help medical professionals to rationalize its supplementation in the said population.

8.2 Prevalence of Vitamin D Deficiency in the Elderly

The elderly population are more susceptible for hypovitaminosis D because of the reduced capacity of skin to synthesize vitamin D. The global prevalence of vitamin D deficiency in the elderly is estimated to be 50% which underlines its public health significance [2]. Levels of 25(OH)D below 50 nmol/l 04 20 ng/ml is considered as vitamin D deficiency with levels between 51 and 74 nmol/l or 21 to 29 ng/ml is considered insufficiency. Sufficient vitamin D levels are considered to be above 75 nmol/l or 30 ng/ml. Correspondingly, the levels of parathormone (PTH) remain within the normal range if the levels of 25(OH)D range between 75 and 100 nmol/l or 31 and 40 ng/ml [3]. Independent of these cut-off values, the elderly are more prone due to reduced synthesis of 7-dehydrocholesterol in the skin, reduced exposure to sunlight, poor dietary intake, chronic diseases, drug interactions, and reduced mobility [4]. Even though vitamin D deficiency has been thought to be common in population at higher altitude and latitude, it has also been noted in tropical climate and at lower altitudes [3, 5]. Although the levels decline in both the genders, it is more steep and starts early in the females in the perimenopausal period whereas in the males the steeper descent starts later after the age of 70 years [6]. Among the susceptible population, 25% are believed to be vitamin D deficient while 50% are believed to be vitamin D insufficient [7].

8.3 Ageing on Vitamin D Production and Levels

The metabolism of vitamin D declines with ageing because of reduced sun exposure as well as reduced capacity of skin to synthesize it. The precursor, 7-dehydrocholesterol, reduces by 50% from the age of 20 to 80 years. It is estimated that with every passing decade, the ability of skin to synthesize previtamin D reduces by 13% [8]. Other factors responsible are relative insensitivity of vitamin D to calcium absorption in the gastrointestinal tract as well as the general decline in renal health. This leads to reduced formation of the active 1,25di(OH)D [9].

Despite the abovementioned factors, a single 15-min exposure of more than 40% of body area to sun will result in sufficient vitamin D production [8]. Hence, although cutaneous synthesis is affected, exposure to sunlight is an important factor in vitamin D metabolism even in the elderly. Exposure of 1000 sqcm of body area daily to artificial UV light is akin to daily supplementation of 800 IU of vitamin D in persons above the age of 70 years [10]. Hence the potential for adequate vitamin D exists since the area of skin in the human body greatly exceeds 1000 sqcm. In the absence of sunlight exposure, artificial UV light can be an important source initiating vitamin D synthesis in the elderly who are in nursing homes or home bound due to lack of mobility [11]. Various factors like time of day, altitude, latitude, use of sunscreen, season, cloudiness, and air pollution are confounding variables which affect the role of sunlight in cutaneous synthesis of pre-vitamin D [12].

Various other factors have a link to reduced metabolism of vitamin D in the elderly. Smoking is known to reduce vitamin D levels, although the mechanism is clearly not known [13]. Higher fat content of the body is related to reduced production as well as higher volume of distribution for vitamin D [14]. The global increase in obesity compounded with reduced physical activity will lead to reduced vitamin D levels [15].

Hence, despite the reduction in vitamin D production with aging, cutaneous reserve capacity for production should be ample. It is, therefore, possible that most older people can produce sufficient vitamin D from the sun.

8.4 Skeletal Effects of Vitamin D Deficiency in the Elderly

Deficiency of vitamin D has direct as well as indirect effects on bone health. There is reduced recruitment and activation of precursor osteoclasts into activated osteoclasts which is mediated by factors secreted by osteoblasts. The indirect effects include reduced mineralization of the osteoid seams due to reduced calcium levels due to vitamin D deficiency. Reports suggest that more than 80% of ambulatory population above the age of 55 years have osteopenia or osteoporosis along with low levels vitamin D [16]. The Osteoporotic Fractures in Men (MrOS) study concluded that vitamin D levels less than 40 nmol/l is associated with greater bone loss around the hip joint which is even greater at ages of 75 years and above [16]. Hence, it can be concluded that low levels of vitamin D have a deleterious effect on bone health of the elderly.

A recent meta-analysis concluded that daily supplementation of vitamin D by 700–800 IU reduces the risk of hip fractures by 26% and other nonvertebral fractures by 23% [17]. This evidence proves the results of previous randomized trials [18]. The supplementing of vitamin D and calcium, and not vitamin D alone, helps in reducing the rate of fragility fractures in the elderly who are institutionalized or reside in nursing homes [19].

Regarding the health of skeletal muscle, vitamin D deficiency has been noted to cause proximal muscle weakness with increased swaying which exacerbates the chances of falls [20]. As mentioned earlier, the defect in osteoid mineralization leads to lesser structural support for the periosteum and muscle covering. Such individuals have chronic pains along the bones which may be misdiagnosed as fibromyalgia, chronic fatigue syndrome, or degenerative disease [21].

Walking speed and improved balance has been noted when vitamin D levels increase from 10 nmol/l to 40 nmol/l which shows an improving trend till 100 nmol/l [22]. Individuals with vitamin D levels less than 25 nmol/l have 2.57 times the chance of experiencing sarcopenia as compared to levels above 50 nmol/l [23]. Low levels of vitamin D have shown to impair physical performance and cause frailty which has been defined as "a biologic syndrome of decreased reserve and resistance to stressors, resulting from cumulative declines across multiple physiologic systems and causing vulnerability to adverse outcomes" [24]. The low levels of vitamin D affects muscle health by reducing its synthesis as well as altering the contractile properties. The contractile properties are affected due to the effect of vitamin D on calcium metabolism and homeostasis. Literature suggests that a 6-min walking distance may be an indicator of healthy levels of vitamin D [25]. Individuals with vitamin D levels below 25 nmol/l have higher odds of decline in physical performance over a 3-year period as compared to levels of above 50 nmol/l [26]. Although a recent meta-analysis concluded that the association between vitamin D and physical performance is controversial and studies have reported divergent results pointing to the complex relationship of vitamin D and its metabolism with physical health [27].

Suboptimal levels of vitamin D are associated with increased risk of falls in the population aged 65 years and above. It has been suggested that supplementation of 800 IU of vitamin D along with calcium daily helps in minimizing falls in the at-risk population [22]. The reduction of falls by 72% was noted as per the supplementation regimen mentioned [28]. In contrast, a recent Cochrane review concluded aerobic exercises as the only significant intervening factor helpful in reduction of falls with the authors suggesting that a multifactorial assessment of the problem is necessary [29].

8.5 Extraskeletal Effects of Vitamin D Deficiency in the Elderly

Numerous tissues in the body with no direct relationship with calcium or phosphorous metabolism express vitamin D receptor, 1-alpha-hydroxylase, and 24-hydroxylase which leads to formation of 1,25di(OH)D when 25(OH)D enters these tissues. This seems to be not influenced by calciotropic hormones like PTH but rather influenced by serum levels of vitamin D, and these actions are responsible for the nonclassical genomic effects of vitamin D. It is known that active vitamin D has similar functions in skeletal muscles as mentioned previously by binding to proteins similar to vitamin D receptor proteins [30]. A few of the reported associations of vitamin D with extraskeletal health which holds particular significance for the elderly are mentioned below:

- Increased risk of colorectal and breast cancers [31].
- Increased risk of autoimmune diseases [32].
- Increased risk of major cardiovascular events along with mortality [33].
- Increased risk of albuminuria and development of chronic kidney disease [34].
- Deterioration of Alzheimer's disease [35].

It is prudent to discuss the evidence that suggests the abovementioned associations since "association" does not mean "causality," and the reported studies are either observational or experimental in nature but not interventional in nature. It needs to be proven whether such associations have any relationship to vitamin D status or simply indicative of poor health status in general. The studies referenced here are few of the many such studies, the majority of which report no association between vitamin D and extraskeletal health status.

8.6 Drug Interactions of Vitamin D in Polypharmacy

With multiple comorbidities being common in the elderly, the population usually are under multiple other medications apart from supplements. Drug interactions of vitamin D can be broadly categorized into two: those that affect intestinal absorption of vitamin D and those that influence metabolism of vitamin D [36].

(a) Lipase inhibitors reduce the hydrolysis of triglycerides in the intestine. This leads to increase in fats in the stools which reduces the pool of vitamin D available for absorption in the intestine.
(b) Statins are widely used in primary and secondary prevention of cardiovascular disorders. It functions by inhibition of HMG-CoA reductase, which is a rate-limiting enzyme in the synthesis of cholesterol. Although the pathways are similar, there is conflicting evidence regarding their synergistic or antagonistic actions [37]. Both metabolism of statins and catabolism of vitamin D involves the actions of CYP3A4. Statins may compete for the active centers of the enzyme leading to decreased vitamin D catabolism and its subsequent elevated levels.
(c) Antiepileptic drugs like barbiturates, benzodiazepines, hydantoins, and aliphatic carboxylic acids are known to influence vitamin D metabolism [38]. These drugs have the potential to induce the catabolic enzymes of vitamin D. This leads to reduced levels in the blood with increased fracture risk. They can also influence intestinal calcium absorption.
(d) Glucocorticoids have also been reported to alter vitamin D metabolism. They are known to induce enzymes which catabolize vitamin D. It has been noted that up to 30% of individuals on glucocorticoids sustain fragility fractures. This can be prevented by vitamin D supplementation.

8.7 Strategies to Prevent Vitamin D Deficiency

8.7.1 Sunlight Exposure

Although higher UV ray exposure can lead to increased synthesis of vitamin D, its risks outweigh the benefits as per the World Health Organization. This strategy can be debated, as seasonal variations in vitamin D levels have been reported. Tropical and sunny countries have better vitamin D profile and lesser incidence of rickets and osteomalacia if diet is taken care of. A previous study from Brazil reports that there is improved concentration of vitamin D after the summers in persons leading a healthy lifestyle of adequate diet and physical activity, but the exceptions noted were females, men over the age of 70 years, and those of African descent. The other group of study population who were institutionalized were noted to have persistent low levels of vitamin D [39]. A similar study from Sweden which is located at a higher latitude did not notice any significant change in the vitamin D levels after summers [40].

Outdoor physical activity can be considered a health improving measure which apart from its importance in vitamin D metabolism also has other benefits relating to overall general health of the being. Education regarding sunburns, use of sunscreens, and early detection of skin lesions can help in achieving balance between the risks and benefits of sunlight exposure.

8.7.2 Diet

Since most foods are scare in vitamin D, fortification of foods commonly used can be a useful strategy to achieve optimum vitamin D levels [41]. The European Calcified Tissue Society has laid out a policy guideline which aims to improve vitamin D levels above 50 nmol/l and reduce the incidence of rickets, osteomalacia, and osteoporosis [42]. The guideline suggests food fortification as an effective strategy for children, adolescents, ethnic minorities, and institutionalized persons. Finland introduced fortification of milk products from 2003 onwards which has helped in improving vitamin D levels across board [43]. Similar reports from Canada, the USA, the UK, and Germany suggest the same.

Fortification of animal feed shall help in improving vitamin D content of the meat products. This program may be limited by factors such as lack of universal dosing for fortification, regulatory policies of each country, and control over industrial production of such foods.

8.7.3 *Vitamin D Supplementation*

Considering the above two factors, supplementation of vitamin D seems to be an easier and achievable way to obtain optimum vitamin D levels. Daily or weekly doses are usually considered to improve compliance among the at-risk population. A meta-regression study suggests a daily dose of 400 IU to prevent levels below 25 nmol/l [44].

Even though attractive, supplementation may be difficult to achieve in persons belonging to lower income group, migrants, and refugees. In contrast, food fortification would cover the whole population, and hence a mixed strategy of food fortification and supplementation shall help achieve optimum vitamin D level in the entire population.

8.8 Treatment of Vitamin D Deficiency in the Elderly

The choice of drug for vitamin D deficiency is cholecalciferol or vitamin D_3. Supplementation with 1000 IU daily shall lead to an increase of 25 nmol/l over the treatment period of 3–4 months. For primary treatment of vitamin D deficiency, daily dosing of 1000 IU vitamin D along with 1 gram of calcium is recommended. 25(OH)D is the drug of choice for patients with chronic liver disease. Similarly, calcitriol (0.5 mg per day) along with calcium (1 g per day) is recommended for patients with chronic kidney disease.

References

1. Weaver CM, Alexander DD, Boushey CJ, Dawson-Hughes B, Lappe JM, LeBoff MS, et al. Calcium plus vitamin D supplementation and risk of fractures: an updated meta-analysis from the National Osteoporosis Foundation. Osteoporos Int. 2016;27(1):367–76.
2. Holick MF. Vitamin D deficiency. N Engl J Med. 2007;357(3):266–81.
3. Holick MF, Chen TC. Vitamin D deficiency: a worldwide problem with health consequences. Am J Clin Nutr. 2008;87(4):1080S–6S.
4. Oudshoorn C, van der Cammen TJM, McMurdo MET, van Leeuwen JPTM, Colin EM. Ageing and vitamin D deficiency: effects on calcium homeostasis and considerations for vitamin D supplementation. Br J Nutr. 2009;101(11):1597–606.
5. Raj V, Barik S, Raj M. Attention to the bone health of a neglected rural-tribal population in India: a pilot study. Indian J Community Med. 2023;48(3):501–4.
6. Maggio D, Cherubini A, Lauretani F, Russo RC, Bartali B, Pierandrei M, et al. 25(OH)D serum levels decline with age earlier in women than in men and less efficiently prevent compensatory hyperparathyroidism in older adults. J Gerontol A Biol Sci Med Sci. 2005;60(11):1414–9.
7. Orwoll E, Nielson CM, Marshall LM, Lambert L, Holton KF, Hoffman AR, et al. Vitamin D deficiency in older men. J Clin Endocrinol Metab. 2009;94(4):1214–22.

8. Chalcraft JR, Cardinal LM, Wechsler PJ, Hollis BW, Gerow KG, Alexander BM, et al. Vitamin D synthesis following a single bout of sun exposure in older and younger men and women. Nutrients. 2020;12(8):2237.
9. de Jongh RT, van Schoor NM, Lips P. Changes in vitamin D endocrinology during aging in adults. Mol Cell Endocrinol. 2017;453:144–50.
10. Chel VG, Ooms ME, Popp-Snijders C, Pavel S, Schothorst AA, Meulemans CC, et al. Ultraviolet irradiation corrects vitamin D deficiency and suppresses secondary hyperparathyroidism in the elderly. J Bone Miner Res. 1998;13(8):1238–42.
11. Chel VGM, Ooms ME, Pavel S, de Gruijl F, Brand A, Lips P. Prevention and treatment of vitamin D deficiency in Dutch psychogeriatric nursing home residents by weekly half-body UVB exposure after showering: a pilot study. Age Ageing. 2011;40(2):211–4.
12. Mousavi SE, Amini H, Heydarpour P, Amini Chermahini F, Godderis L. Air pollution, environmental chemicals, and smoking may trigger vitamin D deficiency: evidence and potential mechanisms. Environ Int. 2019;122:67–90.
13. Cutillas-Marco E, Fuertes-Prosper A, Grant WB, Morales-Suárez-Varela M. Vitamin D deficiency in South Europe: effect of smoking and aging. Photodermatol Photoimmunol Photomed. 2012;28(3):159–61.
14. Snijder MB, van Dam RM, Visser M, Deeg DJH, Dekker JM, Bouter LM, et al. Adiposity in relation to vitamin D status and parathyroid hormone levels: a population-based study in older men and women. J Clin Endocrinol Metab. 2005;90(7):4119–23.
15. Bixby H, Bentham J, Zhou B, Di Cesare M, Paciorek CJ, Bennett JE, et al. Rising rural body-mass index is the main driver of the global obesity epidemic in adults. Nature. 2019;569(7755):260–4.
16. Seton M, Jackson V, Lasser KE, Doppelt S, Pierre-Jacques M, Connelly M. Low 25-hydroxyvitamin D and osteopenia are prevalent in persons > or =55 yr with fracture at any site: a prospective, observational study of persons fracturing in the community. J Clin Densitom. 2005;8(4):454–60.
17. Bischoff-Ferrari HA, Dawson-Hughes B, Baron JA, Burckhardt P, Li R, Spiegelman D, et al. Calcium intake and hip fracture risk in men and women: a meta-analysis of prospective cohort studies and randomized controlled trials. Am J Clin Nutr. 2007;86(6):1780–90.
18. Dawson-Hughes B, Harris SS, Krall EA, Dallal GE. Effect of calcium and vitamin D supplementation on bone density in men and women 65 years of age or older. N Engl J Med. 1997;337(10):670–6.
19. Avenell A, Gillespie WJ, Gillespie LD, O'Connell DL. Vitamin D and vitamin D analogues for preventing fractures associated with involutional and post-menopausal osteoporosis. Cochrane Database Syst Rev. 2005;3:CD000227.
20. Dam TTL, von Mühlen D, Barrett-Connor EL. Sex-specific association of serum vitamin D levels with physical function in older adults. Osteoporos Int. 2009;20(5):751–60.
21. Hicks GE, Shardell M, Miller RR, Bandinelli S, Guralnik J, Cherubini A, et al. Associations between vitamin D status and pain in older adults: the Invecchiare in Chianti study. J Am Geriatr Soc. 2008;56(5):785–91.
22. Bischoff-Ferrari HA, Dietrich T, Orav EJ, Hu FB, Zhang Y, Karlson EW, et al. Higher 25-hydroxyvitamin D concentrations are associated with better lower-extremity function in both active and inactive persons aged > or =60 y. Am J Clin Nutr. 2004;80(3):752–8.
23. Visser M, Deeg DJH, Lips P. Longitudinal aging study Amsterdam. Low vitamin D and high parathyroid hormone levels as determinants of loss of muscle strength and muscle mass (sarcopenia): the longitudinal aging study Amsterdam. J Clin Endocrinol Metab. 2003;88(12):5766–72.
24. Shardell M, Hicks GE, Miller RR, Kritchevsky S, Andersen D, Bandinelli S, et al. Association of low vitamin D levels with the frailty syndrome in men and women. J Gerontol A Biol Sci Med Sci. 2009;64(1):69–75.

25. Boxer RS, Dauser DA, Walsh SJ, Hager WD, Kenny AM. The association between vitamin D and inflammation with the 6-minute walk and frailty in patients with heart failure. J Am Geriatr Soc. 2008;56(3):454–61.
26. Wicherts IS, van Schoor NM, Boeke AJP, Visser M, Deeg DJH, Smit J, et al. Vitamin D status predicts physical performance and its decline in older persons. J Clin Endocrinol Metab. 2007;92(6):2058–65.
27. Annweiler C, Schott AM, Berrut G, Fantino B, Beauchet O. Vitamin D-related changes in physical performance: a systematic review. J Nutr Health Aging. 2009;13(10):893–8.
28. Bischoff-Ferrari HA, Willett WC, Wong JB, Giovannucci E, Dietrich T, Dawson-Hughes B. Fracture prevention with vitamin D supplementation: a meta-analysis of randomized controlled trials. JAMA. 2005;293(18):2257–64.
29. Gillespie LD, Robertson MC, Gillespie WJ, Sherrington C, Gates S, Clemson LM, et al. Interventions for preventing falls in older people living in the community. Cochrane Database Syst Rev. 2012;2012(9):CD007146.
30. Nemere I, Schwartz Z, Pedrozo H, Sylvia VL, Dean DD, Boyan BD. Identification of a membrane receptor for 1,25-dihydroxyvitamin D3 which mediates rapid activation of protein kinase C. J Bone Miner Res. 1998;13(9):1353–9.
31. Ma Y, Zhang P, Wang F, Yang J, Liu Z, Qin H. Association between vitamin D and risk of colorectal cancer: a systematic review of prospective studies. J Clin Oncol. 2011;29(28):3775–82.
32. Zanetti M, Harris SS, Dawson-Hughes B. Ability of vitamin D to reduce inflammation in adults without acute illness. Nutr Rev. 2014;72(2):95–8.
33. Leu M, Giovannucci E. Vitamin D: epidemiology of cardiovascular risks and events. Best Pract Res Clin Endocrinol Metab. 2011;25(4):633–46.
34. de Boer IH, Katz R, Chonchol M, Ix JH, Sarnak MJ, Shlipak MG, et al. Serum 25-hydroxyvitamin D and change in estimated glomerular filtration rate. Clin J Am Soc Nephrol. 2011;6(9):2141–9.
35. Annweiler C, Llewellyn DJ, Beauchet O. Low serum vitamin D concentrations in Alzheimer's disease: a systematic review and meta-analysis. J Alzheimers Dis. 2013;33(3):659–74.
36. Robien K, Oppeneer SJ, Kelly JA, Hamilton-Reeves JM. Drug-vitamin D interactions: a systematic review of the literature. Nutr Clin Pract. 2013;28(2):194–208.
37. Mazidi M, Rezaie P, Vatanparast H, Kengne AP. Effect of statins on serum vitamin D concentrations: a systematic review and meta-analysis. Eur J Clin Investig. 2017;47(1):93–101.
38. Siniscalchi A, Murphy S, Cione E, Piro L, Sarro GD, Gallelli L. Antiepileptic drugs and bone health: current concepts. Psychopharmacol Bull. 2020;50(2):36–44.
39. Maeda SS, Kunii IS, Hayashi LF, Lazaretti-Castro M. Increases in summer serum 25-hydroxyvitamin D (25OHD) concentrations in elderly subjects in São Paulo, Brazil vary with age, gender and ethnicity. BMC Endocr Disord. 2010;10(1):12.
40. Samefors M, Tengblad A, Östgren CJ. Sunlight exposure and vitamin D levels in older people-an intervention study in Swedish nursing homes. J Nutr Health Aging. 2020;24(10):1047–52.
41. Pinheiro MM, Schuch NJ, Genaro PS, Ciconelli RM, Ferraz MB, Martini LA. Nutrient intakes related to osteoporotic fractures in men and women–the Brazilian Osteoporosis Study (BRAZOS). Nutr J. 2009;8:6.
42. Lips P, Cashman KD, Lamberg-Allardt C, Bischoff-Ferrari HA, Obermayer-Pietsch B, Bianchi ML, et al. Current vitamin D status in European and Middle East countries and strategies to prevent vitamin D deficiency: a position statement of the European calcified tissue society. Eur J Endocrinol. 2019;180(4):P23–54.
43. Cashman KD, Vitamin D. Deficiency: defining, prevalence, causes, and strategies of addressing. Calcif Tissue Int. 2020;106(1):14–29.
44. Cashman KD, Ritz C, Kiely M, Odin Collaborators null. Improved dietary guidelines for vitamin D: application of individual participant data (IPD)-level meta-regression analyses. Nutrients. 2017;9(5):469.

Chapter 9
Vitamin D and Musculoskeletal System

Sandeep Patel and Tarkik Thami

Abstract Vitamin D plays an imperative role in maintaining musculoskeletal health by regulating calcium homeostasis and bone metabolism. It enhances intestinal calcium absorption, promoting optimal bone mineralization and preventing disorders like rickets and osteomalacia. Beyond skeletal health, vitamin D influences muscle function, strength, and coordination, reducing the risk of falls and fragility fractures in the elderly. Deficiency in vitamin D is prevalent globally and is linked to impaired bone density and increased fracture risk. Emerging evidence also highlights its role in managing chronic musculoskeletal disorders, such as fibromyalgia, chronic fatigue syndrome, and osteoarthritis. Chronic vitamin D deficiency has been implicated in exacerbating symptoms of fibromyalgia through its effects on pain modulation, inflammation, and muscle function. Among other effects, vitamin D has also been associated with improved muscle strength, reduced injury risk, and enhanced performance in athletes. Conversely, its deficiency can impair physical function and predispose individuals to stress fractures and muscle injuries. Factors such as limited sun exposure, dietary insufficiencies, and aging contribute to suboptimal vitamin D levels, necessitating supplementation in at-risk populations. This chapter underscores the importance of vitamin D in musculoskeletal health and advocates for its adequate intake to improve quality of life.

Keywords Vitamin-D · Fibromyalgia · Musculoskeletal health · Sports medicine

---

S. Patel · T. Thami (✉)
Department of Orthopaedics, PGIMER, Chandigarh, India

V. Kumar, S. Patel (eds.), *Vitamin D: A Comprehensive Guide for Medical Professionals*, https://doi.org/10.1007/978-981-96-6491-7_9

Learning Objectives

- Understanding the functions of vitamin D (1,25$(OH)_2$_D_3) in relation to bone health, bone mineralization, and regulation of bone mineral density (BMD).
- Explaining the pathways through which 1,25$(OH)_2$_D_3 impacts muscle health, including its role in muscle function, strength, and performance.
- Studying the association between common encountered orthopedic outpatient ailments and vitamin D deficiency.
- Analyzing the potential benefits and risks of vitamin D supplementation for bone health, considering factors such as age, dietary habits, and sunlight exposure.

9.1 Introduction

Vitamin D, aka the "sunshine vitamin," is a crucial nutrient that plays a vital role in human bodily functions. Unlike other vitamins, vitamin D behaves more like a hormone, exerting its effects on multiple organ systems. The principal role of vitamin D is regulation of calcium and phosphorus metabolism which is necessary for muscle function and ionic equilibrium inside our body [1]. Vitamin D is a fat-soluble vitamin and can be produced in the skin upon exposure to ultraviolet rays (from natural sunlight) (Fig. 9.1). Alternatively, it can be obtained through dietary sources such as eggs, red meats, and fish [2]. Vitamin D accrued from sun exposure or diet is biologically inert and needs to be converted to its active metabolites before it can function like a hormone. The kidney and the liver play an important role in double hydroxylation of vitamin D as a two-step function. First hydroxylation of vitamin D to 25-hydroxyvitamin D occurs in the liver, and second hydroxylation to 1–25-dihydroxyvitamin D takes place in the kidneys. Vitamin D is intricately related to the functioning of the musculoskeletal system, and its impact on bone health is of supreme importance [1].

Today, vitamin D is recognized as a multifunctional hormone that plays crucial roles in calcium homeostasis, bone health, muscle function, immune regulation, and many other physiological processes [1, 2]. Ongoing research continues to uncover new lights into the diverse functions of 1,25$(OH)_2$_D_3 and its implications for health and disease [3].

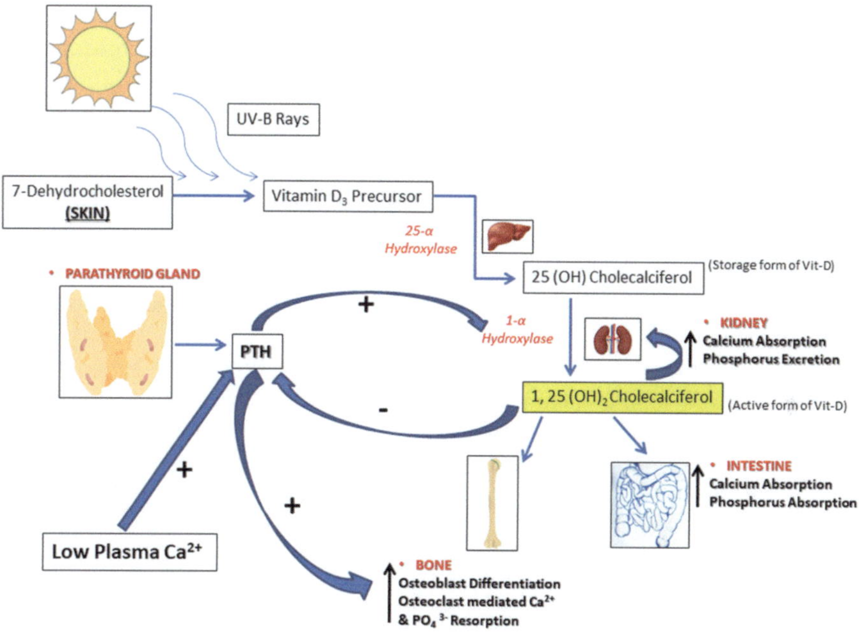

Fig. 9.1 Flowchart depicting vitamin-D homeostasis and calcium metabolism

9.2 Vitamin D and Bone Health

A. **Physiology and Metabolism of Vitamin D.**

Vitamin D is a precursor chemical that gives rise to its metabolites, which function as prohormones, exerting crucial endocrine effects. In order to maintain serum Ca^{2+} levels, the biologically active metabolite $1,25(OH)_2$-D-3 coordinates and facilitates the mobilization of calcium in diet through absorption via gut, reabsorption of filtered calcium in the kidney, and also helps transport calcium to the bone for maturation and mineralization (Fig. 9.1). It also suppresses parathormone secretion and prevents demineralization of bones while its deficiency secretes parathormone and mobilizes calcium from bones to maintain serum calcium levels with consequent demineralization of bones leading to osteoporosis, osteomalacia, and pathological fractures in adults and rickets in children. Calcium homeostasis is important for routine cell metabolism, cellular functions like signal transduction, and neuromuscular stimulus transmission [3, 4].

B. **Role in Musculoskeletal Health.**

Low serum concentrations of vitamin D are associated with improper mineralization of long bones of the growing skeleton in human beings [4]. Various studies have proved that both calcium and vitamin D are needed in adequate amounts to maintain bone health and muscle function. Most of the meta-analyses have looked into the effect of vitamin D on bone mineral density (BMD) and fracture incidence [2, 5]. Low vitamin D status triggers secondary hyperparathyroidism, increases bone resorption, and leads to myalgias. Based on recently published data, it has been recommended to consume 800 IU of vitamin D per day, combined with 1200 mg of calcium per day, to prevent fractures and falls in older adults [6, 7].

Multiple recent research articles have indicated a correlation between deficient levels of serum vitamin D (<80 nmol/L) and diminished muscle strength, susceptibility to muscle-tendon unit injuries, infections, and even impaired training efficacy among young athletes [7–10]. A recent meta-analysis aimed to assess the link between vitamin D status and sports-related injuries; it was found that serum vitamin D levels below 75 nmol/L were linked to a heightened risk of stress fractures, with a higher odds ratio observed for stress fractures in cases of vitamin D insufficiency. Girgis et al. [11] have underscored the multifaceted role of vitamin D beyond bone health, highlighting its influence on muscle strength, hastened muscle aging, and hindered muscle repair and function. 1,25$(OH)_2$_D-3 exerts its effect on muscle function through the vitamin D receptor (VDR); muscle function can also vary in individuals based on the type of genome sequencing of VDR [11]. It has been recommended that maintaining 25(OH)D levels above 20 nmol/L is crucial for optimal calcium balance, thereby preventing muscle dysfunction. Figure 9.2 highlights the effects of vitamin D deficiency on the human musculoskeletal system.

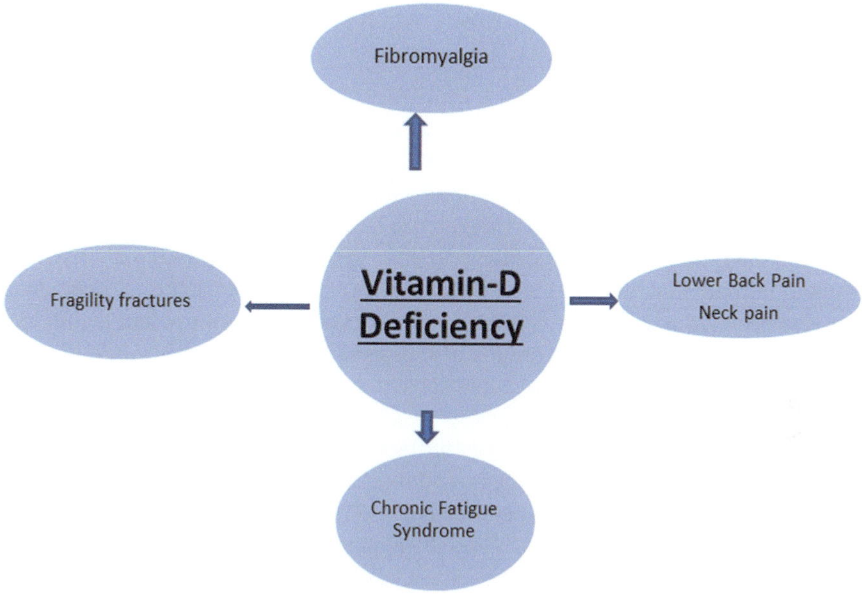

Fig. 9.2 Effects of vitamin D deficiency on the musculoskeletal system

9.3 Vitamin D Deficiency and Fibromyalgia

The biological mechanisms underlying the association between fibromyalgia and vitamin D insufficiency are not fully elucidated yet. However, vitamin D receptors are present in various tissues, including those involved in pain perception and immune system regulation. It is hypothesized that vitamin D may modulate pain pathways and immune system function, thereby influencing the development and progression of fibromyalgia symptoms.

The potential association between fibromyalgia and vitamin D deficiency has important clinical implications for the management and treatment of the condition. Healthcare providers may consider assessing vitamin D levels in patients with fibromyalgia, particularly those with persistent symptoms or risk factors for deficiency. Supplementation with vitamin D can be advocated for patients with lower levels, although further research is needed to determine the optimal dosage and its impact on fibromyalgia symptoms [12]. The vitamin D receptor exhibits extensive expression in muscles, hence exerting a direct regulatory influence. 25OHD also has a potential role in myogenesis, cell proliferation, differentiation, protein synthesis, and mitochondrial function [3, 6]. Continued research in this area is essential to better understand the role of vitamin D in fibromyalgia and improve treatment outcomes for affected individuals.

9.4 Vitamin D Deficiency and Chronic Fatigue Syndrome (CFS)

Vitamin D has been the subject of interest in the context of chronic fatigue syndrome (CFS) due to its *role in immune function, inflammation regulation*, and overall health [13, 14]. Certain research findings indicate a potential correlation between inadequate vitamin D levels and a heightened susceptibility to chronic fatigue syndrome (CFS), while some indications propose that supplementation of vitamin D might offer potential benefits for individuals afflicted by CFS. Vitamin D deficiency has also been associated with weakness of proximal muscle groups and mainly the type II muscle fibers in these groups of muscles. Another possible mechanism of muscle weakness is the serum hypophosphatemia caused by secondary hyperparathyroidism due to vitamin D deficiency.

It's imperative to note that CFS is an incompletely elucidated condition with multiple potential contributing factors, and there is unlikely to be a one-size-fits-all approach to treatment. While some individuals with CFS may improve clinically by optimizing their vitamin D levels through supplementation or other means, others may not experience significant improvements. A higher incidence of vitamin D deficiency/ insufficiency is commonly found in individuals experiencing fatigue alongside stable chronic medical conditions. Administering ergocalciferol therapy to restore normal levels of vitamin D has been shown to notably alleviate the severity of fatigue symptoms in these patients [15].

Nonetheless, the research on the association between vitamin D and CFS is still limited and inconclusive. Some studies have found a correlation between low vitamin D levels and CFS symptoms, while others have not. Additionally, the effectiveness of vitamin D supplementation in improving symptoms of CFS remains uncertain, as results from clinical trials have been mixed.

9.5 Vitamin D Deficiency and the Spine

Vitamin-D deficiency in adults can cause osteomalacia, a disorder characterized by weakened bones and increased susceptibility to fractures, which may manifest as back pain. Furthermore, vitamin D has anti-inflammatory properties and may modulate pain signaling pathways, thereby influencing back pain perception and severity. A few studies have reported a positive association between lower back pain and vitamin D deficiency, while some others have reported that there is no statistically significant association between them. A recent meta-analysis [16] reports that ethnicity, gender, duration of symptoms, and severity of vitamin D insufficiency may contribute to pathogenesis of chronic lower back pain in these patients. Supplementation with vitamin D should be advocated for patients with decreased serum levels, usually in combination with other treatments for management of back pain.

Serum vitamin D is frequently low even in postmenopausal females, predisposing them to osteopenia and osteoporosis which places them at a higher risk of fragility fractures of the spine [17]. Osteoporotic fractures of the spine usually present as chronic dull aching pain usually associated with paraspinal muscle spasm with occasional episodes of acute exacerbation of pain. As per data from the National Osteoporosis Foundation, osteopenia or osteoporosis affects approximately 30 million women and 14 million men in the United States [5, 18]. The decline in physical function among patients following vertebral fractures is significant and akin to that observed in cases of hip fractures.

Several mechanisms explain how vitamin D deficiency predisposes individuals to fragility fractures. Firstly, inadequate vitamin D levels lead to decreased calcium absorption, resulting in impaired bone mineralization and reduced bone strength. Secondly, vitamin D insufficiency compromises muscle function, balance, and coordination, increasing the risk of falls and subsequent fractures. Moreover, vitamin D plays a role in muscle–bone crosstalk, influencing bone remodeling and fracture healing processes [11, 19]. Strategies aimed at improving vitamin D status, such as supplementation, dietary modifications, and sunlight exposure, are crucial for reducing fracture risk, especially among high-risk populations, including older adults, institutionalized individuals, and those with limited sun exposure. Addressing vitamin D insufficiency through targeted interventions holds promise for reducing the burden of fragility fractures, improving the quality of life, and promoting healthy aging.

Vitamin D deficiency can indirectly predispose an individual to neck pain since it is closely associated with calcium metabolism. Both extracellular and intracellular calcium levels are essential for muscle function and recovery. Some recent studies [20] have studied the associations between neck pain and serum vitamin D levels, but more studies are required to study the association between these two variables.

9.6 Vitamin D Deficiency and Sports Medicine

Adequate vitamin D levels are essential for maintaining musculoskeletal health, crucial for athletes' performance and injury prevention. Vitamin D regulates calcium absorption, supporting bone mineralization and strength, which is vital for withstanding the physical demands of sports activities. Moreover, vitamin D deficiency is associated with an increased risk of stress fractures, muscle injuries, and impaired muscle function, highlighting its significance in athlete health [9, 10].

Emerging evidence suggests that optimizing vitamin D status may positively influence athletic performance [9]. Vitamin D receptors are present in skeletal muscles, indicating its potential role in muscle function and strength especially in calcium intracellular signaling pathways [20]. Studies have shown associations between vitamin D status and measures of muscle strength, power, and endurance, suggesting that adequate vitamin D levels may enhance athletic performance, particularly in activities requiring explosive movements and endurance [21]. Vitamin D supplementation is particularly essential for athletes engaging in indoor sports since they have poor exposure to sunlight hence compromising their endogenous vitamin D production. High-demand athletes should prioritize maintaining optimal vitamin D levels through sun exposure, dietary sources, and supplementation as needed to support their training, performance, and overall well-being.

9.7 Prevention of Vitamin D Deficiency

The best option in prevention of vitamin D deficiency is to enhance our own body's production of vitamin D from exposure of skin to natural sun. It is estimated that 15–20 minutes of sun exposure on a daily basis is sufficient to meet the needs of vitamin D. It is pertinent to remember here that foods which contain a good amount of vitamin D are just a few. Milk and dairy products like cheese, soya, tofu and curd, eggs, meat (beef), citrus fruit juices, mushrooms, and freshwater fish are good sources of natural vitamin D.

9.8 Conclusion

Vitamin D plays a pivotal role for optimizing the bone health and muscle function. Extensive evidence underscores the importance of maintaining the minimum required serum $1,25(OH)_2\_D_3$ levels for the upkeep of bone health and for effective prevention and treatment of osteoporosis when coupled with sufficient Ca^{2+} intake. A deeper understanding of the distinct contributions of dietary vitamin D intake and sunlight exposure, alongside various influential factors, on clinical outcomes, will greatly aid in formulating meaningful and tailored recommendations for different demographic groups. This information will be instrumental in shaping public health policies aimed at efficiently addressing vitamin D deficiency. Presently, there is consensus that adequate sun exposure or vitamin D supplementation is advisable for institutionalized individuals, the frail elderly, and those with serum $1,25(OH)_2\_D_3$ levels below 25 nmol/L. Additionally, it is increasingly recognized that individuals with limited sun exposure are likely to benefit from vitamin D supplementation to maintain sufficient serum $1,25(OH)_2\_D_3$ levels.

To summarize, vitamin D plays a pivotal role in promoting optimal musculoskeletal health by facilitating proper bone mineralization, supporting muscle function, and mitigating the risk of musculoskeletal disorders.

References

1. Rebelos E, Tentolouris N, Jude E. The role of vitamin D in health and disease: a narrative review on the mechanisms linking vitamin D with disease and the effects of supplementation. Drugs. 2023;83:665–85. https://doi.org/10.1007/s40265-023-01875-8.
2. Mendes MM, Botelho PB, Ribeiro H. Vitamin D and musculoskeletal health: outstanding aspects to be considered in the light of current evidence. Endocr Connect. 2022;11(10):e210596.
3. Raiten DJ, Picciano MF. Vitamin D and health in the 21st century: bone and beyond. Executive summary. Am J Clin Nutr. 2004;80(6 Supplement):1673S–7S. https://doi.org/10.1093/ajcn/80.6.1673S.
4. Bouillon R, Van Schoor NM, Gielen E, Boonen S, Mathieu C, Vanderschueren D, Lips P. Optimal vitamin D status: a critical analysis on the basis of evidence-based medicine. J Clin Endocrinol Metab. 2013;98:E1283–304. https://doi.org/10.1210/jc.2013-1195.
5. Avenell A, Mak JC, O'Connell D. Vitamin D and vitamin D analogues for preventing fractures in post-menopausal women and older men. Cochrane Database Syst Rev. 2014;4:CD000227. https://doi.org/10.1002/14651858.CD000227.pub4.
6. Chapuy MC, Pamphile R, Paris E, et al. Combined calcium and vitamin D3 supplementation in elderly women: confirmation of reversal of secondary hyperparathyroidism and hip fracture risk: the Decalyos II study. Osteoporosis Int. 2002;13:257–64.
7. Bevilacqua G, Laskou F, Clynes MA, Jameson KA, Boucher BJ, Noonan K, Cooper C, Dennison EM. Determinants of circulating 25-hydroxyvitamin D concentration and its association with musculoskeletal health in midlife: findings from the Hertfordshire Cohort Study. Metabolism Open. 2021;12:100143. https://doi.org/10.1016/j.metop.2021.100143.
8. Yoon S, Kwon O, Kim J. Vitamin D in athletes: focus on physical performance and musculoskeletal injuries. Phys Act Nutr. 2021;25:20–5. https://doi.org/10.20463/pan.2021.0011.
9. Ammerman BM, Ling D, Callahan LR, Hannafin JA, Goolsby MA. Prevalence of vitamin D insufficiency and deficiency in young, female patients with lower extremity musculoskeletal complaints. Sports Health. 2021;13:173–80. https://doi.org/10.1177/1941738120953414.
10. Jakobsen MM, Nygaard RH, Hojbjerg JA, Larsen JB. The association between vitamin D status and overuse sport injuries: a systematic review and meta-analysis. Transl Sports Med. 2021;4:553–64. https://doi.org/10.1002/tsm2.269.
11. Girgis CM, Cha KM, So B. Mice with myocyte deletion of vitamin D receptor have sarcopenia and impaired muscle function: myocyte VDR mice have impaired muscle function. J Cachexia Sarcopenia Muscle. 2019; https://doi.org/10.1002/jcsm.12460.
12. Ellis SD, Kelly ST, Shurlock JH, Hepburn ALN. The role of vitamin D testing and replacement in fibromyalgia: a systematic literature review. BMC Rheumatol. 2018;2:28. https://doi.org/10.1186/s41927-018-0035-6.
13. Nowak A, Boesch L, Andres E, Battegay E, Hornemann T, Schmid C, Bischoff-Ferrari HA, Suter PM, Krayenbuehl P-A. Effect of vitamin D3 on self-perceived fatigue: a double-blind randomized placebo-controlled trial. Medicine. 2016;95(52):e5353. https://doi.org/10.1097/MD.0000000000005353.
14. Gloth FM, Lindsay JM, Zelesnick LB, Greenough WB. Can vitamin D deficiency produce an unusual pain syndrome? Arch Intern Med. 1991;151(8):1662–4. https://doi.org/10.1001/archinte.1991.00400080144029.
15. Roy S, Sherman A, Monari-Sparks MJ, Schweiker O, Hunter K. Correction of low vitamin D improves fatigue: effect of correction of low vitamin D in fatigue study (EViDiF Study). N Am J Med Sci. 2014;6(8):396–402. https://doi.org/10.4103/1947-2714.139291.

16. Zadro J, Shirley D, Ferreira M, Carvalho-Silva AP, Lamb SE, Cooper C, Ferreira PH. Mapping the association between Vitamin D and low back pain: a systematic review and meta-analysis of observational studies. Pain Physician. 2017;20(7):611–40.
17. Suganthan N, Kumanan T, Kesavan V, Aravinthan N, Rajeshkannan N. Vitamin D status among postmenopausal osteoporotic women: a hospital based cross-sectional study from Northern Sri Lanka. BMC Nutr. 2020;6:15. https://doi.org/10.1186/s40795-020-00341-y.
18. Sözen T, Özışık L, Başaran NÇ. An overview and management of osteoporosis. Eur J Rheumatol. 2017;4(1):46–56. https://doi.org/10.5152/eurjrheum.2016.048.
19. Crescioli C. Vitamin D restores skeletal muscle cell remodeling and myogenic program: potential impact on human health. Int J Mol Sci. 2021;22(4):1760. https://doi.org/10.3390/ijms22041760.
20. Cai C. Treating vitamin D deficiency and insufficiency in chronic neck and back pain and muscle spasm: a case series. Perm J. 2019;23:18–241. https://doi.org/10.7812/TPP/18.241.
21. de la Puente YM, Collado Yurrita L, Ciudad Cabañas MJ, Cuadrado Cenzual MA. Role of vitamin D in athletes and their performance: current concepts and new trends. Nutrients. 2020;12(2):579. https://doi.org/10.3390/nu12020579.

Chapter 10
Vitamin D and the Endocrine System

Trupti N. Prasad and Rimesh Pal

Abstract Vitamin D is critical for bone and mineral metabolism. The two major sources of vitamin D in the human body are sun exposure and diet accounting for 80% and 20%, respectively. In the skin, vitamin D synthesis occurs in the Malpighian layer of the epidermis; exposure of 7-dehydrocholestrol to ultraviolet B rays produces previtamin D which is then converted to vitamin D. Vitamin D then undergoes a series of hydroxylations in the liver and kidney, respectively, to generate the active form. The main physiological role of vitamin D is to regulate calcium and phosphorus metabolism and maintain bone health. This constitutes the classical action of vitamin D. In addition, vitamin D also plays some putative role in immune regulation, insulin secretion and glucose metabolism, reproductive function, pituitary hormonal secretion, and other endocrine systems that comprise the nonclassical actions of the hormone. The exact mechanisms of many of these extraskeletal actions are still unclear and need further investigation.

Keywords Vitamin D · Endocrine system · Bone and mineral metabolism · Extraskeletal effects

Learning Objectives
- Production, activation, and metabolism of vitamin D.
- Classic and nonclassic actions of vitamin D.
- Interaction of vitamin D with other hormones.
- Disorders related to vitamin D.

T. N. Prasad · R. Pal (✉)
Department of Endocrinology, Postgraduate Institute of Medical Education and Research (PGIMER), Chandigarh, India

V. Kumar, S. Patel (eds.), *Vitamin D: A Comprehensive Guide for Medical Professionals*, https://doi.org/10.1007/978-981-96-6491-7_10

10.1 Introduction

Vitamin D is critical for bone and mineral metabolism; however, in addition, it interacts with other endocrine systems as well. This chapter aims to summarize the interaction of vitamin D with the endocrine systems of the human body.

10.2 Vitamin D Metabolism in Humans

The two major sources of vitamin D in the human body are sun exposure and diet accounting for 80% and 20%, respectively. Vitamin D obtained from diet exists as either vitamin D_2 (ergocalciferol) which is of plant origin or as vitamin D_3 (cholecalciferol) which is obtained from animal sources. Both forms have equivalent potencies. In the skin, vitamin D synthesis occurs in the Malpighian layer of the epidermis; exposure of 7-dehydrocholestrol to ultraviolet B rays produces previtamin D which is then converted to vitamin D.

In circulation, vitamin D is bound to vitamin D binding protein (VDBP) (85–88%) while around 12–15% is bound to albumin and only less than 0.1% is in free/unbound form. Vitamin D then undergoes a series of hydroxylations to generate the active form. In the liver, vitamin D undergoes 25-hydroxylation by the enzyme CYP27A1, which is not tightly regulated. 25(OH)D undergoes 1α-hydroxylation to yield calcitriol [1,25$(OH)_2$D] in the proximal convoluted tubule of kidneys by the enzyme CYP2B1 which is under tight regulation. Parathyroid hormone (PTH) and low serum phosphate are major inducers of 1α-hydroxylation. 1,25$(OH)_2$D, FGF23, and raised serum calcium, in turn, suppress 1α-hydroxylation. The active form of vitamin D then acts on the vitamin D receptor (VDR) which is a nuclear hormone receptor. The classic target organs for the action of vitamin D include the bone, kidney, and intestine which express VDR in high concentrations. VDR is also widely expressed throughout the body. 1,25$(OH)_2$D binds to VDR which then heterodimerizes with retinoid X receptors and interacts with vitamin D response element (VDRE). Inactivation of vitamin D occurs mainly via 24-hydroxylation by the enzyme CYP24A1 (24-hydroxylase). 1,25$(OH)_2$D and fibroblast growth factor-23 (FGF23) are the major inducers of this enzyme.

Functions of Vitamin D The main physiological role of vitamin D is to regulate calcium and phosphorus metabolism and maintain bone health. This constitutes the classical action of vitamin D. In addition, vitamin D also plays some putative role in immune regulation, insulin secretion and glucose metabolism, reproductive function, pituitary hormonal secretion, and other endocrine systems that comprise the nonclassical actions of the hormone. The exact mechanisms of many of these extraskeletal actions are still unclear and need further investigation.

Vitamin D in Calcium and Phosphate Homeostasis Vitamin D along with PTH and FGF-23 regulates calcium and phosphate metabolism. Vitamin D promotes intestinal calcium absorption via ion channels TRPV5 and TRPV6, accounting for 20–70% of ingested calcium absorption. It also increases renal calcium reabsorption, particularly in the distal convoluted tubules. In the bones, VDR is present on the osteoblasts where $1,25(OH)_2D$ promotes the differentiation of osteoblasts and increases RANKL expression which stimulates osteoclastic activity. $1,25(OH)_2D$ is also required for the production of collagen and markers of bone formation such as osteocalcin and alkaline phosphatase.

$1,25(OH)_2D$ inhibits PTH secretion and decreases the proliferation of PTH-producing cells in the parathyroid gland. $1,25(OH)_2D$ also increases the sensitivity of the calcium-sensing receptor (CaSR) on the parathyroid glands to serum calcium which further inhibits PTH secretion. In this regard, active vitamin D analogues are thus used for the management of secondary hyperparathyroidism especially in the setting of chronic kidney disease.

1α-hydroxylase is also expressed in macrophages, placenta, and various other cells. The extrarenal secretion of $1,25(OH)_2D$ is not under strict hormonal regulation. Extrarenal 1α-hydroxylase is stimulated by cytokines such as interferon-gamma (IFN-γ) and tumor necrosis factor-alpha (TNF-α) and to a lesser extent by PTH. Its formation is upregulated in granulomatous diseases like tuberculosis, sarcoidosis, and lymphoma and results in PTH-independent hypercalcemia.

Vitamin D also regulates FGF23 secretion. FGF23 is produced by osteocytes and osteoblasts. $1,25(OH)_2D$ promotes secretion of FGF23, whereas FGF23 downregulates renal 1α-hydroxylase and thereby inhibits $1,25(OH)_2D$ synthesis. Diseases associated with an excess of FGF23 either due to mutations in phosphate regulating genes with homologies to endopeptidases on the X chromosome (*PHEX*) or *FGF23* gene results in low levels of phosphate and $1,25(OH)_2D$ resulting in hypophosphatemic osteomalacia. Likewise, excessive secretion of FGF23 by bony fibrous dysplastic lesions (either sporadic or in association with McCune–Albright syndrome) or by mesenchymal tumors as in tumor-induced osteomalacia results in very low levels of serum phosphate and active vitamin D. In contrast, in tumoral calcinosis, mutations in UDP-N-acetyl-α-D galactosamine: polypeptide N-acetylgalactosaminyltransferase (GALNT3), which is essential for glycosylation of FGF23 leads to reduced levels of FGF23 and subsequently hyperphosphatemia and increased $1,25(OH)_2D$.

10.3 Nonclassical Actions of Vitamin D

1. **Role in Cellular Proliferation and Differentiation**: Vitamin D regulates functions of several genes that control cell proliferation, differentiation, and apoptosis, where it inhibits cell proliferation and promotes differentiation. Vitamin D also decreases angiogenesis and limits metastasis. The proposed mechanisms for its antiproliferative effects are stimulation of cell cycle inhibitors p21 and p27 as well as E-cadherin and other cell adhesion molecules. Low levels of vitamin D

have been implicated in the development and progression of various cancers. Evidence is limited to define any optimal level of vitamin D for its proposed anticancer benefit. Nevertheless, evidence suggests that vitamin D may not prevent malignancy, but adequate vitamin D levels may improve survival and decrease mortality in cancer patients.

2. **Role in Immune Regulation**: Vitamin D plays a role in both innate and adaptive immune responses. VDR is expressed on the surface of activated T and B cells, dendritic cells (DCs), and macrophages. The role vitamin D as an immune modulator came to the forefront during the COVID-19 pandemic when observational studies suggested that hypovitaminosis D increases the severity of COVID-19 and predicts poor prognosis. Besides, vitamin D supplementation was proposed to improve clinical outcomes in COVID-19.
3. **Role in Glucose Homeostasis**: $1,25(OH)_2D$ stimulates insulin secretion; however, its precise mechanism is unknown. Vitamin D by regulating intracellular calcium levels favors depolarization of the pancreatic β-cells and insulin release. These findings have been further supported by demonstrating the presence of VDR, CYP27B1, and calbindin-D in β-cells of the pancreas. Further vitamin D might also confer protection from cytokine-mediated damage to the β-cells. A systemic review and meta-analysis suggested a possible association between vitamin D insufficiency and diabetes mellitus. However, evidence is restricted to observational studies. Several randomized controlled trials (RCTs) have been conducted to evaluate the effect of vitamin D supplementation to prevent diabetes. These studies show no benefit in reducing the risk of diabetes mellitus. However, subgroup analysis observed that patients with vitamin D deficiency benefitted more in terms of prevention of incident diabetes compared to those with sufficient vitamin D status.

10.4 Endocrine Disorders Related to Vitamin D

1. **Rickets and Osteomalacia**

Vitamin D Deficiency Rickets: Vitamin D deficiency is the most common cause of rickets in India and other developing nations. Vitamin D deficiency rickets is particularly more common in exclusively breastfed infants as breast milk is a poor source of vitamin D. Other risk factors for vitamin D deficiency rickets include dark skin, rampant use of sunscreens and skin-protective barriers, minimal sun exposure, and malabsorptive disorders.

The recommended daily intake of vitamin D varies according to different societal guidelines. As per the Institute of Medicine (IOM) recommendation is an intake of 600 IU/day for individuals 1–70 years of age and 800 IU/day for individuals older than 70 years.

As per the Endocrine Society, a vitamin D level of less than 30 ng/ml is regarded as vitamin D deficiency. It recommends an intake of 400–1000 IU/day for infants,

600–1000 IU/day for children 1–18 years of age, and 1500–2000 IU/day for all adults aged more than 18 years. Usually, manifestations of vitamin D deficiency rickets are seen when 25(OH)D levels are less than 10 ng/ml.

Vitamin D-dependent Rickets (VDDR): VDDR is a group of genetic disorders in which there is a defect in vitamin D activation or action. These disorders usually present at 2 months to 2 years of age. The child can present with hypocalcemic seizures or tetany, hypotonia, irritability, and delayed motor milestones which later progress to gross deformities of the limbs.

VDDR are broadly of three types with type 1 being further classified into type 1A and type 1B. VDDR type 1A is due to a mutation in the gene encoding for 1α-hydroxylase (*CYP27B1*). This results in deficient formation of $1,25(OH)_2D$. VDDR type 1B is due to mutations in *CYP2R1* which encodes for the enzyme 25-hydroxylase.

VDDR type 2 results from resistance to the action of vitamin D. This could be due to either mutation in VDR (type 2A) or a defect in vitamin D response element-binding protein (type 2B). As VDR is essential for a normal hair cycle, VDDR type 2A is associated with alopecia. Nevertheless, alopecia is seen in only around 50% of the cases. The presence of alopecia is associated with the most severe forms of VDDR type 2A and thus has a very early presentation with hypocalcemia and also predicts poor response to therapy. Lately, a new type of VDDR has been described. This is due to increased inactivation of active vitamin D.

Clinical Presentation: Symptoms are more or less similar irrespective of the underlying etiology of rickets and osteomalacia. The common clinical presentations include proximal muscle weakness, bone pains, backache, limb deformities, and fractures. Muscle weakness is due to altered activity of muscle protein and reduced ATP generation in the muscle due to hypocalcemia and hypophosphatemia. Studies have also shown expression of VDR on the myoblasts and a possible role of the direct action of vitamin D on muscle function cannot be ruled out. Hypocalcemic symptoms like seizures, tetany, and carpopedal spasms are not uncommon in VDDR.

Skeletal deformities depend on the age of presentation. Presentation in first year of life can be with open fontanelles, craniotabes, frontal bossing, dolichocephaly, widening of wrists and ankles, double malleoli, rachitic rosary, Harrison sulcus, and deformities of the elbow. As the child begins to walk, lower limb deformities emerge in the form of genu valgum (knock knees), genu varum (bow legs), windswept deformity, and anterior bowing of the tibia (saber shin). Fractures are not very common as the bones are usually soft.

Biochemistry: Table 10.1 summarizes the biochemical parameters of the different types of rickets.

Imaging: Radiographs reveal generalized reduced bone density with thinning of cortices, looser zones, and biconcavity of vertebrae, known as "codfish vertebrae." In severe cases, due to secondary hyperparathyroidism subperiosteal resorption of bones and osteitis fibrosa cystica (commonly known as brown tumors) may be seen.

Management: Treatment depends on the cause of rickets (Table 10.1) and varies as per different guidelines and recommendations. Either D_2 or D_3 is used for treatment of VDD. Active forms of vitamin D (calcitriol) are the mainstay of the management of VDDR. VDDR type 2 requires a very high dose of calcitriol as there is end-organ resistance.

2. **Diabetes Mellitus:** Adequate levels of vitamin D have been shown to improve insulin secretion as well as sensitivity. Studies have shown that vitamin D deficiency is associated with an increased risk of both type 1 and type 2 diabetes mellitus. Thus, vitamin D supplementation has been tried for the prevention of

Table 10.1 Summary of biochemical parameters and management of different types of vitamin D-related rickets

	Vitamin D deficiency	Vitamin D-dependent type 1A	Vitamin D-dependent type 1B	Vitamin D-dependent type 2A and 2B
Basic pathogenesis	Nutritional/ malabsorption	Loss of function of renal 1α-hydroxylase	Loss of function of hepatic 25-hydroxylase	Mutation in vitamin D receptor or vitamin D response element-binding protein
Enzyme defect	–	CYP27B1	CYP2R1	–
Serum calcium	Low/normal	Low	Low	Low
Serum phosphate	Normal/low	Normal/low	Normal/low	Normal/low
Parathyroid hormone	High	High	High	High
Alkaline phosphatase	High	High	High	High
25-hydroxyvitamin D [25(OH)D]	Low	Normal	Low	High
1,25-dihydroxy vitamin D [1,25(OH)$_2$D]	Low/normal	Low	Low/low normal	High
Treatment	50,000 IU of either D2 or D3 weekly for 8–12 weeks followed by 1000–2000 IU daily as maintenance dose In India, vitamin D preparations are available as 60,000 IU rather than 50,000 IU	Calcitriol 0.3–2 μg/day	Calcifediol 20–50 μg/day; Calcitriol 0.3–2 μg/day	Elemental calcium 1–3 g/day and Calcitriol 5–60 μg/day

diabetes. In the D2D study, vitamin D supplementation at a dose of 4000 IU/day did not prevent the progression of prediabetes to diabetes. In a subgroup analysis of D2D, the benefit of vitamin D supplementation in achieving normoglycemia was seen especially in those who had baseline vitamin D levels of <12 ng/ml.

3. **Polycystic Ovarian Syndrome (PCOS):** Vitamin D stimulates ovarian steroidogenesis, decreases serum anti-Mullerian hormone (AMH) and improves folliculogenesis. PCOS is associated with low vitamin D levels. Vitamin D levels correlate inversely with insulin resistance, fasting and stimulated glucose concentration, and homeostasis model assessment-estimated insulin resistance (HOMA-IR) in patients with PCOS. The severity of vitamin D deficiency has also been correlated with higher body mass index (BMI) and waist circumference and vice versa. Supplementation of vitamin D in obese PCOS patients leads to improvement in HOMA-IR and Matsuda insulin sensitivity index. Vitamin D has also been shown to reduce fasting blood glucose levels in women with PCOS. Apart from surrogate markers of insulin resistance, few studies have shown that vitamin D supplementation, particularly in those deficient in vitamin D, regularized the menstrual cycles.
4. **Thyroid Disorders:** Vitamin D regulates thyroid stimulating hormone (TSH) secretion as well as exerts modulatory action on thyrocytes and increases thyroid hormone release. Vitamin D replacement has been shown to reduce TSH levels in patients with hypothyroidism. There exists a negative correlation between serum TSH and vitamin D levels in euthyroid nonpregnant people. However, during pregnancy, there is a positive correlation seen which points towards both direct and indirect effects of vitamin D on TSH release which varies across the hormonal status of the person. Vitamin D deficiency also increases the risk of autoimmune thyroid disorders. A study by Kivity et al. showed that patients with autoimmune thyroid disease had a significantly higher prevalence of vitamin D deficiency compared to normal healthy people (30.6% vs. 72%, $P < 0.001$), nevertheless, causality cannot be established.
5. **Pituitary Hormone Secretion:** Vitamin D decreases *Pit-1* transcription even in the absence of RXR. Thus it might have some role in the regulation of pituitary hormone secretion such as prolactin, growth hormone (GH), TSH, corticotropin, and gonadotropins. It has been observed that patients with pituitary tumors have lower levels of vitamin D. In a study by Ozcelik et al., out of 88 pituitary adenomas, vitamin D deficiency was present in 70.4%, whereas 21.6% had insufficient levels of vitamin D.

$1,25(OH)_2D$ increases GH secretion directly acting on VDR in the pituitary. $1,25(OH)_2D$ also regulates insulin-like growth factor-1 (IGF-1) synthesis in the liver. In turn, GH and IGF-1 (independent of GH) increase production of $1,25(OH)_2D$ in the renal tubules. In active acromegaly, a condition characterized by excessive GH production by a pituitary adenoma, there is an alteration in VDBP and VDR, which may contribute to poor bone quality.

Lastly, vitamin D deficiency or resistance leads to an increase in the secretion of prolactin. Moreover, cases of prolactinoma are associated with lower vitamin D levels.

6. **Reproductive Functions**: VDR and its enzymes are present in the testicular tissues where it exerts autocrine and/or paracrine functions and regulates testosterone secretion and spermatogenesis. How vitamin D regulates testosterone secretion is poorly understood, but one possible mechanism is alteration in sex hormone binding globulin (SHBG) levels and thus the bioavailability of testosterone. Some studies have shown an association of vitamin D deficiency with hypogonadism in males. A more definite benefit of vitamin D has been in improving semen quality by modulating intracellular calcium and lipids in spermatozoa and thus improving sperm motility, capacitation of sperm, and acrosome reaction.

   Vitamin D stimulates the production of ovarian steroids estrogen, estrone, and progesterone in the ovaries. In those undergoing intrauterine insemination, vitamin D replacement improved the success rate.
7. **Adrenal Function:** 1,25$(OH)_2$D may downregulate 21α-hydroxylase (CYP21A2) and may affect adrenal steroidogenesis. Furthermore, some role of vitamin D in mediating immune-mediated destruction of adrenal tissue has been postulated.
8. **Vitamin D Toxicity:** It is defined as markedly raised 25(OH)D concentrations (usually >150 ng/ml) associated with raised serum calcium levels and hypercalciuria and very low to undetectable PTH. This usually occurs after excessive and prolonged intake of mega doses of vitamin D usually administered parenterally. Vitamin D toxicity does not occur after excessive sun exposure as the body regulates the production of previtamin D in the skin. Management involves discontinuation of further vitamin D doses and management of hypercalcemia with diuresis and antiresorptive drugs like bisphosphonates or denosumab. Glucocorticoids reduce intestinal absorption of calcium, increase renal calcium excretion, and also lead to the formation of inactive vitamin D metabolites in the liver and are thus useful in this setting.

10.5 Conclusions

Vitamin D acts as a hormone rather than a vitamin and regulates calcium and phosphate metabolism in concert with PTH and FGF23. Apart from the classic actions of vitamin D, vitamin D receptor is ubiquitously distributed and interacts with endocrine systems and has been linked to various endocrine and metabolic disorders. These nonclassic actions of vitamin D are future areas of research to explore the further clinical applications of vitamin D.

Chapter 11
Vitamin D and Cardiovascular Disease

Basant Kumar, Ritika Mahajan, and Ramandeep Kaur

11.1 Vitamin D Metabolism

Vitamin D, a type of fat-soluble vitamin, comprises five categories identified as 2–6. The main variants are cholecalciferol, also known as vitamin D3, and ergocalciferol, also known as vitamin D2. They have structural differences primarily in the side chain; vitamin D2 has a C22–C23 double bond and an extra methyl group at C24 [1]. Plants and fungi produce vitamin D2 from ergosterol, whereas human skin synthesizes vitamin D3 from 7-dehydrocholesterol (7-DHC) when exposed to UVB radiation [2]. A photochemical reaction initiated by UV photons transforms 7-DHC into previtamin D3, subsequently converting to vitamin D3 through thermal isomerizations [3]. Vitamin D3 is carried in the bloodstream by Vitamin D binding protein (DBP) to the liver, where it undergoes conversion by CYP2R1 to 25-hydroxyvitamin D3 [25(OH)D3], which is the main circulating form of vitamin D3. In the kidneys, CYP27B1 transforms 25(OH)D3 into its active state, known as 1,25-dihydroxyvitamin D3 [1,25(OH)2D3] [4–6].

Cells such as cardiomyocytes, smooth muscle cells, fibroblasts, and vascular endothelial cells contain Vitamin D Receptor (VDR) and 1α-hydroxylase, allowing for the synthesis of the active form of vitamin D [7–9]. Experimental models have

B. Kumar (✉)
Department of Cardiology, Advanced Cardiac Centre, PGIMER, Chandigarh, India

R. Mahajan
Centre for Innovation and BioDesign (CIBioD), PGIMER, Chandigarh, India

R. Kaur
Department of Endocrinology, PGIMER, Chandigarh, India

V. Kumar, S. Patel (eds.), *Vitamin D: A Comprehensive Guide for Medical Professionals*, https://doi.org/10.1007/978-981-96-6491-7_11

highlighted several cardiovascular impacts of vitamin D. These include anti-hypertrophic properties, inhibition of cardiomyocyte proliferation, stimulation of vascular smooth muscle cell proliferation, induction of vascular endothelial growth factor expression, and suppression of the renin–angiotensin–aldosterone system (RAAS) and natriuretic peptide secretion [10, 11]. Noteworthy is the Vitamin D Receptor (VDR) activation to suppress the expression of angiotensin I along with the local generation of angiotensin II within myocardial, renal arteries, and kidney tissue [12]. These mechanisms underscore the multifaceted regulatory role of vitamin D in cardiovascular physiology, implicating its potential significance as a modulator of cardiovascular health and pathology. Vitamin D enhances the contractility of myocytes by increasing the levels of calcium within the cells [13] and regulates metalloproteinases and their inhibitors, impacting heart failure development. Apart from its direct influences, vitamin D exerts indirect effects on the cardiovascular system through its involvement in various cardiovascular risk factors. A lack of vitamin D has been associated with the onset of systemic hypertension, dyslipidemia, and type 2 diabetes mellitus [14]. Furthermore, recent findings indicate that vitamin D may possess anti-inflammatory properties, as it can hinder the activation of tumor necrosis factor alpha and nuclear factor kappa B, while simultaneously encouraging the activation of interleukin 10. These cytokines are recognized as pivotal mediators in the pathogenesis of cardiovascular diseases [15, 16]. This highlights the broader impact of vitamin D deficiency on cardiovascular health, extending beyond its direct physiological actions to encompass modulation of inflammatory pathways and systemic metabolic processes implicated in cardiovascular pathology (Fig. 11.1).

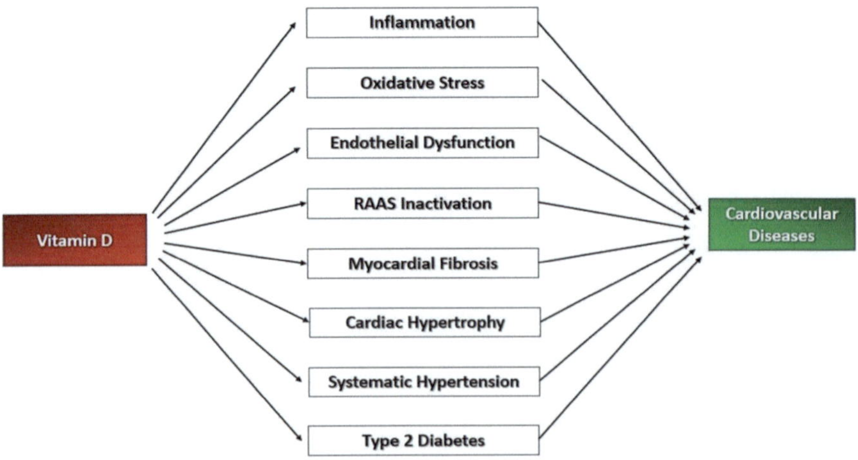

Fig. 11.1 Proposed mechanism of CVD due to deficiency of vitamin D

11.2 Vitamin D and Coronary Artery Disease

Many studies have investigated the relationship between insufficient levels of vitamin D and coronary artery disease (CAD) [17–19]. The initial findings from Denmark in 1978 highlighted a significant connection between lower levels of vitamin D and both stable angina and acute myocardial infarction (AMI) when compared to control subjects [20]. Subsequent research, including a case–control study in 1990, reaffirmed the lower levels of vitamin D in AMI patients compared to controls, with a noticeable decrease in the risk of AMI observed with higher quartiles of vitamin D [21]. Investigations into the association between vitamin D deficiency and coronary artery disease (CAD) have continued in subsequent studies [19, 22, 23]. Notably, the comprehensive Health Professionals Follow-up Study, where participants with vitamin D levels ≥30 ng/mL exhibited an approximately 50% reduction in the adjusted risk of AMI over a 10-year follow-up period [24]. Supporting these findings, a meta-analysis indicated a relative risk of 1.5 for cardiovascular events and 1.4 for CAD when comparing the lowest to the highest categories of vitamin D levels [25].

Even after accounting for major cardiovascular risk factors, the connection between reduced levels of vitamin D and an increased risk of acute myocardial infarction (AMI) seems to persist. It is noteworthy that vitamin D deficiency not only correlates with the risk of coronary artery disease (CAD) but also with its severity. In individuals who opt for elective coronary angiography, vitamin D deficiency has been associated with the presence of multi-vessel and more diffuse CAD, as well as with more severe coronary artery stenosis [26, 27]. Furthermore, variations in the Vitamin D Receptor (VDR) gene have been implicated in the occurrence of stent restenosis following percutaneous coronary intervention [28]. Interestingly, studies have shown that vitamin D can stimulate the expression of miR-145, a critical microRNA in vascular smooth muscle cells involved in regulating their contractile phenotype [29]. Reduced levels of miR-145 have been associated with adverse vascular alterations, including stenosis and vascular calcification [30].

Correia et al. provided initial evidence of an independent link between vitamin D deficiency and heightened in-hospital mortality among patients with acute myocardial infarction (AMI) [31]. Their research demonstrated significantly higher in-hospital mortality among patients with vitamin D levels ≤10 ng/mL compared to those with >10 ng/mL (24% vs. 5%). Despite these observations, the current literature acks conclusive evidence affirming the efficacy of vitamin D supplementation as a cardiovascular protection strategy in CAD [32–34]. Information regarding vitamin D supplementation in primary prevention is scarce and subject to debate, and the potential advantages of administering vitamin D during the early stages of acute myocardial infarction (AMI) have yet to be investigated.

11.3 Vitamin D and Heart Failure

Heart failure (HF) pathogenesis involves multifaceted mechanisms encompassing hemodynamic dysfunction, activation of neurohormonal pathways, increased inflammatory responses, and availability of micronutrients [35].

Groundbreaking research conducted by Shane et al. [36] showed that around 25% of patients with advanced heart failure undergoing evaluation for heart transplant had insufficient levels of vitamin D in their serum. Subsequent investigations have consistently observed diminished levels of vitamin D among patients diagnosed with heart failure (HF) compared to control cohorts, with prevalence estimates varying between 20% and 90% depending on the threshold used to define vitamin D insufficiency and the specific characteristics of the study population. Notably, a study conducted within the Intermountain Healthcare system, which included a large cohort of over 40,000 individuals from the general population, each with at least one measurement of vitamin D, demonstrated a gradual rise in the prevalence of heart failure as levels of vitamin D decreased.

Lower concentrations of vitamin D were found to be inversely associated with the risk of developing HF over an average follow-up period of 1.3 years, showing adjusted hazard ratios of 2.0 for very low levels (≤15 ng/mL) and 1.3 for low levels (16–30 ng/mL), respectively [37]. Additionally, findings from the Ludwigshafen Risk and Cardiovascular Health (LURIC) study revealed a an independent and inverse correlation between vitamin D levels and markers such as N-terminal pro-brain natriuretic peptide (NT pro-BNP) levels and New York Heart Association (NYHA) functional class. Additionally, it exhibited a direct correlation with left ventricular ejection fraction values [37]. Recent investigations have extended the spectrum, associating low vitamin D levels with diminished functional capacity and heightened hospitalization rates in HF patients with preserved ejection fraction [38, 39]. The Vitamin D Treating Patients with Chronic Heart Failure (VINDICATE) study reported favorable effects of vitamin D on left ventricular structure and function, indicating its potential therapeutic role in chronic HF [40]. Moreover, short-term vitamin D supplementation in chronic HF patients demonstrated decreased renin-angiotensin-aldosterone system (RAAS) activity [40, 42]. The VINDICATE trial, specifically, illustrated that one-year high-dose vitamin D supplementation in chronic HF patients with vitamin D deficiency, alongside optimal medical therapy, significantly improved left ventricular function [40]. Nevertheless, clinical trials evaluating the supplementation of vitamin D in HF risk groups and chronic HF patients yielded inconclusive results regarding cardiovascular mortality and morbidity improvement [41–44].

Notably, supplementation of vitamin D in patients with chronic heart failure appeared to decrease inflammatory marker levels and enhance quality of life without translating into reduced mortality or improved left ventricular function [44, 45].

The divergent findings observed across these trials may be ascribed to the utilization of varying vitamin D supplementation protocols, encompassing different dosage levels (both high and low), frequencies of administration (daily, weekly, and monthly), and durations of treatment (ranging from short-term to long-term), across heterogeneous study populations [46].

11.4 Vitamin D and Arterial Fibrillation

Recent investigations have demonstrated an association between diminished levels of vitamin D and atrial fibrillation (AF), suggesting a potential role for vitamin D as a therapeutic target in this clinical context [47]. The plausible mechanisms underlying the heightened AF risk associated with low vitamin D levels encompass direct effects on the atrium and the indirect modulation of cardiovascular risk factors. Notably, reduced vitamin D levels correlate with heightened inflammation, a pivotal factor in AF pathogenesis [48]. Indeed, individuals with elevated C-reactive protein values exhibit a reported twofold increase in AF risk [49]. Furthermore, the diminished levels of vitamin D may contribute to augmented AF risk through the activation of the renin–angiotensin–aldosterone system (RAAS), a regulator of atrial remodeling and inflammatory responses, thereby influencing both structural and electrical alterations in the atrium [50]. Consistent with this, lower vitamin D levels are associated with more extensive left atrial fibrosis [51]. Despite these associations, robust evidence supporting the efficacy of vitamin D supplementation in reducing AF risk remains lacking.

11.5 Vitamin D and Its Interaction with Antiplatelet Drugs

Verdoia et al. [52] conducted an initial investigation into the correlation between vitamin D status and platelet function in individuals undergoing dual antiplatelet therapy for acute coronary syndrome (ACS) or post-percutaneous coronary intervention in stable coronary artery disease (CAD). Their results indicated an impaired inhibition of adenosine-di-phosphate (ADP)-induced platelet aggregation by clopidogrel and ticagrelor in individuals with lower vitamin D levels. Subsequently, Lu et al. [53] provided additional evidence, demonstrating that patients with ischemic stroke who were resistant to clopidogrel, as defined by the maximum platelet aggregation rate induced by ADP, had notably lower levels of vitamin D compared to those who responded to clopidogrel. The inhibitory effect of Vitamin D on platelets may arise from either collagen or ADP-mediated mechanisms [54].

11.6 Future Perspectives

A consistent correlation has been observed between diminished levels of vitamin D and cardiovascular disease (CVD); however, the available information regarding its causal relationship remains limited. Conflicting results have arisen from epidemiological and interventional studies aimed at determining whether vitamin D assessment provides incremental information beyond traditional risk factors. Significantly, Mendelian randomization investigations have produced inconclusive findings, lacking definitive evidence to establish a causal relationship between genetic variations linked to vitamin D concentration and cardiovascular mortality. This suggests that while low vitamin D levels are correlated with heightened cardiovascular risk, they may not inherently contribute causally to such risk [55]. It is imperative to recognize that vitamin D status is significantly influenced by lifestyle, diet, comorbidities, and seasonal fluctuations. Therefore, a singular measurement of vitamin D offers limited insight into its long-term cardiovascular implications. Consequently, an exploration of alternative biomarkers, such as 1,25-dihydroxyvitamin D and 24,25-dihydroxyvitamin D, which more accurately reflect vitamin D status [56], is warranted. Validating these biomarkers as cardiovascular risk indicators and subsequently considering them for the identification of patients most likely to benefit from vitamin D supplementation represents a prudent avenue for further investigation.

11.7 Conclusion

While epidemiological investigations consistently highlight a strong correlation between diminished vitamin D levels and heightened cardiovascular risk, ongoing debates persist concerning the precise involvement of vitamin D in cardiovascular disease (CVD). The unresolved question revolves around whether vitamin D serves merely as an incidental indicator or functions as a causative agent susceptible to therapeutic intervention. Similarly, uncertainty prevails regarding the potential benefits of supplementation in individuals at elevated CVD risk with insufficient vitamin D levels, necessitating rigorous and well-designed investigations to elucidate the impact on outcomes. The intricate nature of these inquiries underscores the need for comprehensive exploration to clarify the role of vitamin D in CVD and to ascertain its potential as a modifiable factor through targeted interventions.

References

1. Jäpelt RB, Jakobsen J. Vitamin D in plants: a review of occurrence, analysis, and biosynthesis. Front Plant Sci. 2013;4:136. https://doi.org/10.3389/fpls.2013.00136. PMID: 23717318; PMCID: PMC3651966

2. Bikle DD. Vitamin D and the skin: physiology and pathophysiology. Rev Endocr Metab Disord. 2012;13(1):3–19. https://doi.org/10.1007/s11154-011-9194-0. PMID: 21845365; PMCID: PMC3687803
3. Aburjai T, Bernasconi S, Manzocchi L, Pelizzoni F. Isolation of 7-dehydrocholesterol from cell cultures of *Solanum malacoxylon*. Phytochemistry. 1996;43:773–6.
4. Andrews JW, Murai T, Page JW. Effects of dietary cholecalciferol and ergocalciferol on catfish. Aquaculture. 1980;19:49–54.
5. Armas LAG, Hollis BW, Heaney RP. Vitamin D_2 is much less effective than vitamin D_3 in humans. J Clin Endocrinol Metab. 2004;89:5387–91. https://doi.org/10.1210/jc.2004-0360.
6. Arnqvist L, Dutta PC, Jonsson L, Sitbon F. Reduction of cholesterol and glycoalkaloid levels in transgenic potato plants by overexpression of a type 1 sterol methyltransferase cDNA. Plant Physiol. 2003;131:1792. https://doi.org/10.1104/pp.102.018788.
7. Franczyk A, Stolarz-Skrzypek K, Wesołowska A, Czarnecka D. Vitamin D and vitamin D receptor activators in treatment of hypertension and cardiovascular disease. Cardiovasc Hematol Disord Drug Targets. 2014;14:34–44.
8. Ni W, Watts SW, Ng M, Chen S, Glenn DJ, Gardner DG. Elimination of vitamin D receptor in vascular endothelial cells alters vascular function. Hypertension. 2014;64:1290–8.
9. Khammissa RAG, Fourie J, Motswaledi MH, Ballyram R, Lemmer J, Feller L. The biological activities of vitamin D and its receptor in relation to calcium and bone homeostasis, cancer, immune and cardiovascular systems, skin biology, and oral health. Biomed Res Int. 2018;2018:9276380.
10. Pilz S, Tomaschitz A. Role of vitamin D in arterial hypertension. Expert Rev Cardiovasc Ther. 2010;8:1599–608.
11. Hiemstra T, Lim K, Thadhani R, Manson JE. Vitamin D and atherosclerotic cardiovascular disease. J Clin Endocrinol Metab. 2019;104:4033–50.
12. Freundlich M, Li YC, Quiroz Y, Bravo Y, Seeherunvong W, Faul C, Weisinger JR, Rodriguez-Iturbe B. Paricalcitol downregulates myocardial renin-angiotensin and fibroblast growth factor expression and attenuates cardiac hypertrophy in uremic rats. Am J Hypertens. 2014;27:720–6.
13. Polly P, Tan TC. The role of vitamin D in skeletal and cardiac muscle function. Front Physiol. 2014;5:145.
14. Norman PE, Powell JT. Vitamin D and cardiovascular disease. Circ Res. 2014;114:379–93.
15. Roffe-Vazquez DN, Huerta-Delgado AS, Castillo EC, Villarreal-Calderón JR, Gonzalez-Gil AM, Enriquez C, Garcia-Rivas G, Elizondo-Montemayor L. Correlation of vitamin D with inflammatory cytokines, atherosclerotic parameters, and lifestyle factors in the setting of heart failure: a 12-month follow-up study. Int J Mol Sci. 2019;20:5811.
16. Wöbke TK, Sorg BL, Steinhilber D. Vitamin D in inflammatory diseases. Front Physiol. 2014;5:244.
17. Siadat ZD, Kiani K, Sadeghi M, Shariat AS, Farajzadegan Z, Kheirmand M. Association of vitamin D deficiency andcoronary artery disease with cardiovascular risk factors. J Res Med Sci. 2012;17:1052–5.
18. Raslan E, Soliman SSA, Nour ZA, Ahmed D, Saad NES. Association of Vitamin D deficiency with chronic stable angina: a case control study. High Blood Press Cardiovasc Prev. 2019;26:77–80.
19. Xu R, Li YY, Ma LL, Yang HN. Association of vitamin D status with coronary artery disease in postmenopausal women. Medicine. 2020;99:e19544.
20. Lund B, Badskjaer J, Lund B, Soerensen OH. Vitamin D and ischaemic heart disease. Horm Metab Res. 1978;10:553–6.
21. Scragg R, Jackson R, Holdaway IM, Lim T, Beaglehole R. Myocardial infarction is inversely associated with plasma 25-hydroxyvitamin D3 levels: a community-based study. Int J Epidemiol. 1990;19:559–63.
22. Siadat ZD, Kiani K, Sadeghi M, Shariat AS, Farajzadegan Z, Kheirmand M. Association of vitamin D deficiency and coronary artery disease with cardiovascular risk factors. J Res Med Sci. 2012;17:1052–5.

23. Raslan E, Soliman SSA, Nour ZA, Ahmed D, Saad NES. Association of vitamin D deficiency with chronic stable angina: a case control study. Cardiovasc Prev. 2019;26:77–80.
24. Giovannucci E, Liu Y, Hollis BW, Rimm EB. 25-hydroxyvitamin D and risk of myocardial infarction in men: a prospective study. Arch Int Med. 2008;168:1174–80.
25. Wang L, Song Y, Manson JE, Pilz S, März W, Michaëlsson K, Lundqvist A, Jassal SK, Barrett-Connor E, Zhang C, et al. Circulating 25-hydroxy-vitamin D and risk of cardiovascular disease: a meta-analysis of prospective studies. Circ Cardiovasc Qual Outcomes. 2012;5:819–29.
26. Dziedzic EA, Gasior JS, Pawłowski M, Wodejko-Kucharska B, Saniewski T, Marcisz A, Dabrowski MJ. Vitamin D level is associated with severity of coronary artery atherosclerosis and incidence of acute coronary syndromes in non-diabetic cardiac patients. Arch Med Sci. 2019;15:359–68.
27. Dziedzic EA, Przychodzen S, Dabrowski M. The effects of vitamin D on severity of coronary artery atherosclerosis and lipid profile of cardiac patients. Arch Med Sci. 2016;12:1199–206.
28. Monraats PS, Fang Y, Pons D, Pires NM, Pols HA, Zwinderman AH, de Maat MP, Doevendans PA, DeWinter RJ, Tio RA, et al. Vitamin D receptor: a new risk marker for clinical restenosis after percutaneous coronary intervention. Expert Opin Ther Targets. 2010;14:243–51.
29. Chang S, Gao L, Yang Y, Tong D, Guo B, Liu L, Li Z, Song T, Huang C. miR-145 mediates the antiproliferative and gene regulatory effects of vitamin D3 by directly targeting E2F3 in gastric cancer cells. Oncotarget. 2015;6:7675–85.
30. Liu X, Cheng Y, Yang J, Qin S, Chen X, Tang X, Zhou X, Krall TJ, Zhang C. Flank sequences of miR-145/143 and their aberrant expression in vascular disease: mechanism and therapeutic application. J Am Heart Assoc. 2013;2:e000407.
31. Correia LC, Sodré F, Garcia G, Sabino M, Brito M, Kalil F, Barreto B, Lima JC, Noya-Rabelo MM. Relation of severe deficiency of vitamin D to cardiovascular mortality during acute coronary syndromes. Am J Cardiol. 2013;111:324–7.
32. Bahrami LS, Ranjbar G, Norouzy A, Arabi SM. Vitamin D supplementation effects on the clinical outcomes of patients with coronary artery disease: a systematic review and meta-analysis. Sci Rep. 2020;10:12923.
33. Legarth C, Grimm D, Krüger M, Infanger M, Wehland M. Potential beneficial effects of vitamin D in coronary artery disease. Nutrients. 2019;12:99.
34. De la Guía-Galipienso F, Martínez-Ferran M, Vallecillo N, Lavie CJ, Sanchis-Gomar F, Pareja-Galeano H. Vitamin D and cardiovascular health. Clin Nutr. 2021;40:2946–57.
35. Jessup M, Brozena S. Heart failure. N Engl J Med. 2003;348:2007–18.
36. Zittermann A, Schleithoff SS, Tenderich G, Berthold HK, Körfer R, Stehle P. Low vitamin D status: a contributing factor in the pathogenesis of congestive heart failure? J Am Coll Cardiol. 2003;41:105–12.
37. Anderson JL, May HT, Horne BD, Bair TL, Hall NL, Carlquist JF, Lappé DL, Muhlestein JB, Intermountain Heart Collaborative (IHC) Study Group. Relation of vitamin D deficiency to cardiovascular risk factors, disease status, and incident events in a general healthcare population. Am J Cardiol. 2010;106:963–8.
38. Pilz S, März W, Wellnitz B, Seelhorst U, Fahrleitner-Pammer A, Dimai HP, Boehm BO, Dobnig H. Association of vitamin D deficiency with heart failure and sudden cardiac death in a large cross-sectional study of patients referred for coronary angiography. J Clin Endocrinol Metab. 2008;93:3927–35.
39. Nolte K, Herrmann-Lingen C, Platschek L, Holzendorf V, Pilz S, Tomaschitz A, Düngen HD, Angermann CE, Hasenfuß G, Pieske B, et al. Vitamin D deficiency in patients with diastolic dysfunction or heart failure with preserved ejection fraction. ESC Heart Fail. 2019;6:262–70.
40. Witte KK, Byrom R, Gierula J, Paton MF, Jamil HA, Lowry JE, Gillott RG, Barnes SA, Chumun H, Kearney LC, et al. Effects of vitamin D on cardiac function in patients with chronic HF: the VINDICATE study. J Am Coll Cardiol. 2016;67:2593–603.
41. Zittermann A, Ernst JB, Prokop S, Fuchs U, Dreier J, Kuhn J, Knabbe C, Birschmann I, Schulz U, Berthold HK, et al. Effect of vitamin D on all-cause mortality in heart failure (EVITA): a 3-year randomized clinical trial with 4000 IU vitamin D daily. Eur Heart J. 2017;38:2279–86.

42. Schroten NF, Ruifrok WP, Kleijn L, Dokter MM, Silljé HH, Lambers Heerspink HJ, Bakker SJ, Kema IP, van Gilst WH, van Veldhuisen DJ, et al. Short-term vitamin D3 supplementation lowers plasma renin activity in patients with stable chronic heart failure: an open-label, blinded end point, randomized prospective trial (VitD-CHF trial). Am Heart J. 2013;166:357–64.
43. Dalbeni A, Scaturro G, Degan M, Minuz P, Delva P. Effects of six months of vitamin D supplementation in patients with heart failure: a randomized double-blind controlled trial. Nutr Metab Cardiovasc Dis. 2014;24:861–8.
44. Jiang WL, Gu HB, Zhang YF, Xia QQ, Qi J, Chen JC. Vitamin D supplementation in the treatment of chronic heart failure: a meta-analysis of randomized controlled trials. Clin Cardiol. 2016;39:56–61.
45. Wang T, Liu Z, Fu J, Min Z. Meta-analysis of vitamin D supplementation in the treatment of chronic heart failure. Scand Cardiovasc J. 2019;53:110–6.
46. D'Amore C, Marsico F, Parente A, Paolillo S, De Martino F, Gargiulo P, Ferrazzano F, De Roberto AM, La Mura L, Marciano C, et al. Vitamin D deficiency and clinical outcome in patients with chronic heart failure: a review. Nutr Metab Cardiovasc Dis. 2017;27:837–49.
47. Huang WL, Yang J, Yang J, Wang HB, Yang CJ, Yang Y. Vitamin D and new-onset atrial fibrillation: a meta-analysis of randomized controlled trials. Hell J Cardiol. 2018;59:72–7.
48. Hu YF, Chen YJ, Lin YJ, Chen SA. Inflammation and the pathogenesis of atrial fibrillation. Nat Rev Cardiol. 2015;12:230–43.
49. Galea R, Cardillo MT, Caroli A, Marini MG, Sonnino C, Narducci ML, Biasucci LM. Inflammation and C-reactive protein in atrial fibrillation: cause or effect? Tex Heart Inst J. 2014;41:461–8.
50. Patel D, Druck A, Hoppensteadt D, Bansal V, Brailovsky Y, Syed M, Fareed J. Relationship between 25-hydroxyvitamin, D, renin, and collagen remodeling biomarkers in atrial fibrillation. Clin Appl Thromb Hemost. 2020;26:1076029619899702.
51. Zhang Z, Yang Y, Ng CY, Wang D, Wang J, Li G, Liu T. Meta-analysis of vitamin D deficiency and risk of atrial fibrillation. Clin Cardiol. 2016;39:537–43.
52. Verdoia M, Pergolini P, Rolla R, Sartori C, Nardin M, Schaffer A, Barbieri L, Daffara V, Marino P, Bellomo G, et al. Novara Atherosclerosis Study Group (NAS). Vitamin D levels and high-residual platelet reactivity in patients receiving dual antiplatelet therapy with clopidogrel or ticagrelor. Platelets. 2016;27:576–82.
53. Lu BC, Shi XJ, Liang L, Dong N, Liu ZZ. Platelet surface CD62p and serum vitamin D levels are associated with clopidogrel resistance in chinese patients with ischemic stroke. J Stroke Cerebrovasc Dis. 2019;28:1323–8.
54. Sultan M, Twito O, Tohami T, Ramati E, Neumark E, Rashid G. Vitamin D diminishes the high platelet aggregation of type2 diabetes mellitus patients. Platelets. 2019;30:120–5.
55. Trummer O, Pilz S, Hoffmann MM, Winkelmann BR, Boehm BO, März W, Pieber TR, Obermayer-Pietsch B, Renner W. Vitamin D and mortality: a Mendelian randomization study. Clin Chem. 2013;59:793–7.
56. Tang JCY, Nicholls H, Piec I, Washbourne CJ, Dutton JJ, Jackson S, Greeves J, Fraser WD. Reference intervals for serum 24,25-dihydroxyvitamin D and the ratio with 25-hydroxyvitamin D established using a newly developed LC-MS/MS method. J Nutr Biochem. 2017;46:21–9.

Chapter 12
Vitamin D and Integumentary System

Ghazal Ahmed

Abstract The integumentary system is the largest organ in the human body. It performs several functions, such as protecting the body from various injuries and hazards, controlling temperature, immunity, excretion, metabolism, sensation, and vitamin D production. Vitamin D is vital in maintaining bone health, muscle and nerve functions, immune and anti-inflammatory actions, etc. Bone health depends on calcium homeostasis, and vitamin D is critical for the absorption of the same. It also prevents osteoporosis. Skin (keratinocytes) initiates the biochemical process for synthesizing vitamin D. However, vitamin D has also been investigated in skin health and diseases. Current evidence indicates that vitamin D is essential for skin and hair health. For specific diseases like psoriasis and atopic dermatitis, there has been evidence for their positive correlation with serum vitamin D deficiency and evidence negating any such correlation. Nevertheless, oral vitamin D use in skin disease treatment and prevention is limited due to its multifactorial origin and the risk of other adverse effects. The present chapter briefly describes the relation and evidence of the integumentary system, skin health, and vitamin D.

Keywords Skin · Vitamin D · Psoriasis · Immunity · Dermatitis · Vitamin D supplementation

---

G. Ahmed (✉)
Department of Dermatology, Venerology and Leprosy, All India Institute of Medical Sciences, Deoghar, Jharkhand, India

V. Kumar, S. Patel (eds.), *Vitamin D: A Comprehensive Guide for Medical Professionals*, https://doi.org/10.1007/978-981-96-6491-7_12

Learning Objectives
1. Vitamin D Synthesis.
2. Vitamin D in Skin Physiology.
3. Vitamin D in Dermatological Diseases.
4. Topical Vitamin D and Its Analogs.

12.1 Introduction

Integument means the outer covering of the body. It comprises skin, hair, nails, and the associated glands for humans, making it a complex integumentary system. Skin cells are called keratinocytes because they are loaded with the unique protein keratin. Vitamin D is synthesized mainly in keratinocytes and maintains its supply to all other body organs. Thus, it is involved in keeping the skeletal and many other systems in a healthy condition.

12.2 Vitamin D Synthesis

Vitamin D is synthesized in the skin. According to the Commission Internationale de l'Eclairage (CIE) [1], the most efficient wavelength for the action spectrum is 295 nm (ultraviolet-B), with a spectral range of 255–330 nm.

7-dehydrocholesterol works as a precursor. It is present in the plasma membranes of suprabasal keratinocytes and epidermal basal and dermal fibroblasts. It is converted to pre-calciferol (pre-vitamin D_3), then to cholecalciferol, i.e., vitamin D_3. Post-synthesis, the plasma membrane releases the vitamin D3 bound to vitamin D-binding protein (DBP), which enters the systemic circulation. It reaches the peak serum concentrations of vitamin D_3 within 24–48 h after exposure to UV-B radiation. The half-life of the plasma vitamin D_3 ranges from 36 to 78 h, and their levels decline exponentially [2]. As vitamin D3 is a lipid-soluble molecule, adipocytes can take up, which can be stored in subcutaneous tissue and omental fat for later use. The distribution of vitamin D_3 into adipose tissue prolongs its total-body half-life to approximately 2 months, as first detected in experiments on submarine personnel.[3, 4] Vitamin D synthesis is schematically depicted in Fig. 12.1.

Once in circulation, vitamin D is converted by a hepatic hydroxylase into 25-hydroxyvitamin D (25(OH)D; calcidiol). The circulating 25(OH)D level indicates the vitamin D status. However, there are some dietary sources, and the level reflects both dietary vitamin D intake and is synthesized from ultraviolet exposure. The serum half-life of 25(OH)D is approximately 15 days [5]. 25(OH)D is biologically inactive except at very high non-physiological levels [6], and the kidney converts it to its active hormonal form 1,25-dihydroxy vitamin D ($1,25(OH)_2D$;

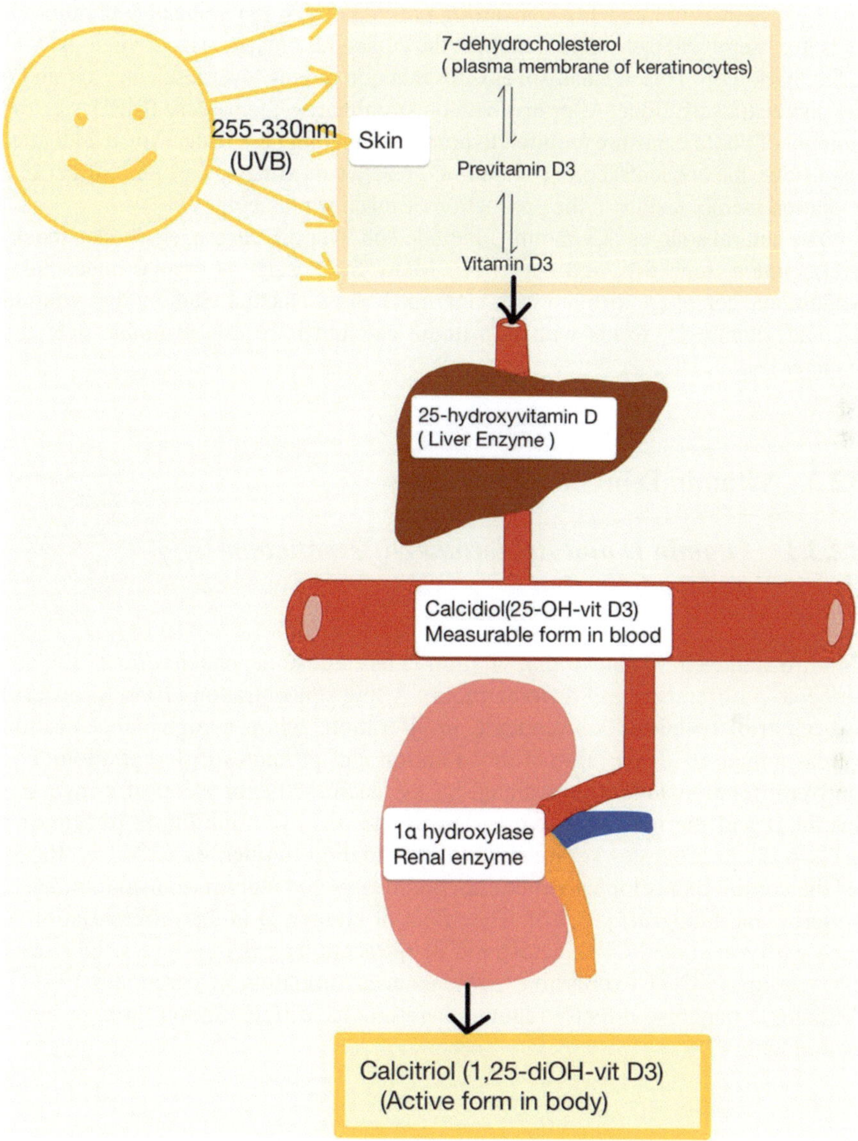

Fig. 12.1 Pathway for vitamin D synthesis

calcitriol) in a process that is usually tightly controlled by the parathyroid hormone. The parathyroid hormone level starts rising when the plasma vitamin D level is 75 nmol/L or lower. Despite this, inadequate vitamin D supply lowers the circulating level of calcitriol [4]. Circulating calcitriol is affected by viable nephron numbers, high serum fibroblast growth factor-23, and inflammatory cytokines such as interleukin (IL)-1, IL-6, and tumor necrosis factor-alpha (TNF-α).

It is essential to know that converting pre-vitamin D_3 to the inactive photoproducts lumisterol and tachysterol balances the cutaneous biosynthesis of vitamin D_3 as a feedback loop. This mechanism ensures that one cannot "overdose" on vitamin D_3 by photoexposure alone. After less than one minimal erythema dose (MED; i.e., the amount of photo exposure required to produce faint pinkness in the skin at 24 h after exposure), the concentration of vitamin D_3 reaches maximal levels, and further UV radiation merely results in the production of inactive metabolites [5].

The natural sources of vitamin D include cod liver oil, cheese, egg yolks, mackerel, salmon, tuna fish, and beef liver. Many foods such as orange juice, milk, yogurt, and cereal are fortified with vitamin D. Supplemental vitamin D as vitamin D_3 and vitamin D_2 forms with or without calcium is readily available over the counter [7].

12.3 Vitamin D in Skin Physiology

12.3.1 Vitamin D and Epidermal Differentiation and Proliferation

In vitro studies have shown that vitamin D has a dose-dependent effect on keratinocyte proliferation and differentiation. A low concentration (10–9 M or less) of calcitriol enhances keratinocyte proliferation, while a high concentration (greater than 10–8 M) inhibits proliferation and promotes differentiation. The anti-proliferative action is mediated by the decreased expression of c-myc and cyclin D and the increased expression of the cell cycle inhibitors p21cip and p27kip [8]. It promotes keratinocyte differentiation through increased synthesis of the cornified envelope's structural components (involucrin, transglutaminase, loricrin, and filaggrin) [15, 16]. The effect of vitamin D in the differentiation is also partly mediated by the elevation of intracellular calcium levels, increased phospholipase C-γ1 expression, and enhanced formation of ceramides [9–11]. Vitamin D may also directly regulate keratinocyte differentiation through interaction with VDR.

12.3.2 Vitamin D and Keratinocyte Apoptosis

It prevents apoptosis at physiological levels mediated by sphingosine-1-phosphate and by activation of MEK/ERK and PI3K/Akt signaling pathways, and the increased ratio of antiapoptotic protein (Bcl-2) to pro-apoptotic protein (Bad and Bax) [12]. However, at high concentrations, it induces apoptosis in keratinocytes [13].

12.3.3 Vitamin D and Skin Immune Functions

T and B lymphocytes, monocytes, and Langerhans cells express both vitamin D receptors and 25-hydroxy-vitamin D-1α-hydroxylase, [14] implicating a vital role of vitamin D in the control and regulation of immune mechanisms. It is a significant regulator of anti-microbial peptides (AMPs). Cathelicidin and β-defensin are direct transcriptional targets of vitamin D. Skin insult increases the local synthesis of active vitamin D via enhanced CYP27B1 expression; also, TLR-2 is increased, which in turn increases the level of cathelicidin through a vitamin D-dependent mechanism. After calcitriol treatment in keratinocytes and sebocytes, an increased expression of hCAP18/ LL-37 and defensin has been seen.

Calcitriol and calcipotriol (vitamin D analog) mediate an immunosuppressive action in the skin through decreased antigen presentation directly by affecting Langerhans cells or indirectly by modulating keratinocyte cytokine production [15, 16]. It has been proposed that this is mediated through the induction of CD4 + CD25+ T regulatory (Treg) cells, which prevents subsequent antigen-specific CD8+ T-cell proliferation and IFN-γ production [17, 18].

Vitamin D suppresses the expression of MHC class II molecules and co-stimulatory molecules like CD40, CD80, and CD86, thus influencing dendritic cells. It stimulates the production of IL-10 and inhibits the production of IL-12, causing suppression of T-cell activation.

12.3.4 Antioxidant and Photoprotection

In vitro studies show that vitamin D generates antioxidants in keratinocytes. These antioxidants probably combat ultraviolet ray-induced harmful free radicals, thus offering photoprotection.

12.3.5 Hair

The importance of vitamin D in hair health is evidenced by the alopecia seen in vitamin D resistance rickets type II, where there is a mutation in the VDR gene [19].

12.4 Vitamin D in Dermatological Diseases

12.4.1 Psoriasis

Vitamin D has a vital role in psoriasis. Serum vitamin D levels in psoriasis patients are significantly deficient compared to non-psoriatic patients in many studies [20–22]. The severity of psoriasis has also been associated with vitamin D serum levels, and psoriatic patients display a higher risk of vitamin D deficiency [23, 24]. However, pooled analysis of vitamin D supplementation did not improve clinical manifestations [25]. PsA patients have lower vitamin D levels than the general population [26]. Although a study found a significant relationship between low 25(OH) D levels and psoriasis, it did not establish a causal relationship [27, 28]. The currently available data fails to support a robust association between VDR rs7975232, rs1544410, rs2228570, and rs731236 polymorphisms and psoriasis susceptibility, which still required the support of more case-control studies [29]. A meta-analysis showed that VDR ApaI, BsmI, and FokI polymorphisms are not associated with psoriasis susceptibility [30]. On the contrary, a population-based study has shown that the mean 25(OH)D levels and deficiency prevalences were not different between those with and without psoriasis. Therefore, the current literature is uncertain about the efficacy and safety of supplemental vitamin D in treating psoriasis.

12.4.2 Atopic Dermatitis

Epidemiological studies have shown an increased AD prevalence in populations living in higher geographic latitudes, with lower sun exposure and less vitamin D production [31]. Cheng HM et al. found an increased likelihood of developing AD in individuals with deficient or insufficient vitamin D levels [32]. The severity of atopic dermatitis by SCORAD score was seen to be inversely related to serum levels of vitamin D. A recent meta-analysis has shown lower serum 25 (OH) D levels in pediatric AD patients compared with healthy controls. There was also an increased risk of vitamin D deficiency among pediatric AD patients; the serum 25 (OH) D level in severe AD patients was significantly lower than in mild AD patients. The SCORAD and EASI scores improved after vitamin D supplementation, suggesting its beneficial effect on AD patients [33]. Another meta-analysis showed that lower serum 25(OH)D was associated with more severe AD and vitamin D supplementation may help lower AD severity [34]. Nevertheless, some contradictory reports suggest no role for vitamin D or a positive association between vitamin D levels and the risk of developing AD [35].

12.4.3 *Vitiligo*

Finamor et al., in their study, included 16 vitiligo patients, treating them with high doses, i.e., 35,000 IU of vitamin D3 once daily for 6 months with a low-calcium diet and hydration. The study found that 14 (87.5%) patients had 25–75% repigmentation without significant metabolic parameter changes [36]. A recent meta-analysis has shown that vitiligo patients showed significantly decreased serum vitamin D levels compared with controls. In vitiligo, patients with indoor/urban work had a significantly lower vitamin D level than their outdoor/rural counterparts [37]· Thus, a 25(OH)D deficiency may contribute to developing vitiligo. However, the exact mechanism remains an enigma.

12.4.4 *Ichthyosis*

Calcitriol inhibits keratinocyte proliferation and contributes to the mineralization of new bones by increasing intestinal calcium and phosphorus absorption. Ichthyosis is caused by abnormal keratinization and, thus, might be associated with an alteration in vitamin D metabolism, leading to rickets and osteomalacia [38]. Nevertheless, whether this association is causal or coincidental still needs to be determined. It is suggested that children in developing countries with severe ichthyosis, especially those with pigmented skin, should be evaluated for rickets. A recent double-blinded randomized controlled trial in congenital non-syndromic ichthyosis treated with oral vitamin D and Acitretin showed no significant difference, proposing that oral vitamin D can be a good alternative [39].

12.4.5 *Skin Cancers*

Exposure to sunlight, specifically to ultraviolet radiation, is a risk factor for skin cancer. On the other hand, vitamin D exerts anti-proliferative and pro-apoptotic effects on melanocytes and keratinocytes in vitro against UV rays [40]. An association between increased incidence of cutaneous melanoma and vitamin D insufficiency, less favorable Breslow tumor depth with lower vitamin D levels, and the presence of vitamin D insufficiency was noted in a meta-analysis [41]. Data suggested an inverse association between vitamin D blood levels and CM thickness at diagnosis [42].

A meta-analysis suggests high vitamin D status was associated with increased keratinocyte cancers and melanoma risks. As 25-(OH)D level is mainly related to sun exposure, a higher risk of skin cancer might be confounded; data for which is lacking in most studies. However, while it was seen that intakes of supplemental or dietary vitamin D were not associated with the risk of melanoma and squamous cell carcinoma (SCC), high intakes of vitamin D from supplements and diet were slightly associated with basal cell carcinoma (BCC) risk, albeit with no heterogeneity across skin cancer [43]. Overall, the evidence suggests that unprotected sun exposure should be avoided to achieve high vitamin D status. Adequate amount of vitamin D should be obtained from a healthy diet [43].

12.5 Topical Vitamin D and Its Analogs

12.5.1 Mechanism of Action

Vitamin D mainly binds to the vitamin D receptor protein, which belongs to the family of thyroid hormone, corticosteroid, and retinoic acid nuclear receptor gene family. When the ligand binds to VDR, the complex, in association with retinoid X receptor alpha (RXR-α), binds to VDR elements, which are deoxyribose nucleic acid (DNA) binding sites specific to vitamin D or its analogs (Fig. 12.2). The therapeutic effects of topical vitamin D occur via a *VDR*-mediated genomic mechanism, resulting in the inhibition of keratinocyte proliferation and mediated non-genomic mechanisms inducing keratinocyte differentiation by increasing intracellular calcium levels [44]. The anti-inflammatory effects may also result from inhibiting the production of IL-2, IL-6, and interferon-gamma (IFN-γ). Further, topical

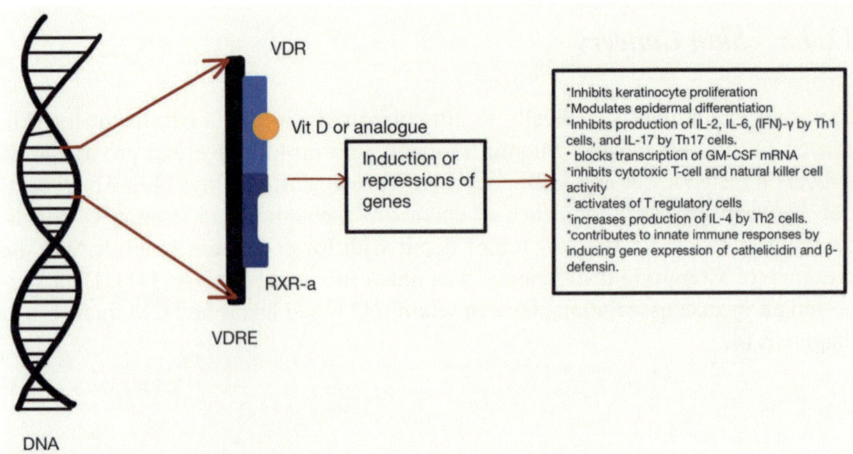

Fig. 12.2 Topical vitamin D's mechanism of action and analogs. (*VDR* itamin D receptor, *VDRE* vitamin D response elements, *RXR-a* retinoid X receptor alpha)

calcipotriol inhibits proinflammatory cytokines and human beta-defensin, which are found in increased levels in psoriatic lesions [45]. The following table shows the commonly available vitamin D analogs and their strengths in the topical forms.

Serial no.	Vitamin D analog	Strength of the topical form
1.	Calcitriol	3 μg/g
2.	Calcipotriol	50 μg/g
3.	Tacalcitol	4 μg/g
4.	Maxacalcitol	25 μg/g

Topical calcitriol has minimal effects on serum calcium and is approved by the United States Food and Drug Administration (USFDA) for treating psoriasis. Compared to systemic steroids, lack of skin atrophy or adrenal suppression is why they are included in the therapeutic armamentarium for psoriasis. Topical formulations of calcitriol and its analogs, like calcipotriol and maxacalcitol, have been shown to improve and have good safety profiles in psoriatic plaques. Other off-label uses are vitiligo, [44] ichthyosis, morphea, lichen sclerosis et atrophicus, and pityriasis rubra pilaris, to name a few.

12.6 Conclusion

Current evidence indicates that vitamin D is essential for skin and hair health. For specific diseases like psoriasis and atopic dermatitis, there has been evidence for their positive correlation with serum vitamin D deficiency and evidence negating any such correlation. It brings us to the point that serum vitamin D level is affected by several factors, such as sunlight /UV light exposure, ethnicity, and dietary intake, emphasizing that results of studies on vitamin D should be interpreted with caution. Another probable reason for the limited use of oral vitamin D is the risk of adverse effects. Oral vitamin D can cause hypercalcemia, hypercalciuria, nephrocalcinosis, and nephrolithiasis. Bone mineral density can also be reduced secondary to its calcemic effect.

While the importance of oral vitamin D supplementation is still an enigma in dermatological diseases, the topical form of vitamin D has an established role in various skin diseases. It is FDA-approved for use in psoriasis. Thus, the skin plays a vital role in vitamin D synthesis, and its health should be maintained to minimize any disease related to vitamin D deficiency.

References

1. CIE. Action spectrum for the production of pre-vitamin D3 in human skin. Technical report 174; 2006.
2. Stamp TC, Haddad JG, Twigg CA. Comparison of oral 25- hydroxycholecalciferol, vitamin D, and ultraviolet light as determinants of circulating 25-hydroxyvitamin D. Lancet. 1977;1:1341–3.

3. Dlugos DJ, Perrotta PL, Horn WG. Effects of the submarine environment on renal-stone risk factors and vitamin D metabolism. Undersea Hyperb Med. 1995;22:145–52.
4. Zittermann A, Frisch S, Berthold HK, et al. Vitamin D supplementation enhances the beneficial effects of weight loss on cardiovascular disease risk markers. Am J Clin Nutr. 2009;89:1321–7.
5. Vanchinathan V, Lim HW. A dermatologist's perspective on vitamin D. Mayo Clin Proc. 2012;87:372–80.
6. Shahriari M, Kerr PE, Slade K, et al. Vitamin D and the skin. Clin Dermatol. 2010;28:663–8.
7. Holick MF, Binkley NC, Bischoff-Ferrari HA, et al. Endocrine Society. Evaluation, treatment, and prevention of vitamin D deficiency: an Endocrine Society clinical practice guideline. J Clin Endocrinol Metab. 2011;96:1911–30.
8. Bikle DD. Vitamin D metabolism and function in the skin. Mol Cell Endocrinol. 2011;347:80–9.
9. Bikle DD, Ng D, Tu CL, et al. Calcium- and vitamin D-regulated keratinocyte differentiation. Mol Cell Endocrinol. 2001;177:161–71.
10. Ratnam AV, Bikle DD, Cho JK. 1,25 dihydroxyvitamin D3 enhances the calcium response of keratinocytes. J Cell Physiol. 1999;178:188–96.
11. Pillai S, Mahajan M, Carlomusto M. Ceramide potentiates, but sphingomyelin inhibits, vitamin D-induced keratinocyte differentiation: comparison between keratinocytes and HL-60 cells. Arch Dermatol Res. 1999;291:284–9.
12. De Haes P, Garmyn M, Carmeliet G, et al. Molecular pathways involved in the antiapoptotic effect of 1,25-dihydroxyvitamin D3 in primary human keratinocytes. J Cell Biochem. 2004;93:951–67.
13. Manggau M, Kim DS, Ruwisch L, et al. 1Alpha,25-dihydroxyvitamin D3 protects human keratinocytes from apoptosis by the formation of sphingosine-1-phosphate. J Invest Dermatol. 2001;117:1241–9.
14. van Etten E, Branisteanu DD, Verstuyf A, et al. Analogs of 1,25-dihydroxyvitamin D3 as dose-reducing agents for classical immunosuppressants. Transplantation. 2000;69:1932–42.
15. Bagot M, Charue D, Lescs MC, et al. Immunosuppressive effects of 1,25-dihydroxyvitamin D3 and its analogue calcipotriol on epidermal cells. Br J Dermatol. 1994;130:424–31.
16. Dam TN, Møller B, Hindkjaer J, et al. The vitamin D3 analog calcipotriol suppresses the number and antigen-presenting function of Langerhans cells in normal human skin. J Investig Dermatol Symp Proc. 1996;1:72–7.
17. Ghoreishi M, Bach P, Obst J, et al. Expansion of antigen-specific regulatory T cells with the topical vitamin d analog calcipotriol. J Immunol. 2009;182:6071–8.
18. Gorman S, Geldenhuys S, Judge M, et al. Dietary vitamin D increases percentages and function of regulatory T cells in the skin-draining lymph nodes and suppresses dermal inflammation. J Immunol Res. 2016;2016:1426503.
19. Malloy PJ, Feldman D. The role of vitamin D receptor mutations in the development of alopecia. Mol Cell Endocrinol. 2011;347:90–6.
20. Maleki M, Nahidi Y, Azizahari S, et al. Serum 25-OH vitamin D level in psoriatic patients and comparison with control subjects. J Cutan Med Surg. 2016;20:207–10.
21. Chandrashekar L, Kumarit GR, Rajappa M, et al. 25-hydroxy vitamin D and ischaemia-modified albumin levels in psoriasis and their association with disease severity. Br J Biomed Sci. 2015;72:56–60.
22. Orgaz-Molina J, Magro-Checa C, Rosales-Alexander JL, et al. Vitamin D insufficiency is associated with higher carotid intima-media thickness in psoriatic patients. Eur J Dermatol. 2014;24:53–62.
23. Ricceri F, Pescitelli L, Tripo L, et al. Deficiency of serum concentration of 25-hydroxyvitamin D correlates with severity of disease in chronic plaque psoriasis. J Am Acad Dermatol. 2013;68:511–2.
24. Moosazadeh M, Damiani G, Khademloo M, et al. Comparing vitamin D level between patients with psoriasis and healthy individuals: a systematic review and meta-analysis. J Evid Based Integr Med. 2023;28:2515690X231211663.

25. Formisano E, Proietti E, Borgarelli C, et al. Psoriasis and vitamin D: a systematic review and meta-analysis. Nutrients. 2023;15:3387.
26. Radić M, Đogaš H, Kolak E, et al. Vitamin D in psoriatic arthritis—a systematic review and meta-analysis. Semin Arthritis Rheum. 2023;60:152200.
27. Theodoridis X, Grammatikopoulou MG, Stamouli EM, et al. Effectiveness of oral vitamin D supplementation in lessening disease severity among patients with psoriasis: a systematic review and meta-analysis of randomized controlled trials. Nutrition. 2021;82:111024.
28. Pitukweerakul S, Thavaraputta S, Prachuapthunyachart S, et al. Hypovitaminosis D is associated with psoriasis: a systematic review and meta-analysis. Kans J Med. 2019;12:103–8.
29. Li J, Sun L, Sun J, et al. Pooling analysis regarding the impact of human vitamin D receptor variants on the odds of psoriasis. BMC Med Genet. 2019;20:161.
30. Lee YH. Vitamin D receptor ApaI, TaqI, BsmI, and FokI polymorphisms and psoriasis susceptibility: an updated meta-analysis. Clin Exp Dermatol. 2019;44:498–505.
31. Weiland SK, Hüsing A, Strachan DP, et al. ISAAC phase one study group. Climate and the prevalence of symptoms of asthma, allergic rhinitis, and atopic eczema in children. Occup Environ Med. 2004;61:609–15.
32. Cheng HM, Kim S, Park GH, et al. Low vitamin D levels are associated with atopic dermatitis, but not allergic rhinitis, asthma, or IgE sensitization, in the adult Korean population. J Allergy Clin Immunol. 2014;133:1048–55.
33. Fu H, Li Y, Huang H, et al. Serum vitamin D level and efficacy of vitamin D supplementation in children with atopic dermatitis: a systematic review and meta-analysis. Comput Math Methods Med. 2022;2022:9407888.
34. Ng JC, Yew YW. Effect of vitamin D serum levels and supplementation on atopic dermatitis: a systematic review and meta-analysis. Am J Clin Dermatol. 2022;23:267–75.
35. Thuesen BH, Heede NG, Tang L, et al. No association between vitamin D and atopy, asthma, lung function or atopic dermatitis: a prospective study in adults. Allergy. 2015;70:1501–4.
36. Finamor DC, Sinigaglia-Coimbra R, Neves LC, et al. A pilot study assessing the effect of prolonged administration of high daily doses of vitamin D on the clinical course of vitiligo and psoriasis. Dermatoendocrinol. 2013;5:222–34.
37. Varikasuvu SR, Aloori S, Varshney S, et al. Decreased circulatory levels of vitamin D in vitiligo: a meta-analysis. An Bras Dermatol. 2021;96:284–94.
38. Litaiem N, Chabchoub I, Bacha T, et al. Rickets in association with skin diseases and conditions: a review with emphasis on screening and prevention. Photodermatol Photoimmunol Photomed. 2020;36:339–50.
39. Bakshi S, Mahajan R, Karim A, et al. Oral vitamin D versus acitretin in congenital non-syndromic ichthyosis: double blinded, randomized controlled trial. J Dtsch Dermatol Ges. 2022;20:297–304.
40. Langberg M, Rotem C, Fenig E, et al. Vitamin D protects keratinocytes from deleterious effects of ionizing radiation. Br J Dermatol. 2009;160:151–61.
41. Shellenberger RA, Gowda S, Kurn H, et al. Vitamin D insufficiency and serum levels related to the incidence and stage of cutaneous melanoma: a systematic review and meta-analysis. Melanoma Res. 2023;33:265–74.
42. Caini S, Boniol M, Tosti G, et al. Vitamin D and melanoma and non-melanoma skin cancer risk and prognosis: a comprehensive review and meta-analysis. Eur J Cancer. 2014;50:2649–58.
43. Mahamat-Saleh Y, Aune D, Schlesinger S. 25-Hydroxyvitamin D status, vitamin D intake, and skin cancer risk: a systematic review and dose-response meta-analysis of prospective studies. Sci Rep. 2020;10:13151.
44. Al-Smadi K, Ali M, Alavi SE, et al. Using a topical formulation of vitamin D for the treatment of vitiligo: a systematic review. Cells. 2023;12:2387.
45. Barrea L, Savanelli MC, Di Somma C, et al. Vitamin D and its role in psoriasis: an overview of the dermatologist and nutritionist. Rev Endocr Metab Disord. 2017;18:195–205.

Chapter 13
Vitamin D and Central Nervous System Across the Lifespan

Riya Sharma and Kamalesh Chakravarty

Abstract Vitamin D is primarily synthesized in human epithelial cells by photochemical synthesis and also acquired from dietary sources. Recent evidence indicate that apart from the classical effect of regulation in calcium hemostasis and bone metabolism, vitamin D plays a crucial role in nervous system development and function. It has been associated with regulating neurotrophin production, neural differentiation, maturation, and various neuromodulator syntheses. Here, we present an overview of different neurological and psychiatric disorders associated with vitamin D deficiency and current evidence regarding the utility of vitamin D supplementation in them.

Keywords Vitamin D · Neurotrophin · Alzheimer's disease · Parkinson's disease · Multiple sclerosis · Schizophrenia · Depression

13.1 Introduction

Vitamin D is one of the most ancient hormones, produced by early life forms for over 750 million years. Until the late 1970s, it was believed that $1,25(OH)_2D_3$ had biologic functions only on calcium and phosphate metabolism [1]. Though the classic target organs have been bone, kidney, parathyroid gland, and intestine, the past few decades have recognized newer roles for vitamin D. Its impact on neurogenesis, neuroprotection, and immune regulation highlights its upcoming influence and potential therapeutic goals in neurological diseases. As the time evolved, many

R. Sharma · K. Chakravarty (✉)
Department of Neurology, PGIMER, Chandigarh, India

V. Kumar, S. Patel (eds.), *Vitamin D: A Comprehensive Guide for Medical Professionals*, https://doi.org/10.1007/978-981-96-6491-7_13

neuronal receptors for vitamin D have been identified as well. Lack of this vitamin as well as the presence of it has important implications.

About 90–100% of the vitamin D needed by humans is obtained through exposure to sunlight [1]. A UVB-mediated, photolytic, nonenzymatic reaction converts 7-dehydrocholesterol into previtamin D3. This previtamin D3 then undergoes a thermal isomerization process, a nonenzymatic reaction, to form vitamin D3 (cholecalciferol) [2]. Once synthesized in the skin, vitamin D3 is transported to the liver, where it is hydroxylated by cytochrome P450 enzymes in hepatic parenchyma to produce 25-hydroxyvitamin D3 ($25OHD_3$). The biologically active form of vitamin D, calcitriol (1,25-dihydroxyvitamin D3), is subsequently generated in the kidneys through the action of 1α-hydroxylase [3].

13.2 Vitamin D and the Brain

Both human and rodent brains express 1α-hydroxylase, facilitating the conversion of $25(OH)\text{-}D_3$ to the active form $1,25(OH)_2D_3$. Nuclear vitamin D receptors (VDRs) in the brain regulate its genomic functions, affecting approximately 3% of the human genome, which includes around 700 genes. This influence has contributed to its recognition as a neurosteroid in recent decades [4]. Studies show that the distribution of these receptors in rodent brains closely mirrors that of humans. Spanning in the central nervous system from cortex, cerebellum, diencephalon, spinal cord, amygdala, hypothalamus, hippocampus to the Schwann cells of the peripheral nervous system, VDRs are distributed extensively suggesting a great functional diversity [4].

The recent development of a genome-wide map of VDR binding has revealed a significant enrichment of VDR-binding sites near genes linked to autoimmune diseases and cancer, as identified through genome-wide association studies [5]. Moreover, both vitamin D metabolites, $25(OH)D_3$ and $1,25(OH)_2D_3$, freely cross the blood–brain barrier.

Neurogenesis occurring in both fetal development and adulthood emerged as the earliest evidence supporting the genetic regulatory effects of vitamin D in the brain. The discovery of adult stem cells in the subgranular zone of the hippocampus, dentate gyrus, and olfactory bulb [6] revealed an intriguing correlation: these regions also express vitamin D receptors (VDRs) in the adult brain, suggesting a potential link with cell growth and differentiation.

Various neurotrophins (NGF, GDNF, BDNF, TGF-α, NT-3, NT-4/5) support the survival and differentiation of neurons (Table 13.1). They can modulate synaptic efficiency and neurotransmitter release enhancing the functional connectivity [7].

The gene expression and production of all these neurotrophins are enhanced by calcitriol. Vitamin D deficiency in fetuses lead to reduced neurotrophin factors linking it to various neurodevelopmental diseases. Furthermore, various calcium-binding proteins distributed in the developing brain act as a calcium buffer that protects the brain from neurotoxins, ischemia, and aging. These are modulated by vitamin D, which further forms the basis for its role in neurodegenerative diseases [8, 9].

Table 13.1 Summary of various neurotrophins and their role

Neurotrophins	Targets	Functions
Nerve growth factor (NGF)	Sympathetic neurons, sensory neurons (dorsal root ganglia), basal forebrain cholinergic neurons, striatal cholinergic neurons, Purkinje cells	Synaptic plasticity and neurotransmission
Brain-derived neurotrophic factor (BDNF) Growth-derived neurotrophic factor (GDNF)	Sensory neurons (nodose dorsal root neurons), spinal motor neurons, basal forebrain cholinergic neurons, substantia nigra dopaminergic neurons, facial motor neurons, retinal ganglion cells	Survival and differentiation of dopaminergic neurons
Neurotrophin 3 (NT-3)	Sympathetic neurons, sensory neurons, basal forebrain cholinergic neurons, locus coeruleus neurons	Neuronal and glial development in precursor cells as well as survival in mature cells
Neurotrophin 4/5 (NT-4/5)	Sympathetic neurons, sensory neurons (nodose dorsal root ganglia), basal forebrain cholinergic neurons, locus coeruleus neurons, motor neurons, retinal ganglion cells	Calcium signaling

Activated T and B lymphocytes, as well as monocytes, express VDR, and $1,25(OH)_2D_3$ functions as an effective immune regulator, contributing to its involvement in autoimmune diseases. The lower incidence of common autoimmune diseases in individuals living closer to the equator is believed to be due to their higher presumed levels of vitamin D synthesis.

In terms of neurotransmitter function, vitamin D treatment has been shown to increase dopamine levels in the brainstem and homovanillic acid in the striatum and hypothalamus. This suggests a potential role for vitamin D in modulating dopaminergic neurotransmission, which is involved in various cognitive and motor functions [10].

In summary, the accumulating evidence over the past few decades and the revelation that numerous tissues and cells throughout the body harbor vitamin D receptors, along with the capacity of certain tissues to convert the primary circulating form of vitamin D, into its active form, especially in the nervous system, has brought forth fresh understandings regarding its extraskeletal functions and its role in various chronic diseases.

13.3 Vitamin D and Neurological Disorders

The rapid accumulation of experimental evidence over the past two decades bringing forth vitamin D as a neuroactive steroid has exponentially increased its association with a wide array of neurological diseases. From neurodevelopmental and neurodegenerative disorders to neuroinflammation, stroke, epilepsy, and neuropsychiatric conditions, the role of vitamin D as a potential therapeutic target has been extensively studied.

13.3.1 Neurocognition and Aging

Extensive epidemiological research has been done to examine the possible link between vitamin D levels and cognitive function. These studies aim to determine whether vitamin D deficiency influences cognitive decline, memory impairment, or neurodegenerative diseases. Establishing a causal relationship remains challenging, as most research on these associations relies on observational and cross-sectional studies, similar to those investigating brain volume and various neurodegenerative disorders [11]. A 2017 meta-analysis, which included 26 observational studies and three randomized controlled trials (RCTs), found that while both cross-sectional and longitudinal observational studies suggested an association between low vitamin D levels and cognitive decline, the RCTs did not demonstrate significant cognitive improvement with vitamin D supplementation [12].

13.3.2 Alzheimer's Disease

Alzheimer's disease (AD) is the most common cause of dementia characterized by the extracellular buildup of amyloid-beta deposits and the formation of neurofibrillary tangles, which result from the intracellular accumulation of hyperphosphorylated tau (p-tau) protein. Vitamin D facilitates clearance of amyloid plaques and may interact with the TGF-beta signaling pathway, which regulates amyloid precursor protein expression. However, there is insufficient evidence to establish a definitive link between vitamin D deficiency and Alzheimer's disease. Research into VDR polymorphisms as potential risk factors for Alzheimer's disease (AD) and the correlation between the ApoE4 genotype and increased vitamin D levels in both human and mouse studies are emerging as areas of growing interest. However, current evidence is limited, indicating a need for further research [11].

13.3.3 Parkinson's Disease

The presence of both vitamin D receptors (VDR) and enzymes involved in vitamin D synthesis in the hypothalamus and substantia nigra [13] suggests the potential for autocrine/paracrine effects of vitamin D in metabolism of acetylcholine and dopamine. Vitamin D is shown to increase dopamine synthesis in CNS making a strong argument for the presence of an interrelation between vitamin D deficiency and Parkinson's disease (PD). Though causality has yet to be determined and vitamin D deficiency may be a risk factor for PD, having PD may also increase the risk of vitamin D deficiency, with immobility reducing sunlight exposure or leading to malnutrition [14].

Research consistently indicates an inverse relationship between serum 25(OH)D levels and motor symptom severity in Parkinson's disease (PD), as assessed by the Unified Parkinson's Disease Rating Scale (UPDRS) and the Hoehn and Yahr (H&Y) staging system. Additionally, low 25(OH)D levels have been associated with worsening nonmotor symptoms, though further studies are needed to confirm these findings [15]. Studies investigating the effectiveness of vitamin D supplementation in fall prevention have yielded mixed results. Few meta-analyses emphasize the need for further research to clarify the relationship between 25(OH)D levels and fall risk [16]. In conclusion, while evidence supporting vitamin D supplementation in PD remains limited, many clinicians still consider its use due to its low risk and potential benefits.

13.3.4 Amyotrophic Lateral Sclerosis

Amyotrophic lateral sclerosis (ALS) is a progressive, presently incurable, neurodegenerative disorder with multiple pathogenetic mechanisms. Evidence suggesting a strong link between vitamin D deficiency and the risk of ALS remains limited, and studies investigating the relationship between vitamin D deficiency or VDR polymorphisms and ALS are scarce. While research has established that a daily supplementation of 2000 IU is safe for ALS patients, the benefits of vitamin D in slowing disease progression remain uncertain due to the small number of available studies and their methodological limitations [11].

13.3.5 Multiple Sclerosis

Multiple sclerosis (MS) is the most prevalent autoimmune neurological disorder causing disability in young adults, with symptoms typically developing between the ages of 20 and 40 years. Of all the neurodegenerative disorders in which vitamin D deficiency has been implicated, the most research and strongest evidence exists for MS.

In a healthy individual, regulatory T cells (Tregs) are essential for controlling the immune response's intensity and effectiveness, thereby contributing to the regulation of autoimmune diseases. Interestingly, in multiple sclerosis (MS) patients, Treg function is diminished. However, research suggests that vitamin D supplementation can enhance Treg activity and promote a shift in the Th1/Th2 balance toward a Th2-dominant state, potentially exerting beneficial immunomodulatory effects [17].

Environmental risk factors are considered to play a pivotal role in the pathogenesis of multiple sclerosis. A well-documented hypothesis linking MS prevalence to

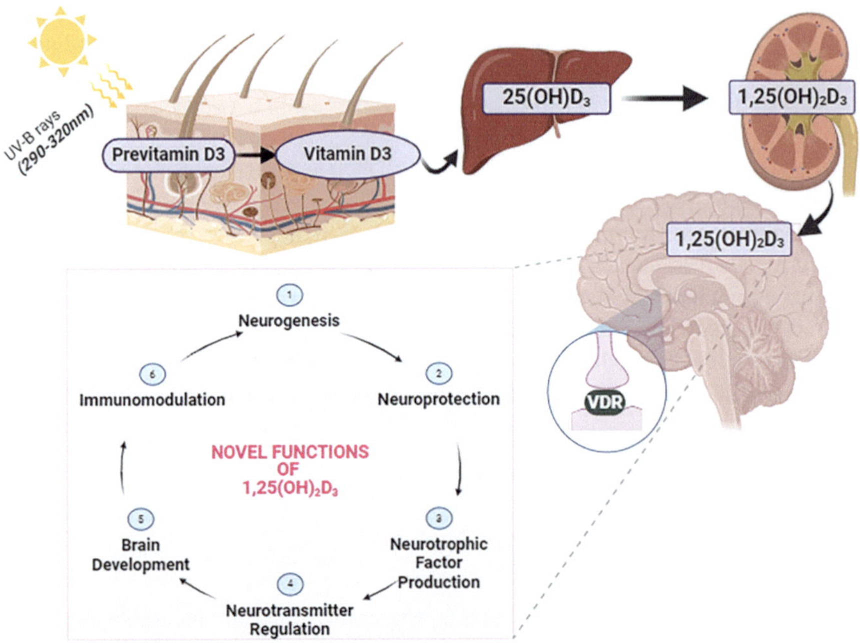

Fig. 13.1 Schematic representation of the synthesis, activation, and novel neurological functions of Vitamin D_3. Ultraviolet B (UVB) radiation (290–320 nm) from sunlight induces the conversion of 7-dehydrocholesterol in the skin to previtamin D_3, which isomerizes to form vitamin D_3. Vitamin D_3 is then hydroxylated in the liver to form 25-hydroxyvitamin D_3 [25(OH)D_3], the major circulating form. Further hydroxylation occurs in the kidneys, producing the biologically active form, 1,25-dihydroxyvitamin D_3 [1,25$(OH)_2D_3$]. This active form crosses the blood-brain barrier and binds to the vitamin D receptor (VDR) in the brain. The inset illustrates novel neurobiological roles of 1,25$(OH)_2D_3$, including neurogenesis, neuroprotection, neurotrophic factor production, neurotransmitter regulation, brain development, and immunomodulation

latitude suggests that sunlight exposure may exert a protective effect, primarily through its role in vitamin D synthesis (Fig. 13.1). Vitamin D is known to modulate immune responses, and its deficiency has been implicated in increased MS susceptibility. Epidemiological studies have consistently demonstrated an inverse correlation between sunlight exposure, serum 25(OH)D levels, and MS disease activity, indicating that lower sunlight exposure may contribute to both disease onset and progression [18].

Small, underpowered randomized trials have found no evidence that vitamin D supplementation reduces brain lesions on MRI or impacts disability and disease progression in patients with multiple sclerosis [19, 20]. Mendelian randomization studies suggest that individuals with elevated 25(OH)D concentrations may experience up to a 28% decrease in the likelihood of developing MS. Among those already diagnosed with the disease, higher vitamin D levels have been linked to a reduced risk of relapse, with some studies indicating up to a 43% decrease in relapse rates [21].

A comprehensive systematic meta-analysis examined clinical trials evaluating the efficacy of vitamin D supplementation in multiple sclerosis (MS) and clinically isolated syndrome (CIS)/optic neuritis. The analysis of these trials found no significant therapeutic benefits of vitamin D in these patients and suggested a potential increased risk of relapse in MS patients receiving high-dose vitamin D supplementation [22]. Interestingly, CHOLINE [23] and SOLAR [24] randomized controlled trials showed positive effects of vitamin D supplementation on secondary endpoints. Moreover, supplementation was associated with a reduction in the number of enlarging or new T2 lesions, new T1 lesions, lower hypointense T1 lesion volume, and slower disability progression [25].

Thus, serum 25(OH)D deficiency should always be corrected when detected, and measures should be taken to prevent vitamin D deficiency in patients with multiple sclerosis.

13.3.6 Diabetic Neuropathy

Vitamin D deficiency has been linked to occurrence and severity of diabetic polyneuropathy. A recent metaanalysis of 26 studies with 6277 participants reported patients with diabetic neuropathy had a significantly lower serum 25 (OH) D levels compared with patients without diabetic neuropathy [26]. The pathogenesis of vitamin D in diabetic neuropathy remains unclear with insulin resistance, atherosclerosis, increased inflammation, and oxidative injury leading to nerve ischemia has been proposed as possible factors. Vitamin D deficiency has also been found to have a role in pathogenesis of small-fiber neuropathy involving the nociceptive fibers.

Vitamin D supplementation has been found beneficial in neuropathic pain and other symptoms of diabetic neuropathy. Multiple studies and meta-analysis report an improved glycaemic control, reduction in pain severity and neuropathy score [26–28]. In view of beneficial effect of vitamin D on bone health, its supplementation of at least 800–2000 IU of vitamin per day is generally recommend in diabetic patients with a target serum level of serum 25(OH)D levels of 75 nmol/L. Patients of diabetic neuropathy should be screened for vitamin D deficiency, and supplementation can be beneficial for the neuropathic symptoms.

13.3.7 Migraine

Migraine is one of the most common neurological disorders encountered by neurologists in routine clinical practice. The sixth survey of the Tromsø Study (Tromsø6), which analyzed data from 11,614 participants, found no significant correlation between serum 25(OH)D levels and migraine. However, lower vitamin D levels were associated with an increased risk of nonmigraine headaches [29].

Clinical trials have evaluated the effects of vitamin D supplementation in migraine patients, reporting a significant reduction in headache frequency and the number of attacks. However, its impact on pain intensity remains less conclusive [30]. Further research is needed to clarify the role of vitamin D in migraine management.

13.3.8 Epilepsy

Severe vitamin D deficiency can cause hypocalcemia, potentially leading to seizures. Additionally, certain antiepileptic drugs, like phenytoin and carbamazepine, accelerate vitamin D metabolism through hepatic enzyme induction. This process can worsen vitamin D deficiency and hypocalcemia, exacerbating seizure disorders [31]. Hypocalcemia resulting from vitamin D deficiency during fetal and infant development can trigger seizures in newborns, sometimes as early as 1 week after birth. This deficiency is prevalent in infants born to mothers with inadequate vitamin D levels, especially when they are exclusively breastfed [32]. Decades of basic research and animal studies provide strong evidence supporting the anticonvulsant effects of vitamin D. However, clinical data remain limited due to small sample sizes and the absence of rigorous study designs, such as randomization and double-blind, placebo-controlled trials [33].

13.3.9 Stroke

In individuals affected by stroke, vitamin D deficiency may contribute to vascular disease by promoting the secretion of atherogenic pro-inflammatory cytokines. These cytokines promote atherosclerotic changes within blood vessels and may increase the risk of plaque instability, potentially leading to ischemic events. Furthermore, vitamin D is actively involved in modulating inflammatory pathways and maintaining endothelial function, both of which are critical in cerebrovascular health. Its deficiency has been identified as an independent risk factor for acute ischemic stroke. The relationship between vitamin D and cerebrovascular as well as cardiovascular health has been extensively studied in both animal and human research. Growing evidence suggests that vitamin D plays a regulatory role in the renin–angiotensin system, with potential adverse effects. Studies on vitamin D receptor (VDR) knockout mice have shown a significant increase in renin expression, whereas wild-type mice exhibited suppressed renin levels following administration of $1,25(OH)_2D$ [34].

Moderate to strong associations between lower serum 25(OH)D levels and stroke were observed across various analytical methods, even after adjusting for traditional demographic and lifestyle factors. These associations were particularly pronounced in young females under 50 years of age. The VITamin D and Omega-3 TriaL

(VITAL) investigated the impact of supplementation on overall cancer incidence and major cardiovascular events, including myocardial infarction, stroke, and cardiovascular-related mortality. However, the study found significant effects only in reducing bone fractures [35]. If vitamin D plays a role in poststroke recovery through pathophysiological mechanisms that remain to be fully elucidated—potentially involving a reduction in cerebral infarct volume and neuroprotective effects—its supplementation may offer a yet-to-be-established therapeutic benefit.

Vitamin D levels should be measured in all stroke patients and should be considered as an independent stroke risk factor. The supplementation of vitamin D could be considered as a fundamental part of stroke therapy, but more studies are awaited [34].

13.4 Psychiatric Disorders

13.4.1 Depression

VDR and 1a-hydroxylase, the enzyme capable of metabolizing 25(OH)D to 1,25$(OH)_2$D, are both present in various parts of the human brain such as the prefrontal cortex, hippocampus, cingulate, thalamus, hypothalamus, and substantia nigra, which are implicated in the pathophysiology of depression. Limited trials have examined the impact of vitamin D supplementation on neuropsychiatric symptoms. A meta-analysis of six studies comparing vitamin D supplementation with a placebo in adults diagnosed with or at risk for depression found no significant effect of vitamin D on depressive symptoms [36]. However, in another recent meta-analysis of 25 studies, including a total of 7534 participants, vitamin D supplementation was found to be effective in individuals with major depressive disorders and serum levels of 25(OH)D levels </50 nmol/L. The beneficial effect was reported by studies where vitamin D supplementation was continued for >/8 weeks at a dosage of </4000 IU/day [37]. Hence, vitamin D supplementation may be considered in deficient patients.

13.4.2 Autism

Gestational hypovitaminosis D has been proposed as a potential contributor to excessive neuronal proliferation in early brain development, possibly linking to autism spectrum disorder (ASD). This theory suggests that vitamin D's role in inhibiting excessive cellular proliferation in early brain development may be relevant to the observed pattern of neural changes in ASD, including early brain overgrowth and subsequent neurodegeneration. Children with insufficient vitamin D intake may require supplementation to meet daily requirements. However, there is no conclusive evidence that vitamin D supplementation improves the core symptoms of ASD [38].

13.4.3 Schizophrenia

The neuropathology of schizophrenia remains complex, but structural abnormalities such as ventricular enlargement and decreased cerebral volume, reminiscent of neurodevelopmental factors, have been consistently observed. Interestingly, these changes mirror findings in developmental vitamin D deficiency models. Additionally, neurotrophic factors like NGF, neurotrophin, and p75NTR, which are influenced by vitamin D, play crucial roles in synaptic regulation and neuronal growth—processes disrupted in schizophrenia. These findings indicate a possible association between vitamin D deficiency and an increased risk of developing schizophrenia [39].

Systematic reviews have consistently demonstrated a higher prevalence of vitamin D deficiency among individuals with schizophrenia. Although clinical evidence linking low vitamin D levels to adverse effects on adult brain health remains unconvincing, preclinical studies in rats and mice—both during development and in adulthood—provide strong evidence that vitamin D deficiency negatively impacts brain structure, delays neuronal maturation, and induces neurochemical and behavioral changes relevant to schizophrenia [40]. However, the causal relationship still remains uncertain.

Role of vitamin D in various neurological disorders and related evidence has been summarized in Table 13.3. Treatment for vitamin D has been summarized in Table 13.2.

Table 13.2 Treatment of vitamin D deficiency [41, 42]

Treatment summary	
Preparation	Two main forms: Cholecalciferol (D3) and ergocalciferol (D2) Cholecalciferol preferred due to higher efficacy. Available in oral capsules and intramuscular injections.
Dosing	1. Severe deficiency (<12 ng/mL or 30 nmol/L) 25,000–50,000 IU of vitamin D2 or D3 orally once per week for 6–8 weeks maintenance—800 IU of vitamin D2 or D3 orally daily 2. Moderate deficiency (12 to <20 ng/mL or 30 to <50 nmol/L) 800–1000 IU of vitamin D2 or D3 orally daily 3. Insufficiency (20–30 ng/mL or 50–75 nmol/L) 600–800 IU of vitamin D2 or D3 orally daily

Table 13.3 Summary of evidence for role of vitamin D in neurological disease

Neurological disease	Role of vitamin D and evidence in literature
Neurocognition and aging	Low vitamin D levels associated with cognitive decline, but limited improvement with supplementation [12]
Alzheimer's disease	Potential role in facilitating amyloid plaque clearance and interacting with the TGF-beta signaling pathway, but evidence limited [11]
Parkinson's disease	Vitamin D deficiency may be a risk factor for PD, but supplementation benefits unclear. Associations found between low vitamin D levels and PD severity, but supplementation benefits inconclusive [15]
Amyotrophic lateral sclerosis	Weak evidence for vitamin D deficiency in ALS risk, supplementation benefits uncertain. Limited evidence suggests vitamin D deficiency may play a role in ALS risk, but supplementation benefits unclear [11]
Multiple sclerosis	Strong evidence for a link between vitamin D deficiency and MS risk, but supplementation effects inconclusive. Significant association between vitamin D deficiency and MS risk, but supplementation effects on reducing brain lesions inconclusive [14–21]
Migraine	Limited evidence for a direct association, potential benefits of supplementation in reducing headache attacks. No significant association found, but supplementation may reduce headache attacks [22, 23]
Diabetic neuropathy	Linked to occurrence/severity of diabetic polyneuropathy, Potential role in small-fiber neuropathy pathogenesis. Supplementation may alleviate symptoms and improve glycemic control, with recommended daily doses aiming for target serum levels [22–24]
Epilepsy	Severe deficiency exacerbates seizures, but clinical evidence for supplementation benefits lacking. Severe deficiency worsens seizures, but limited clinical data support supplementation benefits [26]
Stroke	Vitamin D deficiency may be an independent risk factor for stroke, supplementation effects unclear. Associations observed between low vitamin D levels and stroke severity, but supplementation effects on recovery uncertain [27]
Psychiatric disorders	Limited evidence for direct causal relationship, inconclusive results from supplementation trials. Vitamin D present in brain regions implicated in disorders, but supplementation trials show no significant effect on symptoms [28–30]

13.5 Conclusion

Over the years, research has highlighted the significance of vitamin D in neurogenesis across the lifespan, ranging from early developmental phases to later stages of aging. However, when it comes to understanding the precise relationship between vitamin D levels and CNS disorders, studies to demonstrate this associations are lacking.

Nevertheless, the importance of ensuring adequate vitamin D levels, particularly in vulnerable populations, is important. Routine screening for vitamin D deficiency and targeted supplementation strategies may hold promise for preventing or mitigating the progression of neurological disorders. Moreover, ongoing research into the molecular mechanisms underlying vitamin D's effects on the brain and its interaction with genetic and environmental factors will further improve clinical practice.

References

1. Holick MF. Vitamin D: a millenium perspective. J Cell Biochem. 2003;88(2):296–307.
2. Heaney RP. Vitamin D endocrine physiology. J Bone Miner Res. 2007;22(Suppl 2):V25–7.
3. Heaney RP, Vitamin D. Endocrine physiology*. J Bone Miner Res. 2009;22(S2):V25–V7.
4. Prüfer K, Veenstra TD, Jirikowski GF, Kumar R. Distribution of 1,25-dihydroxyvitamin D3 receptor immunoreactivity in the rat brain and spinal cord. J Chem Neuroanat. 1999;16(2):135–45.
5. Ramagopalan SV, Heger A, Berlanga AJ, Maugeri NJ, Lincoln MR, Burrell A, et al. A ChIP-seq defined genome-wide map of vitamin D receptor binding: associations with disease and evolution. Genome Res. 2010;20(10):1352–60.
6. Taupin P, Gage FH. Adult neurogenesis and neural stem cells of the central nervous system in mammals. J Neurosci Res. 2002;69(6):745–9.
7. Siegel GJ, Albers RW, Katzman R, Agranoff B. Basic neurochemistry. Boston: Little, Brown; 1981.
8. Baimbridge KG, Celio MR, Rogers JH. Calcium-binding proteins in the nervous system. Trends Neurosci. 1992;15(8):303–8.
9. de Viragh PA, Haglid KG, Celio MR. Parvalbumin increases in the caudate putamen of rats with vitamin D hypervitaminosis. Proc Natl Acad Sci. 1989;86(10):3887–90.
10. Tekes K, Gyenge M, Folyovich A, Csaba G. Influence of neonatal vitamin A or vitamin D treatment on the concentration of biogenic amines and their metabolites in the adult rat brain. Horm Metab Res. 2009;41(4):277–80.
11. Roman S, Mowry EM. Vitamin D and the central nervous system: development, protection, and disease. In: Liao EP, editor. Extraskeletal effects of Vitamin D: a clinical guide. Cham: Springer; 2018. p. 227–47.
12. Goodwill AM, Szoeke C. A systematic review and meta-analysis of the effect of low vitamin D on cognition. J Am Geriatr Soc. 2017;65(10):2161–8.
13. Eyles DW, Smith S, Kinobe R, Hewison M, McGrath JJ. Distribution of the vitamin D receptor and 1 alpha-hydroxylase in human brain. J Chem Neuroanat. 2005;29(1):21–30.
14. Knekt P, Kilkkinen A, Rissanen H, Marniemi J, Sääksjärvi K, Heliövaara M. Serum vitamin D and the risk of Parkinson disease. Arch Neurol. 2010;67(7):808–11.
15. Plantone D, Primiano G, Manco C, Locci S, Servidei S, De Stefano N. Vitamin D in neurological diseases. Int J Mol Sci. 2022;24(1)
16. Wei F-L, Li T, Gao Q-Y, Huang Y, Zhou C-P, Wang W, et al. Association between Vitamin D supplementation and fall prevention. Front Endocrinol. 2022;13
17. Smolders J, Thewissen M, Peelen E, Menheere P, Tervaert JW, Damoiseaux J, et al. Vitamin D status is positively correlated with regulatory T cell function in patients with multiple sclerosis. PLoS One. 2009;4(8):e6635.
18. Vitkova M, Diouf I, Malpas C, Horakova D, Kubala Havrdova E, Patti F, et al. Association of latitude and exposure to ultraviolet B radiation with severity of multiple sclerosis: an international registry study. Neurology. 2022;98(24):e2401–e12.
19. Stein MS, Liu Y, Gray OM, Baker JE, Kolbe SC, Ditchfield MR, et al. A randomized trial of high-dose vitamin D2 in relapsing-remitting multiple sclerosis. Neurology. 2011;77(17):1611–8.
20. Bhargava P, Cassard S, Steele SU, Azevedo C, Pelletier D, Sugar EA, et al. The vitamin D to ameliorate multiple sclerosis (VIDAMS) trial: study design for a multicenter, randomized, double-blind controlled trial of vitamin D in multiple sclerosis. Contemp Clin Trials. 2014;39(2):288–93.
21. Vandebergh M, Dubois B, Goris A. Effects of Vitamin D and body mass index on disease risk and relapse hazard in multiple sclerosis. Neurol Neuroimmunol Neuroinflamm. 2022;9(3):e1165.

22. McLaughlin L, Clarke L, Khalilidehkordi E, Butzkueven H, Taylor B, Broadley SA. Vitamin D for the treatment of multiple sclerosis: a meta-analysis. J Neurol. 2018;265(12):2893–905.
23. Camu W, Lehert P, Pierrot-Deseilligny C, Hautecoeur P, Besserve A, Jean Deleglise A-S, et al. Cholecalciferol in relapsing-remitting MS: a randomized clinical trial (CHOLINE). Neurol Neuroimmunol Neuroinflamm. 2019;6(5):e597.
24. Hupperts R, Smolders J, Vieth R, Holmøy T, Marhardt K, Schluep M, et al. Randomized trial of daily high-dose vitamin D(3) in patients with RRMS receiving subcutaneous interferon β-1a. Neurology. 2019;93(20):e1906–e16.
25. Hupperts R, Smolders J, Vieth R, Holmøy T, Marhardt K, Schluep M, et al. Randomized trial of daily high-dose vitamin D3 in patients with RRMS receiving subcutaneous interferon -1a. Neurology. 2019;93(20):e1906–e16.
26. Yammine K, Abi Kharma J, Kaypekian T, Assi C, Zeeni N. Is diabetic neuropathy associated with vitamin D status? A meta-analysis. Br J Nutr. 2022;127(7):972–81.
27. Reda Halawa M, Zaky Ahmed I, Fawzy Abouelezz N, Roushdy Mohamed N, Khalil NHA, Hendawy LMA. The impact of vitamin D supplementation on peripheral neuropathy in a sample of Egyptian prediabetic individuals. F1000Res. 2021;10:817.
28. Yammine K, Wehbe R, Assi C. A systematic review on the efficacy of vitamin D supplementation on diabetic peripheral neuropathy. Clin Nutr. 2020;39(10):2970–4.
29. Kjærgaard M, Eggen AE, Mathiesen EB, Jorde R. Association between headache and serum 25-Hydroxyvitamin D; the Tromsø study: Tromsø 6. Headache. 2012;52(10):1499–505.
30. Gazerani P, Fuglsang R, Pedersen JG, Sørensen J, Kjeldsen JL, Yassin H, et al. A randomized, double-blinded, placebo-controlled, parallel trial of vitamin D3 supplementation in adult patients with migraine. Curr Med Res Opin. 2019;35(4):715–23.
31. Mintzer S, Boppana P, Toguri J, DeSantis A. Vitamin D levels and bone turnover in epilepsy patients taking carbamazepine or oxcarbazepine. Epilepsia. 2006;47(3):510–5.
32. Holick MF, Lim R, Dighe AS. Case 3-2009. N Engl J Med. 2009;360(4):398–407.
33. Pendo K, DeGiorgio M, Christopher M. Vitamin D for the treatment of epilepsy: basic mechanisms, animal models and clinical trials. Front Neurol. 2016;7
34. Moretti R, Morelli ME, Caruso P. Vitamin D in neurological diseases: a rationale for a pathogenic impact. Int J Mol Sci. 2018;19(8):2245.
35. Manson JE, Bassuk SS, Lee IM, Cook NR, Albert MA, Gordon D, et al. The VITamin D and OmegA-3 TriaL (VITAL): rationale and design of a large randomized controlled trial of vitamin D and marine omega-3 fatty acid supplements for the primary prevention of cancer and cardiovascular disease. Contemp Clin Trials. 2012;33(1):159–71.
36. Li G, Mbuagbaw L, Samaan Z, Falavigna M, Zhang S, Adachi JD, et al. Efficacy of vitamin D supplementation in depression in adults: a systematic review. J Clin Endocrinol Metab. 2014;99(3):757–67.
37. Cheng YC, Huang YC, Huang WL. The effect of vitamin D supplement on negative emotions: a systematic review and meta-analysis. Depress Anxiety. 2020;37(6):549–64.
38. Kerley CP, Power C, Gallagher L, Coghlan D. Lack of effect of vitamin D(3) supplementation in autism: a 20-week, placebo-controlled RCT. Arch Dis Child. 2017;102(11):1030–6.
39. DeLuca GC, Kimball SM, Kolasinski J, Ramagopalan SV, Ebers GC. Review: the role of vitamin D in nervous system health and disease. Neuropathol Appl Neurobiol. 2013;39(5):458–84.
40. Cui X, McGrath JJ, Burne THJ, Eyles DW. Vitamin D and schizophrenia: 20 years on. Mol Psychiatry. 2021;26(7):2708–20.
41. Vieth R. Critique of the considerations for establishing the tolerable upper intake level for vitamin D: critical need for revision upwards. J Nutr. 2006;136(4):1117–22.
42. Bischoff-Ferrari HA, Dawson-Hughes B, Orav EJ, Staehelin HB, Meyer OW, Theiler R, et al. Monthly high-dose Vitamin D treatment for the prevention of functional decline: a randomized clinical trial. JAMA Intern Med. 2016;176(2):175–83.

Chapter 14
Vitamin D and the Gastrointestinal System

Mithu Bhowmick and Jayanta Samanta

Abstract Vitamin D is a prohormone that is primarily responsible for calcium homeostasis and maintaining bone health, but recently the role of vitamin D in cellular proliferation and differentiation, immunomodulatory functions, and gut microbiome modulation has been well established. Several gastrointestinal conditions such as inflammatory bowel disease, celiac disease, pancreatic diseases, chronic liver diseases, malabsorptive bariatric surgery, and gastrointestinal malignancy predispose to vitamin D deficiency with resultant adverse clinical outcomes. Early diagnosis and adequate vitamin D supplementation are imperative to prevent detrimental outcomes. The purpose of this chapter is to comprehensively focus on vitamin D physiology and functions, association of vitamin D with specific gastrointestinal diseases, and evidence-based recommendations for evaluation and treatment of vitamin D deficiency.

Keywords Vitamin D · Gastrointestinal diseases · Deficiency · Vitamin D supplementation

Learning Objectives
- Molecular characteristics and physiology of vitamin D
- Common gastrointestinal conditions and association with vitamin D
- Vitamin D and gastrointestinal malignancy
- Evaluation, prevention, and treatment of vitamin D deficiency

M. Bhowmick · J. Samanta (✉)
Department of Gastroenterology, Postgraduate Institute of Medical Education and Research, Chandigarh, India

V. Kumar, S. Patel (eds.), *Vitamin D: A Comprehensive Guide for Medical Professionals*, https://doi.org/10.1007/978-981-96-6491-7_14

14.1 Introduction

Vitamin D is a prohormone that is primarily responsible for calcium homeostasis and maintaining bone health. Vitamin D deficiency predisposes to low bone mineral density (osteopenia and osteoporosis) and increases fracture risk. Several gastrointestinal conditions such as inflammatory bowel disease (IBD), celiac disease, pancreatic diseases, chronic liver diseases, malabsorptive bariatric surgery, and gastrointestinal malignancy predispose to vitamin D deficiency with resultant adverse clinical outcomes. This chapter aims to comprehensively discuss vitamin D physiology and functions, association of vitamin D with specific gastrointestinal diseases, and evidence-based recommendations for evaluation and treatment of vitamin D deficiency.

14.2 Biomolecular Characteristics of Vitamin D

Vitamin D, also known as sunshine vitamin, is essentially a prohormone that has a primary role in calcium homeostasis and bone metabolism. Vitamin D exists in two major forms: vitamin D2 (ergocalciferol) and vitamin D3 (cholecalciferol). The former is obtained from plant sources, and the latter is obtained either from skin or from animal sources.

In the skin, cholecalciferol is synthesized from the conversion of 7-dehydrocholecalciferol after exposure to ultraviolet-B radiation. Dietary vitamin D is absorbed as fat-soluble vitamin in the proximal small intestine after incorporation into chylomicrons. Dietary deficiency of vitamin D and people living in northern latitude are prone to develop vitamin D deficiency [1].

Vitamin D is initially an inert molecule, which requires activation in both liver and kidney by two consecutive hydroxylation process. The first step of activation occurs in the liver which results in the formation of 25-hydroxy vitamin D; the subsequent hydroxylation process occurs in the kidney yielding the active form 1,25 dihydroxy vitamin D (calcitriol) by the enzyme 1-α hydroxylase. Low plasma calcium or phosphate level regulates parathyroid hormone level leading to increased production of active vitamin D (Fig. 14.1).

The inactivation process of vitamin D occurs in the kidney initiated by enzyme 24-hydroxylase leading to formation of calcitroic acid with no biologic activity. The production of vitamin D is tightly regulated by negative feedback mechanism.

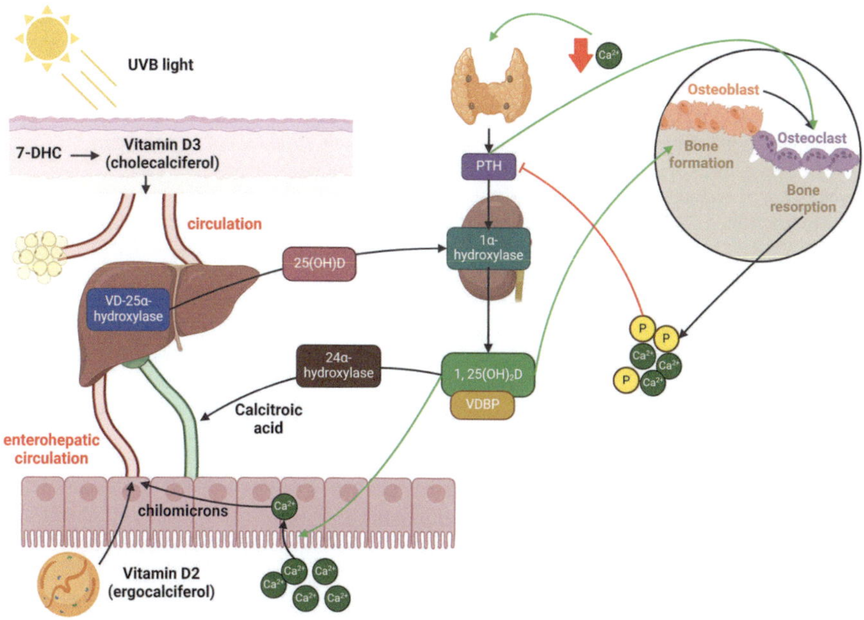

Fig. 14.1 Pathways of vitamin D metabolism

Calcitriol can downregulate 1-α hydroxylase and upregulate 24-hydroxylase enzymes leading to the formation of 24,25 dihydroxy vitamin D. Elevated PTH due to hypocalcaemia can reduce activity of 24 hydroxylase, allowing higher levels of active vitamin D to circulate.

14.3 Physiological Functions of Vitamin D

The active form of vitamin D binds to VDR (vitamin D receptor), a nuclear receptor that heterodimerizes with retinoid X receptor. The heterodimer enters nucleus and binds to vitamin D response element (VDRE) in nucleus resulting in expression of key genes involved in vitamin D action. Vitamin D also exerts nongenomic functions by binding to vitamin D receptor found outside nucleus (Table 14.1) [2].

Table 14.1 Physiological functions of vitamin D [1, 2]

Calcium-regulating organs	Non-calcium-regulating organs
Intestine Promotes calcium absorption by both transcellular and paracellular pathway Active phosphorus absorption by increasing expression of sodium-dependent phosphate transporter (NaPi-IIb) in jejunum *Bone* Promotes bone formation by promoting differentiation of chondrocytes and mineralization of bone Increases bone resorption by increasing number and activity of osteoclast along with PTH by inducing RANKL (receptor activator of nuclear factor kappa-B ligand) *Kidney* Enhances calcium absorption from distal renal tubule by increasing vitamin D receptor (VDR), calbindin, TRPV5, and calcium pump *Parathyroid hormone (PTH)* Regulates PTH production by inducing CASR (calcium-sensing receptor) Low calcium stimulates PTH secretion, high vitamin D, and calcium inhibit PTH secretion Inhibits phosphorus absorption in kidney	*Immune function* Enhances innate immune function by stimulating synthesis of cathelicidin, antimicrobial peptides Inhibits adaptive immune response by inhibiting maturation dendritic cells, inhibiting T cell proliferation and upregulation of Th2 response and T-reg pathways *Cardiovascular system* Decreases renin activity Inhibits vascular smooth muscle proliferation and calcification Reduces oxidative stress, inflammation, and thrombogenesis *Diabetes* Modulates β cell growth and differentiation Increases insulin sensitivity and insulin release by increasing intracellular calcium *Cancer* Antiproliferative action by enhancing cell cycle inhibitors p21 and p27 Inhibits β-catenin activity Increases E-cadherin expression

14.4 Gut Microbiome and Vitamin D

Gut microbiota plays an important role in maintaining mucosal barrier integrity, protection against pathogens, synthesis of vitamins, and regulation of immune response [3]. Gut dysbiosis, a state of disrupted homeostasis, can predispose to several diseases like IBD, diabetes, autoimmune diseases, and malignancies. Vitamin D plays an important role in immune regulation, proliferation, and intestinal homeostasis due to widespread expression of VDR in the gut. Low vitamin D/VDR expression compromises the intestinal barrier and increases intestinal permeability, causing altered microbiota composition and microbial translocation across gut epithelium, which results in heightened autoimmunity [3]. Bellera F et al., in a systematic review of 25 studies, found that vitamin D supplementation was associated with a significant change in major microbial phyla including *Firmicutes, Actinobacteria*, and *Bacteroidetes*, with either decrease or increase in relative abundance. The study also showed that high intake of vitamin D was associated with a change in alpha and beta diversity with resultant shift in bacterial composition and species richness [4]. The clinical implication or the translational significance of these findings is yet to be deciphered. Moreover, the current data are heterogenous, and well-designed randomized controlled trials (RCTs) are required to understand the role of vitamin D in the modulation of gut microbiota.

14.5 Vitamin D and Gastrointestinal Diseases

Gastrointestinal (GI) conditions like IBD, celiac disease, and malabsorptive bariatric surgery can predispose to vitamin D deficiency through several mechanisms including impaired absorption and metabolism leading to hypovitaminosis D. This has a negative impact on bone health with consequent risk of fracture. These can be further compounded by use of glucocorticoids in many GI conditions. Moreover, inadequate vitamin D levels, in a putative bidirectional interaction, may negatively affect the clinical course of the underlying disease.

14.5.1 Inflammatory Bowel Disease

Inflammatory bowel disease, which includes ulcerative colitis (UC) and Crohn's disease (CD), is chronic relapsing and remitting inflammatory disorder of GI tract. The etiopathogenesis involves a complex interaction between the genetic, environmental, microbial factors, and the immune responses. In Indian population with UC, Law et al., in their study, showed prevalence of vitamin D deficiency and insufficiency as 56% and 34%, respectively [5]. In another study from Norway, with limited sun exposure, Frigstad et al. found that vitamin D deficiency is prevalent in 44% of UC patients [6]. Similarly, a small study (34 CD patients) by Joseph et al. in southern India found that prevalence of vitamin D deficiency was 79%, while 12% were insufficient [7]. Vitamin D deficiency is an important risk factor for osteopenia and osteoporosis. In a study of 120 IBD patients in India, 50% of the population had either osteopenia or osteoporosis, with no significant difference between CD and UC [8]. Hypovitaminosis D in IBD can be attributed to malabsorption because of mucosal inflammation and/or surgery, poor dietary intake, decreased UV light exposure, and increased uptake by inflammatory cells in the affected sites [9]. Moreover, osteoporosis/osteopenia in IBD has been attributed to chronic ongoing inflammation with increase in inflammatory markers (IL-1β, TNF-α, IL-6, IL-15, IL-17) and corticosteroid therapy that increases osteoclastogenesis by increasing RANKL (receptor activator of nuclear factor kappa-B ligand) and reducing osteoprotegerin level [10]. Thus, superadded vitamin D deficiency can compound this factor to increase the severity of osteoporosis/osteopenia in these patients.

Vitamin D has also been implicated as a biomarker of disease activity, with one study showing low vitamin D level to be associated with increased disease activity (OR = 1.53), mucosal inflammation (OR = 1.25), clinical relapse (OR = 1.23), and low quality of life scores [11]. In another study from India of 126 UC patients, 81% were vitamin D deficient, and on multivariate analysis, low vitamin D levels (OR = 6.1) and histologically active disease (OR = 4) were found to be predictors of poor quality of life [12]. Hypovitaminosis D is also associated with loss of clinical response to biological therapy, with one retrospective study demonstrating vitamin D $\leq$ 25 ng/mL as an identifier of loss of response to biological drugs [13].

A broadly accepted definition of vitamin D deficiency/insufficiency in IBD remains uncertain, as the current normal range is based on studies of bone metabolism. Therapeutic efficacy of vitamin D in active disease is still unclear. Vitamin D supplementation in IBD may have some direct effects in maintaining integrity of mucosal barrier, modulation of immune system, maintenance of gut microbiota, and reducing relapse and exert some indirect effects in the form of increase response to drugs and biologics. We probably need more data to determine the optimum level necessary for disease prevention and management of IBD, as well as the most effective dosage and duration for replacement.

14.5.2 Celiac Disease

Celiac disease is an autoimmune condition triggered by gluten ingestion in genetically predisposed individual who are positive for HLADQ2 and/or HLADQ8. Celiac disease has a myriad of clinical manifestations, which can range from silent disease detected by serological testing to symptomatic disease, which includes intestinal and/or extraintestinal manifestations. The classic intestinal manifestations include diarrhea, bloating, failure to thrive, weight loss while extraintestinal manifestations include iron deficiency anemia, muscle wasting, joint/muscle pain, reduced bone mineral density, and many others [14]. The prevalence of celiac disease (8.53/1000 population) and latent celiac disease (3.70/1000 population) is highest in northern India [15]. Vitamin D deficiency in newly diagnosed celiac disease had a prevalence of 63.3% in Indian children aged up to 18 years as reported by G A et al., whereas pooled prevalence of osteopenia and osteoporosis in adults were 39.6% and 14.4% [16, 17]. In a retrospective study of 570 patients, Dhar J et al. found that patients with delayed diagnosis (>3 years) of celiac disease had significantly low median vitamin D level compared to patients with early diagnosis (<3 years) of celiac disease [18].

Vitamin D deficiency in celiac disease is primarily caused by malabsorption, while other mechanisms include steatorrhea and secondary lactase deficiency causing avoidance of dairy products. Celiac disease also interferes with the absorption of dietary calcium due to the inflammatory process involving proximal part of intestine, which is the predominant site of calcium absorption as well as reduced calcium-binding protein in damaged enterocytes further contributing to calcium loss and secondary hyperparathyroidism [19]. Hyperparathyroid state increases calcitriol level, which can further accentuate metabolism of 25-OH vitamin D.

Vitamin D supplementation can interfere with the disease course by limiting enterocyte damage and permeability and altering the adaptive immune response by reducing T cell stimulatory capacity and inducing immunological tolerance [20]. The current treatment of celiac disease is lifelong gluten-free diet (GFD). GFD per se is associated with normalization of mucosal structure and increased serum levels of vitamin D. Samanta J et al. in an RCT of 56 newly diagnosed celiac disease cases demonstrated that GFD alone was as effective as GFD with vitamin D (60,000 IU/

week) supplementation in improving bone mineralization at 3 months [21]. Nevertheless, dietary supplementation alone may not be always sufficient to restore vitamin D levels, and the ideal amount to normalize vitamin D has not been established.

Akshaya G et al. showed that weekly vitamin D supplementation of 60,000 IU along with GFD, in celiac patients with low vitamin D levels, can increase vitamin D from a median pretreatment value of 12.55–64.93 ng/ml in 1 week [16]. Another study showed a significant increase from 9.45 ± 0.45 ng/ml to 13.53 ± 1.5 ng/ml after 6 months, although none reached normal levels [22]. Regardless of vitamin D level at onset or during GFD, major society guidelines from the American College of Gastroenterology (ACG), the British Society of Gastroenterology (BSG), and European Society Paediatric Gastroenterology, Hepatology and Nutrition (ESPGHAN) recommend monitoring and supplementation of vitamin D in celiac disease [14].

14.5.3 Pancreatitis

14.5.3.1 Acute Pancreatitis

Acute pancreatitis (AP) is characterized by self-digestion of acinar cells of pancreas by activated trypsin induced by various etiological factors leading to local and systemic inflammation. About 80–85% have interstitial pancreatitis and 15–20% have necrotizing pancreatitis. The clinical course is mild in 80% of cases, the overall mortality is 2%, which, however, approaches 30% in patients with persistent organ failure [23]. Severity and early organ failure in AP are sequelae of the initial cytokine storm while compensatory anti-inflammatory response syndrome (CARS) drives the later infective complication phase [24]. Vitamin D, known for its immunomodulatory function, can downregulate the inflammatory cascade by decreasing activation of Toll-like receptors (TLRs) and inhibiting activation of nuclear factor-κB (NF-κβ) [25]. The role of vitamin D in disease modulation is further established with several studies showing impaired or reduced function of VDR gene being associated with increased risk of acute pancreatitis [26]. Huh JH et al., in a study of 242 patients, showed that vitamin D deficiency on admission was an independent risk factor for severe AP (OR = 5.37) and intensive care unit (ICU) admission (OR = 3.09) [27]. Vitamin D deficiency in AP can be attributed to impaired absorption, poor nutrition, prolonged fasting, and exocrine insufficiency. In a large study of 256,580 patients with AP, AP patients were found to be associated with vitamin D deficiency (OR = 1.25), osteoporosis (OR = 1.89), and fractures (OR = 1.58) [28]. Another multiethnic cohort study by Setiawan VW et al. indicated that vitamin D, particularly from milk, negatively correlated with gallstone-related AP [29]. While anecdotal reports have shown a negative association between vitamin D deficiency and AP, there are no definite guidelines/recommendations for the same, and thus, more definitive studies are needed in this aspect.

14.5.3.2 Chronic Pancreatitis

Chronic pancreatitis (CP) is characterized by extensive destruction of pancreatic parenchyma leading to pancreatic atrophy, calcification, exocrine, and endocrine dysfunction with consequent malabsorption and malnutrition in varying combinations. Pancreatic exocrine and endocrine dysfunction develops in 20–30% of cases following a bout of AP while it can be seen in one-half to one-third of CP cases [23]. Vitamin D deficiency in CP is related to steatorrhea, poor nutritional status, and diabetes mellitus among others. A systematic review and meta-analysis of 12 studies found that the pooled prevalence of vitamin D insufficiency was 57.6%, the highest among fat-soluble vitamins [30]. In an Indian study of 103 patients, with majority (70%) having alcohol-related CP, vitamin D was insufficient (10–30 ng/ml) in 40.8% cases and deficient (<10 ng/ml) in 19.4% of cases, while the prevalence of osteoporosis and osteopenia was 46% and 12%, respectively [31]. A recent meta-analysis demonstrated the substantial effect of both oral and intramuscular administration of vitamin D in CP [32]. Further as vitamin D has a plausible role in insulin secretion and action, supplementation may be beneficial in patients with CP-related diabetes, although data is scanty. The recently published ACG guidelines from the United European Gastroenterology (UEG) and several society guidelines on CP recommend periodic screenings for malnutrition, including evaluation for fat-soluble vitamin deficiencies and osteoporosis [25, 33, 34].

14.5.4 Liver Diseases

14.5.4.1 Chronic Liver Disease

Liver plays a crucial role in the metabolism and pleiotropic function of vitamin D. Vitamin D deficiency in liver diseases has been associated with hepatic decompensation, bacterial infection, increased severity of fibrosis, and mortality [35–37]. The prevalence of vitamin D deficiency and low bone mineral density (osteopenia and osteoporosis) in Indian patients with cirrhosis were 89.4% and 86%, respectively, with higher proportion in decompensated cirrhosis [38]. Low vitamin D levels in chronic liver disease can be attributed to [39]

- impaired hydroxylation in severe disease,
- insufficient intake,
- inadequate sun exposure,
- impaired absorption secondary to bowel congestion from portal hypertension and hypoalbuminemia.

In addition, vitamin D has been shown to improve liver disease by regulating innate immune response by affecting TLR expression and intestinal permeability, as well as modulating adaptive immune response by inhibiting Th1 and Th17 cell responses and inducing Th2 and Treg cell differentiation [40]. Furthermore, vitamin

Table 14.2 Implications of vitamin D in liver diseases

Disease	Role of vitamin D
Alcoholic liver disease [42]	Severe deficiency associated with higher aspartate aminotransferase levels, increased hepatic venous pressure gradient, MELD, and child-Pugh scores. Vitamin D supplementation either in vivo or in vitro can downregulate inflammatory cascade by reducing TNF-alfa production
Autoimmune hepatitis [43]	Severe deficiency associated with treatment nonresponse risk of developing cirrhosis and liver-related mortality or requirement for liver transplantation
Hepatocellular carcinoma (HCC) [40]	Studies have shown inverse relationship between serum levels of vitamin D and HCC Vitamin D may impact the development of HCC by activating apoptosis, decreasing oxidative stress, and inflammation

D has antifibrotic activity by inhibiting activation and proliferation of hepatic stellate cells via VDR signaling [41]. The various implications and outcomes of vitamin D deficiency in liver disease have been outlined in Table 14.2.

14.5.4.2 Metabolic-Associated Steatotic Liver Disease

Metabolic-associated steatotic liver disease (MASLD) is the most common liver disease worldwide. The spectrum of MASLD comprises of bland steatosis, steatohepatitis, fibrosis to cirrhosis, and hepatocellular carcinoma (HCC). The pathophysiology is a complex interlinked process of insulin resistance, lipotoxicity, infiltration of liver by proinflammatory cells, gut dysbiosis with resultant activation of hepatic stellate cells, and fibrogenesis [44]. Several studies had demonstarated the potential role of vitamin D/VDR in MASLD owing to its insulin sensitizing properties, anti-inflammatory, antifibrotic, and its role in homeostasis of adipose tissue and gut barrier [45]. There is conflicting evidence on the exact role of vitamin D supplementation on body weight, BMI, fasting blood glucose, or insulin resistance, and thus, more data is required before definitive conclusions can be made [45–47].

14.6 Gastrointestinal Malignancy

Vitamin D affects the activation and inhibition of cellular pathways through gene expression, which includes cellular growth, apoptosis, cellular differentiation, angiogenesis, metastasis, and immune system modulation [48, 49]. Several of these cellular processes are regulated by vitamin D-responsive genes that include WAF1, cyclin C, C-JUN, transforming growth factor-β (TGF-β) family, and calcium-sensing receptor (CaSR) [50]. Multiple observational studies have demonstrated a potential positive effect of vitamin D supplementation in gastrointestinal

Table 14.3 Potential antitumor actions of vitamin D in gastrointestinal malignancies

Organs	Antitumor actions
Esophagus [54, 55]	Reduced IL-6 and glutathione peroxidase expression (GPX1) VDR expression serves as prognostic marker
Stomach [56, 57]	Induction of apoptosis Regulation of hedgehog signaling (Hh) pathway Synergistic effect with chemotherapeutic agents Paricalcitol, a calcitriol analog, induces apoptosis and suppresses inflammation
Pancreas [58, 59]	Induces cyclin-dependent kinase inhibitor p21, p27 and cell cycle arrest Regulation of hedgehog signaling (Hh) pathway Synergistic effect with chemotherapeutic agents Regulates insulin action and vitamin D deficiency; increases risk of cancer in diabetics
Colorectal malignancy [60, 61]	Mutation of *PIK3CA* and *KRAS* is associated with VDR overexpression Induces cyclin-dependent kinase inhibitor p21, p27 Inhibition of protooncogene c-myc, cyclin D1

malignancies such as oesophageal cancer, gastric cancer, HCC, pancreatic cancer, and colorectal cancer [51–53]. The probable antitumor mechanisms of vitamin D in these malignancies have been highlighted in Table 14.3. However, several studies including randomized trials did not demonstrate any beneficial role in chemoprevention. Moreover, the dosing remains uncertain in different gastrointestinal malignancies [62, 63].

14.7 Obesity and Bariatric Surgery

Globally the prevalence of obesity has increased in the recent years. In 2022, WHO reported 2.5 billion adults 18 years and above were overweight including 890 million obese adults. Vitamin D deficiency is one of the most frequently occurring micronutrient deficiencies, with one comprehensive review showing a 35% higher prevalence in obese patients compared to eutrophic adults [64]. The pathogenesis remains multifaceted, the key mechanisms being decreased consumption, poor sun exposure, and vitamin D sequestration in adipose tissue [65]. Although vitamin D has a role in insulin sensitivity, recent systematic review did not show any significant effect of vitamin D supplementation on the BMI, waist circumference, and waist-to-hip ratio [66].

Bariatric surgery is currently the most efficacious treatment of obesity, and the number of such surgeries has significantly increased in recent years. Nutritional deficiencies including vitamin D deficiency frequently occur following bariatric surgery and more so for malabsorptive bariatric surgery. A recent systematic review and metanalysis by Giustina et al. showed that prevalence of vitamin D insufficiency/deficiency was higher in malabsorptive surgeries [Roux-en Y gastric bypass (RYGB), biliopancreatic diversion (BPD)] than restrictive surgeries [laparoscopic sleeve gastrectomy (SG), adjustable gastric band (AGB)] [67]. The main causes of vitamin D deficiency include inadequate supplementation during rapid weight loss, bile salt deficiency, intestinal bacterial overgrowth, and vitamin D deficiency prior to surgery [67]. This has a substantial influence on bone health, with high turnover bone loss and an increased risk of fracture. The American Association of Clinical Endocrinologists (AACE), The Obesity Society (TSO), and American Society for Metabolic & Bariatric Surgery (ASMBS) have recommended evaluation for parathyroid hormone, vitamin D, serum calcium, 24-h urinary calcium, and vitamin D supplementation following bariatric surgery, ranging from 3000 IU daily to 50,000 IU one to three times weekly to daily in severe malabsorption [68].

14.8 Signs and Symptoms of Vitamin D Deficiency

The symptoms complex ranges from asymptomatic to symptomatic disease secondary to hypocalcemia and hyperparathyroidism. Patients with severe vitamin D deficiency can experience symptoms of myalgia, fatigue, arthralgias, weakness, muscle fasciculations, and osteoporosis with fractures in older individuals. Vitamin D insufficiency is well associated with the risk of hypertension or cardiovascular events, a greater incidence of malignancies, migraine, and neuropsychiatric diseases such as schizophrenia, dementia, or depression [69].

14.9 Evaluation of Vitamin D Deficiency

25(OH) vitamin D is the most abundant circulating form of vitamin D, with a half-life of around 2–3 weeks. Although 1,25(OH)D3 is the physiologically active form of vitamin D, it is not recommended for measurement due to its short half-life of 4–6 h [70]. The Endocrine Society Practice Guidelines describes vitamin D deficiency as serum 25(OH)D levels below 20 ng/ml, insufficiency as 21–29 ng/ml, and sufficient as at least 30 ng/ml [71].

14.10 Treatment

The recommended dietary intake of vitamin D above for adults aged 19–70 years and above is 600–800 IU/day. Both the preparation of vitamin D [D2(ergocalciferol) and D3 (cholecalciferol)] are effective in treatment and prevention of vitamin D deficiency. The Endocrine Society clinical practice guideline suggests 50,000 IU of vitamin D2 or vitamin D3 once a week for 8 weeks, or 6000 IU/daily, to attain a blood level of 25(OH)D greater than 30 ng/ml, followed by 1500–2000 IU/day maintenance treatment [71]. In patients with malabsorption syndrome, two to three times higher dose is recommended at least 6000–10,000 IU/day to maintain blood level of 25(OH)D above 30 ng/ml, followed by 3000–6000 IU daily for maintenance [71]. In individuals who cannot take oral vitamin D supplementation, parenteral vitamin D in the form of intramuscular injection is equally effective as oral supplementation to achieve the target levels [72]. Secondary hyperparathyroidism may remain even after vitamin D replenishment. If calcium absorption is also impaired, calcium supplementation at 1–2 gm/day should be given to prevent secondary hyperparathyroidism [73]. Furthermore, there are limited studies on the appropriate dosing and duration of vitamin D supplementation in various gastrointestinal diseases. Although the current practice is to extrapolate from the standard recommendations, this warrant controlled clinical trials to define optimum vitamin D levels, with the final objective of defining disease-specific requirements of vitamin D supplementation.

14.11 Hypervitaminosis D

Hypervitaminosis D although rare results from excessive vitamin D intake from prescription error or over-the-counter medications, lack of monitoring in patients requiring high doses of vitamin D supplementation like osteoporosis, celiac disease, IBD, and malabsorptive surgeries. The clinical signs and symptoms result from hypercalcemia which can range from nonspecific symptoms like weakness, fatigue, anorexia to more severe symptoms like confusion, irritability, abdominal pain, peptic ulcer, pancreatitis, polyuria, polydipsia, nephrolithiasis, and cardiac arrhythmias [74]. Treatment consists of discontinuation of vitamin D and calcium, intravenous hydration with normal saline, calcitonin, glucocorticoids, and bisphosphonates. Most cases resolve without serious complications; some patients may develop acute renal failure necessitating hemodialysis [74]. Hence, routine measurement of calcium, phosphorus, and vitamin D level is warranted in patients on long-term vitamin D supplementation.

14.12 Conclusion

Vitamin D is a pleiotropic hormone that has a plethora of functions, primarily responsible for calcium homeostasis, although its function extends well beyond bone metabolism. Vitamin D deficiency is common in gastrointestinal diseases, and the pathogenesis remains multifactorial. The impact of vitamin D on the various GI disease outcomes is fairly well established and underlines the role of careful assessment of vitamin D status in the incipient stage of disease. Supplementation with adequate dosages and duration of vitamin D in deficient states may be crucial to prevent adverse health outcomes.

Author Disclosure All the authors have no potential conflicts of interests.

Writing Assistance Nil. No artificial-assisted technologies have been used in the production of the submitted work.

Financial Support Nil.

References

1. Rebelos E, Tentolouris N, Jude E. The role of vitamin D in health and disease: a narrative review on the mechanisms linking vitamin D with disease and the effects of supplementation. Drugs. 2023;83(8):665–85.
2. Bikle DD, Feingold KR, Anawalt B, et al. Vitamin D: production, metabolism and mechanisms of action. In: Endotext. South Dartmouth: MDText.com, Inc.; December 31, 2021; 2021.
3. Yamamoto EA, Jørgensen TN. Relationships between vitamin D, gut microbiome, and systemic autoimmunity. Front Immunol. 2020;10:3141.
4. Bellerba F, Muzio V, Gnagnarella P, et al. The association between vitamin D and gut microbiota: a systematic review of human studies. Nutrients. 2021;13(10):3378.
5. Law AD, Dutta U, Kochhar R, et al. Vitamin D deficiency in adult patients with ulcerative colitis: prevalence and relationship with disease severity, extent, and duration. Indian J Gastroenterol. 2019;38(1):6–14.
6. Frigstad SO, Høivik M, Jahnsen J, et al. Vitamin D deficiency in inflammatory bowel disease: prevalence and predictors in a Norwegian outpatient population. Scand J Gastroenterol. 2017;52(1):100–6.
7. Joseph AJ, George B, Pulimood AB, et al. 25 (OH) vitamin D level in Crohn's disease: association with sun exposure & disease activity. Indian J Med Res. 2009;130(2):133–7.
8. Bandyopadhyay D, Bandyopadhyay S, Ghosh P, et al. Extraintestinal manifestations in inflammatory bowel disease: prevalence and predictors in Indian patients. Indian J Gastroenterol. 2015;34(5):387–94.
9. Del Pinto R, Pietropaoli D, Chandar AK, et al. Association between inflammatory bowel disease and vitamin D deficiency: a systematic review and meta-analysis. Inflamm Bowel Dis. 2015;21(11):2708–17.
10. Vernia F, Valvano M, Longo S, et al. Vitamin D in inflammatory bowel diseases. Mechanisms of action and therapeutic implications. Nutrients. 2022;14(2):269.
11. Gubatan J, Chou ND, Nielsen OH, et al. Systematic review with meta-analysis: association of vitamin D status with clinical outcomes in adult patients with inflammatory bowel disease. Aliment Pharmacol Ther. 2019;50(11–12):1146–58.
12. Sachan A, Thungapathra M, Kaur H, et al. Comprehensive assessment of nutritional and functional status of patients with ulcerative colitis and their impact on quality of life. Indian J Gastroenterol. 2024;43(1):254–63.
13. Valvano M, Magistroni M, Mancusi A, et al. The usefulness of serum vitamin D levels in the assessment of IBD activity and response to biologics. Nutrients. 2021;13(2):323.
14. Raiteri A, Granito A, Giamperoli A, et al. Current guidelines for the management of celiac disease: a systematic review with comparative analysis. World J Gastroenterol. 2022;28(1):154–75.
15. Ramakrishna BS, Makharia GK, Chetri K, et al. Prevalence of adult celiac disease in India: regional variations and associations. Am J Gastroenterol. 2016;111(1):115–23.

16. Akhshayaa G, Seth A, Kumar P, Jain A. Prevalence and management of vitamin D deficiency in children with newly diagnosed coeliac disease: cohort study. Paediatr Int Child Health. 2021;41(4):247–52.
17. Ganji R, Moghbeli M, Sadeghi R, et al. Prevalence of osteoporosis and osteopenia in men and premenopausal women with celiac disease: a systematic review. Nutr J. 2019;18(1):9.
18. Dhar J, Samanta J, Sharma M, et al. Impact of delay in diagnosis in patients with celiac disease: a study of 570 patients at a tertiary care center. Indian J Gastroenterol. 2022;41(1):30–6.
19. Staun M, Jarnum S. Measurement of the 10,000-molecular weight calcium-binding protein in small-intestinal biopsy specimens from patients with malabsorption syndromes. Scand J Gastroenterol. 1988;23(7):827–32.
20. Infantino C, Francavilla R, Vella A, et al. Role of vitamin D in celiac disease and inflammatory bowel diseases. Nutrients. 2022;14(23):5154.
21. Samanta J, Kumar S, Dhar J, et al. Is there any added benefit of vitamin-d supplementation over and above gluten-free diet in newly diagnosed celiac disease patients with baseline vitamin-d deficiency? A randomized controlled trial. Gut. 2022;71:A167.
22. Verma A, Lata K, Khanna A, et al. Study of effect of gluten-free diet on vitamin D levels and bone mineral density in celiac disease patients. J Family Med Prim Care. 2022;11(2):603–7.
23. Forsmark CE, Vege SS, Wilcox CM. Acute pancreatitis. N Engl J Med. 2016;375(20):1972–81.
24. Phillip V, Steiner JM, Algül H. Early phase of acute pancreatitis: assessment and management. World J Gastrointest Pathophysiol. 2014;5(3):158–68.
25. Cai F, Hu C, Chen CJ, Han YP, Lin ZQ, Deng LH, Xia Q. Vitamin D and pancreatitis: a narrative review of current evidence. Nutrients. 2022;14(10):2113.
26. El-Mahdy RI, Ramadan HK, Mohammed HSED, et al. Impact of the etiology and vitamin D receptor TaqI rs731236 gene polymorphism on the severity of acute pancreatitis. J Hepatobiliary Pancreat Sci. 2020;27(11):896–906.
27. Huh JH, Kim JW, Lee KJ. Vitamin D deficiency predicts severe acute pancreatitis. United European Gastroenterol J. 2019;7(1):90–5.
28. Abou Saleh M, Alkhayyat M, Mansoor E, et al. The risk of vitamin D deficiency, osteoporosis, and fractures in acute pancreatitis. Pancreas. 2020;49(5):629–33.
29. Setiawan VW, Pandol SJ, Porcel J, et al. Dietary factors reduce risk of acute pancreatitis in a large multiethnic cohort. Clin Gastroenterol Hepatol. 2017;15(2):257–265.e3.
30. Martínez-Moneo E, Stigliano S, Hedström A, et al. Deficiency of fat-soluble vitamins in chronic pancreatitis: a systematic review and meta-analysis. Pancreatology. 2016;16(6):988–94.
31. Prabhakaran A, Bhasin DK, Rana SS, et al. Bone mineral metabolism and bone mineral density in alcohol related and idiopathic chronic pancreatitis. Trop Gastroenterol. 2014;35:107–12.
32. Wiese M, Gärtner S, Doller J, et al. Nutritional management of chronic pancreatitis: a systematic review and meta-analysis of randomized controlled trials. J Gastroenterol Hepatol. 2021;36(3):588–600.
33. Gardner TB, Adler DG, Forsmark CE, et al. ACG clinical guideline: chronic pancreatitis. Am J Gastroenterol. 2020;115(3):322–39.
34. Löhr JM, Dominguez-Munoz E, Rosendahl J, et al. United European gastroenterology evidence-based guidelines for the diagnosis and therapy of chronic pancreatitis (HaPanEU). United European Gastroenterol J. 2017;5(2):153–99.
35. Putz-Bankuti C, Pilz S, Stojakovic T, et al. Association of 25-hydroxyvitamin D levels with liver dysfunction and mortality in chronic liver disease. Liver Int. 2012;32(5):845–51.
36. Anty R, Tonohouan M, Ferrari-Panaia P, et al. Low levels of 25-hydroxy vitamin D are independently associated with the risk of bacterial infection in cirrhotic patients. Clin Transl Gastroenterol. 2014;5(5):e56.
37. Targher G, Bertolini L, Scala L, et al. Associations between serum 25-hydroxyvitamin D3 concentrations and liver histology in patients with non-alcoholic fatty liver disease. Nutr Metab Cardiovasc Dis. 2007;17(7):517–24.
38. Grover I, Singh N, Gunjan D, et al. Vitamin D status & bone health in patients with liver cirrhosis. Indian J Med Res. 2023;158(3):284–91.

39. Iruzubieta P, Terán Á, Crespo J, et al. Vitamin D deficiency in chronic liver disease. World J Hepatol. 2014;6(12):901–15.
40. Triantos C, Aggeletopoulou I, Thomopoulos K, et al. Vitamin D-liver disease association: biological basis and mechanisms of action. Hepatology. 2021;74(2):1065–73.
41. Udomsinprasert W, Jittikoon J. Vitamin D and liver fibrosis: molecular mechanisms and clinical studies. Biomed Pharmacother. 2019;109:1351–60.
42. Trépo E, Ouziel R, Pradat P, et al. Marked 25-hydroxyvitamin D deficiency is associated with poor prognosis in patients with alcoholic liver disease. J Hepatol. 2013;59(2):344–50.
43. Ebadi M, Bhanji RA, Mazurak VC, et al. Severe vitamin D deficiency is a prognostic biomarker in autoimmune hepatitis. Aliment Pharmacol Ther. 2019;49(2):173–82.
44. Kuchay MS, Choudhary NS, Mishra SK. Pathophysiological mechanisms underlying MAFLD. Diabetes Metab Syndr. 2020;14(6):1875–87.
45. Barchetta I, Cimini FA, Cavallo MG. Vitamin D and metabolic dysfunction-associated fatty liver disease (MAFLD): an update. Nutrients. 2020;12(11):3302.
46. Rezaei S, Tabrizi R, Nowrouzi-Sohrabi P, et al. The effects of vitamin D supplementation on anthropometric and biochemical indices in patients with non-alcoholic fatty liver disease: a systematic review and meta-analysis. Front Pharmacol. 2021;12:732496.
47. Chen X, Zhao Y, Zhang R, et al. The effect of vitamin D supplementation on some metabolic parameters in patients with nonalcoholic fatty liver disease: a systematic review and meta-analysis of 8 RCTs. Medicine (Baltimore). 2023;102(42):e35717.
48. Chung I, Han G, Seshadri M, et al. Role of vitamin D receptor in the antiproliferative effects of calcitriol in tumor-derived endothelial cells and tumor angiogenesis in vivo. Cancer Res. 2009;69:967–75.
49. Lamprecht SA, Lipkin M. Cellular mechanisms of calcium and vitamin D in the inhibition of colorectal carcinogenesis. Ann N Y Acad Sci. 2001;952:73–87.
50. Lamprecht SA, Lipkin M. Chemoprevention of colon cancer by calcium, vitamin D and folate: molecular mechanisms. Nat Rev Cancer. 2003;3:601–14.
51. Wu K, Feskanich D, Fuchs CS, et al. A nested case–control study of plasma 25-hydroxyvitamin D concentrations and risk of colorectal cancer. J Natl Cancer Inst. 2007;99:1120–9.
52. Skinner HG, Michaud DS, Giovannucci E, et al. Vitamin D intake and the risk for pancreatic cancer in two cohort studies. Cancer Epidemiol Biomarkers Prev. 2006;15:1688–95.
53. Goyal H, Perisetti A, Rahman MR, et al. Vitamin D and gastrointestinal cancers: a narrative review. Dig Dis Sci. 2019;64(5):1098–109.
54. Chen PT, Hsieh CC, Wu CT, et al. 1α,25-dihydroxyvitamin D3 inhibits esophageal squamous cell carcinoma progression by reducing IL6 signaling. Mol Cancer Ther. 2015;14(6):1365–75.
55. Gan X, Chen B, Shen Z, Liu Y, et al. High GPX1 expression promotes esophageal squamous cell carcinoma invasion, migration, proliferation and cisplatin-resistance but can be reduced by vitamin D. Int J Clin Exp Med. 2014;7(9):2530–40.
56. Baek S, Lee YS, Shim HE, et al. Vitamin D3 regulates cell viability in gastric cancer and cholangiocarcinoma. Anat Cell Biol. 2011;44(3):204–9.
57. Park MR, Lee JH, Park MS, et al. Suppressive effect of 19-nor-1α-25-dihydroxyvitamin D2 on gastric cancer cells and peritoneal metastasis model. J Korean Med Sci. 2012;27(9):1037–43.
58. Schwartz GG, Eads D, Rao A, et al. Pancreatic cancer cells express 25-hydroxyvitamin D-1 alpha-hydroxylase and their proliferation is inhibited by the prohormone 25-hydroxyvitamin D3. Carcinogenesis. 2004;25(6):1015–26.
59. Yu WD, Ma Y, Flynn G, Muindi JR, et al. Calcitriol enhances gemcitabine anti-tumor activity in vitro and in vivo by promoting apoptosis in a human pancreatic carcinoma model system. Cell Cycle. 2010;9(15):3022–9.
60. Kure S, Nosho K, Baba Y, Irahara N, et al. Vitamin D receptor expression is associated with PIK3CA and KRAS mutations in colorectal cancer. Cancer Epidemiol Biomarkers Prev. 2009;18(10):2765–72.
61. Klampfer L. Vitamin D and colon cancer. World J Gastrointest Oncol. 2014;6:430–7.

62. Wactawski-Wende J, Kotchen JM, Anderson GL, et al. Calcium plus vitamin D supplementation and the risk of colorectal cancer [published correction appears in N Engl J Med. 2006 Mar 9;354(10):1102]. N Engl J Med. 2006;354(7):684–96.
63. Piper MR, Freedman DM, Robien K, et al. Vitamin D-binding protein and pancreatic cancer: a nested case-control study. Am J Clin Nutr. 2015;101(6):1206–15.
64. Pereira-Santos M, Costa PR, Assis AM, et al. Obesity and vitamin D deficiency: a systematic review and meta-analysis. Obes Rev. 2015;16(4):341–9.
65. Lespessailles E, Toumi H. Vitamin D alteration associated with obesity and bariatric surgery. Exp Biol Med (Maywood). 2017;242(10):1086–94.
66. Duan L, Han L, Liu Q, et al. Effects of vitamin D supplementation on general and central obesity: results from 20 randomized controlled trials involving apparently healthy populations. Ann Nutr Metab. 2020;76(3):153–64.
67. Giustina A, di Filippo L, Facciorusso A, et al. Vitamin D status and supplementation before and after bariatric surgery: recommendations based on a systematic review and meta-analysis. Rev Endocr Metab Disord. 2023;24(6):1011–29.
68. Mechanick JI, Youdim A, Jones DB, et al. Clinical practice guidelines for the perioperative nutritional, metabolic, and nonsurgical support of the bariatric surgery patient--2013 update: cosponsored by American Association of Clinical Endocrinologists, the Obesity Society, and American Society for Metabolic & Bariatric Surgery. Endocr Pract. 2013;19(2):337–72.
69. Chang SW, Lee HC. Vitamin D and health – the missing vitamin in humans. Pediatr Neonatol. 2019;60(3):237–44.
70. Holick MF. Vitamin D status: measurement, interpretation, and clinical application. Ann Epidemiol. 2009;19(2):73–8.
71. Holick MF, Binkley NC, Bischoff-Ferrari HA, et al. Evaluation, treatment, and prevention of vitamin D deficiency: an Endocrine Society clinical practice guideline [published correction appears in J Clin Endocrinol Metab. 2011 Dec;96(12):3908]. J Clin Endocrinol Metab. 2011;96(7):1911–30.
72. Gupta N, Farooqui KJ, Batra CM, et al. Effect of oral versus intramuscular vitamin D replacement in apparently healthy adults with vitamin D deficiency. Indian J Endocrinol Metab. 2017;21(1):131–6.
73. Davies M, Heys SE, Selby PL, et al. Increased catabolism of 25-hydroxyvitamin D in patients with partial gastrectomy and elevated 1,25-dihydroxyvitamin D levels. Implications for metabolic bone disease. J Clin Endocrinol Metab. 1997;82(1):209–12.
74. Marcinowska-Suchowierska E, Kupisz-Urbańska M, Łukaszkiewicz J, et al. Vitamin D toxicity-A clinical perspective. Front Endocrinol (Lausanne). 2018;20(9):550.

Chapter 15
Radiodiagnostics in Vitamin D Deficiency and Hypervitaminosis

Anindita Sinha, Madhurima Sharma, and Mahesh Prakash

Abstract Radiological manifestations of vitamin D deficiency can manifest in childhood as rickets and in adults as osteomalacia. Characteristic changes occur at the growth plate and appendicular skeleton. Bone softening in osteomalacia causes characteristic Looser's zones and bone deformities in severe cases. Age of presentation and severity may vary depending on the cause, which ranges from genetic to nutritional or renal.

Hypervitaminosis of vitamin D may present with osteopenia or osteosclerosis with dense metaphyseal bands and visceral and soft tissue calcifications.

Keywords Vitamin D · Rickets · Osteomalacia · Hypophosphatasia · Tumour-induced osteomalacia · Hypervitaminosis

Bone metabolism and bone turnover occur throughout life and are mediated by osteoclastic resorption and osteoblastic deposition of osteoid with osteoid mineralisation, which requires homeostasis of calcium, phosphate and vitamin D and alkaline phosphatase. Vitamin D is responsible for maintaining normal serum calcium levels by calcium absorption from the intestine and osteoclastic activation which leads to bone resorption. Vitamin D also promotes mineralisation of osteoid.

A. Sinha (✉) · M. Sharma · M. Prakash
Department of Radiodiagnosis, Post Graduate Institute of Medical Education & Research, Chandigarh, India

V. Kumar, S. Patel (eds.), *Vitamin D: A Comprehensive Guide for Medical Professionals*, https://doi.org/10.1007/978-981-96-6491-7_15

15.1 Vitamin D Deficiency

Vitamin D deficiency causes defective bone mineralisation leading to bone softening and deformities. Defective endochondral ossification leads to metaphyseal abnormalities [1].

Vitamin D deficiency before the fusion of the growth plate causes rickets and after the fusion of the growth plate causes osteomalacia.

15.1.1 Rickets

Rickets is a manifestation of vitamin D deficiency in children, most commonly nutritional, due to dietary deficiency, or malabsorption or due to lack of sunlight exposure. Deficiency in synthesis of vitamin D in liver due to liver disease (leading to defective 25-hydroxylation of vitamin D) or kidney disease; renal rickets (chronic kidney disease leading to deficiency of the enzyme 1 alpha hydroxylase which converts 25-hydroxyvitamin D to its active metabolite 1,25-dihydroxy vitamin D or calcitriol), and certain medications can also give rise to vitamin D deficiency. Calcitriol acts on intestinal cells to facilitate calcium absorption. These form of calcipenic rickets can be corrected by vitamin D supplementation and are known as vitamin D-dependent rickets.

Rickets can also be caused by phosphate loss (phosphopenic rickets) due to genetic mutations (X-linked hypophosphataemic rickets (XLHR), autosomal dominant hypophosphataemic rickets (ADHR) and autosomal recessive hypophosphataemic rickets (ARHR)), renal tubular defects like Fanconi syndrome or phosphate malabsorption or dietary deficiency. These conditions lead to vitamin D-resistant rickets [2].

Both calcipenia and phosphopenia result in abnormalities of the growth plate or physis.

Physis or growth plate consists of four zones [3]:

1. Resting zone—nearest to the epiphysis, with inactive chondroblasts
2. Proliferative zone—with active chondroblasts which lay extracellular matrix, responsible for most longitudinal growth; in rickets, there is lack of apoptosis and unregulated proliferation of chondrocytes with loss of columnar arrangement of the chondrocytes
3. Hypertrophic zone—with hypertrophic chondrocytes and less extracellular matrix; this zone is further divided into the following:
   (a) Zone of maturation
   (b) Zone of provisional degeneration
   (c) Zone of provisional calcification—with chondrocyte death and calcium release. Deficiency of vitamin D leads to decreased hydroxyapatite formation necessary for matrix mineralisation and decrease in osteoclastic resorption of osteoid laid down by osteoblasts.
4. Zone of calcification

Fig. 15.1 Nutritional rickets in a child. Radiographs of bilateral wrists show widening of the growth plate more prominent in the ulna, with metaphyseal irregularity and fraying; the distal radius shows metaphyseal splaying fraying

Radiological changes—radiology may be normal in early stages despite biochemical changes of vitamin D deficiency.

The earliest changes are seen in areas of the greatest growth: around the wrist (distal radius and ulna), ankle (distal tibia) and knee joint (distal femur, proximal tibia) when the child starts weight bearing.

Bones show coarsened trabeculae; epiphysis are widened and poorly mineralised.

Widening, fraying and cupping of the metaphysis occur due to stress at the sites of ligament attachment [4] (Fig. 15.1).

The epiphyses are poorly mineralised.

Bowing deformities—genu valgum or genu varum of the knees (Fig. 15.2a, b) and coxa vara of the hip.

Skull deformities (infancy)—softening of the skull bones (craniotabes) with flattened occiput, frontal bossing, widened sutures, basilar invagination. Loss of lamina dura commonly described in hyperparathyroidism is also seen with rickets [5].

Chest becomes bell-shaped or pectus carinatum. The anterior ends of the ribs become widened (Rachitic rosary). Uncommonly rib fractures may be seen [6].

Spine may show kyphoscoliosis with ballooned disks.

The pelvic bones in severe disease may show protrusion. Slipped capital femoral epiphysis may occur.

Bones may also show delayed ossification and greenstick fractures.

Scoring System for Severity of Rickets

A quantitative radiographic scoring system known as Rickets Severity Score (RSS) has been validated to assess severity of nutritional rickets, based on assessment of wrist and knee radiographs. It has also been utilised in X-linked hypophosphataemic rickets and in assessment of treatment response in hypophosphatasia.

It is a 10-point scale, with the more severely affected wrist and knee being evaluated.

For the wrist (total 4 points possible)—radius and ulna are scored separately.

1. Widened growth plate, metaphyseal irregularity but no cupping or concavity
2. Metaphyseal concavity with fraying

For the knee (total 6 points possible)—femur and tibia are scored separately.

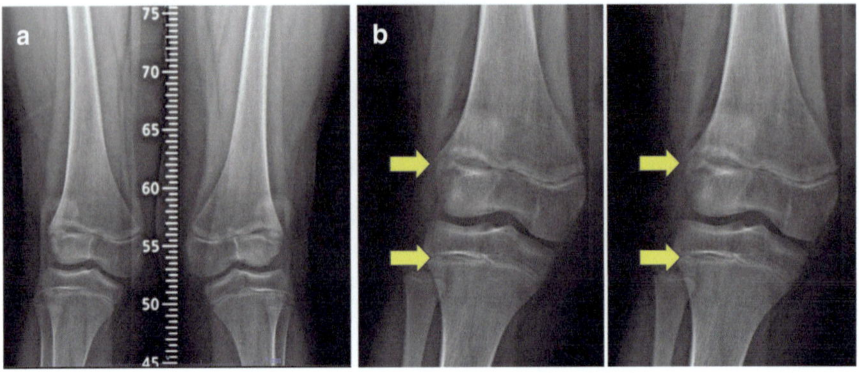

Fig. 15.2 Radiographs of bilateral knees in a young child (**a**) show bilateral knock knees or genu valgum. Enlarged view of the right knee (**b**) shows widening of the physis of both lateral condyles with metaphyseal irregularity (yellow arrows in **b**)

1. Partial lucency with smooth metaphyseal margin visible
2. Partial lucency with smooth metaphyseal margin not visible
3. Complete lucency with wide separation of epiphysis from metaphysis

If both condyles are affected, the score is multiplied by 1, and if only one condyle is affected, the score is multiplied by 0.5 [7, 8].

Healing Rickets

With onset of treatment, radiographic changes commence between 3 and 6 months [1].

The radiolucent widened growth plate becomes gradually mineralised and becomes radiodense, and the width decreases. The radiodensity of the affected growing segment of distal bone may still be less dense (Fig. 15.3a, b).

Harris growth arrest lines may be visible between the period of disease onset to treatment.

Bone remodelling may be incomplete and show persistent bowing deformity of the limbs in cases of severe rickets [1].

15.1.2 Mimicker of Rickets

Hypophosphatasia

Hypophosphatasia is a genetic disorder which can present in perinatal, infantile, childhood or adult age groups.

The disorder is caused by mutations in the gene encoding for tissue non-specific alkaline phosphatase.

Perinatal hypophosphataemia shows grossly unmineralized skull bones and short tubular bones with widened sutures [1, 9]. Children present with rickets like changes in metaphysis and early loss of deciduous teeth and craniosynostosis [1]. Adults

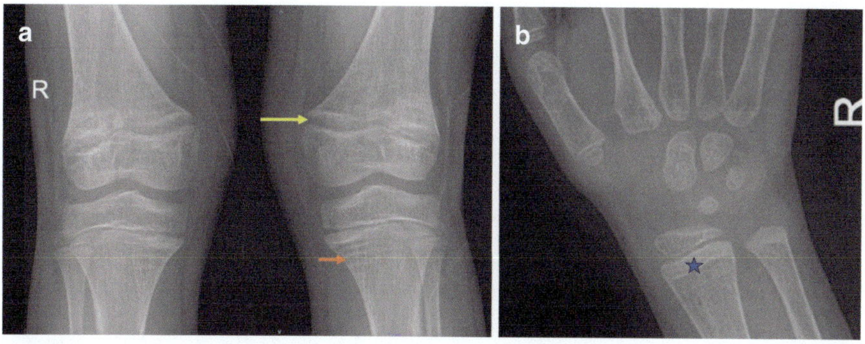

Fig. 15.3 Healing rickets. Radiograph of the bilateral knees (**a**) show increased mineralisation of the metaphysis (long arrow) with residual deformity and growth arrest lines in bilateral tibia (short arrow in **a**). Radiograph of the wrist (**b**) shows increased density of the metaphysis of the distal radius and ulna in healing rickets (star in **b**)

present with hyperostosis and Looser's zones on lateral cortex of femur in contrast to medial cortex in osteomalacia [1].

X-Linked Hypophosphataemic Rickets (Fig. 15.4a, b)
This is the most common hereditary form of rickets and is caused by mutation of PHEX gene and upregulation of FGF23 leading to phosphate wasting.

Clinically, patients present with severe bowing. Serum phosphate levels are low, and calcium and calcitriol and parathyroid hormone levels may be normal.

Radiologically, patients have thickened cortices and coarse trabeculae.

Looser's zones are seen even in paediatric patients unlike in nutritional rickets. Craniosynostosis occurs in children.

By adulthood—generalized increase in bone density, particularly in the axial skeleton.

Older patients show enthesopathic changes with ossification of anulus fibrous, paravertebral ligaments and capsules of appendicular and apophyseal (facet) joints. Ossification of the interosseous membrane can occur. Exuberant extraskeletal enthesial ossification mimics seronegative spondyloarthropathy/fluorosis/diffuse idiopathic skeletal hyperostosis. However, sacroiliac joints are normal [1].

15.1.3 Osteomalacia

In adults, vitamin D deficiency results in osteomalacia, with inadequate or delayed mineralization of osteoid in mature cortical and spongy bone. The characteristic feature is Looser's zones or pseudo-fractures.

Secondary causes of osteomalacia
Conditions associated with osteomalacia

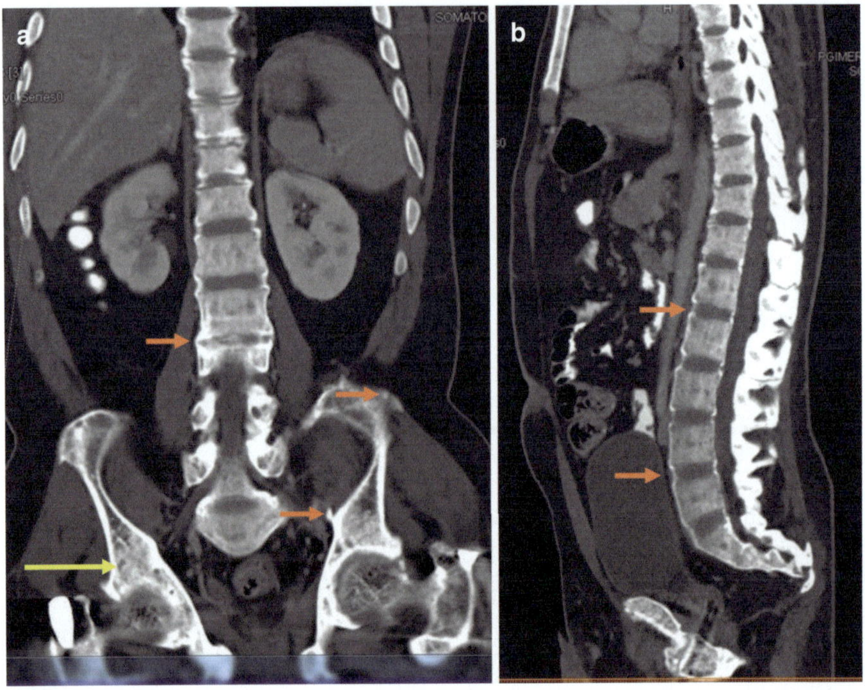

Fig. 15.4 Coronal (**a**) and sagittal (**b**) reformats of CT in a young male X-linked hypophosphatemic rickets. Diffuse osteosclerosis with bone softening with triradiate pelvis (long arrow in **a**) is seen. Extensive entheseal calcification is noted in the vertebrae and iliac bones (short arrows in a & B)

Renal failure
Chronic liver disease
Administration of phenobarbital
Decreased absorption of vitamin D
Bisphosphonates (for the treatment of Paget disease)
Defects in phosphate metabolism
phosphate deficiency, e.g. phosphate binders (some antacids)
familial hypophosphataemia and Fanconi syndrome

Radiological Features of Osteomalacia (Fig. 15.5a–g)

Bones show osteopenia with coarse, irregular trabeculae.

Looser's zone (pseudo-fractures or Milkman's fractures are seen as 2–3 mm transverse lucent band perpendicular to cortex, sometimes with sclerotic margins), caused by stress on weakened bones. They are characteristically seen in pubic rami, medial femoral neck, axillary border of scapula, ribs and may be bilaterally symmetrical.

Less commonly the metatarsals, metacarpals, ilium and acromion may be affected [1].

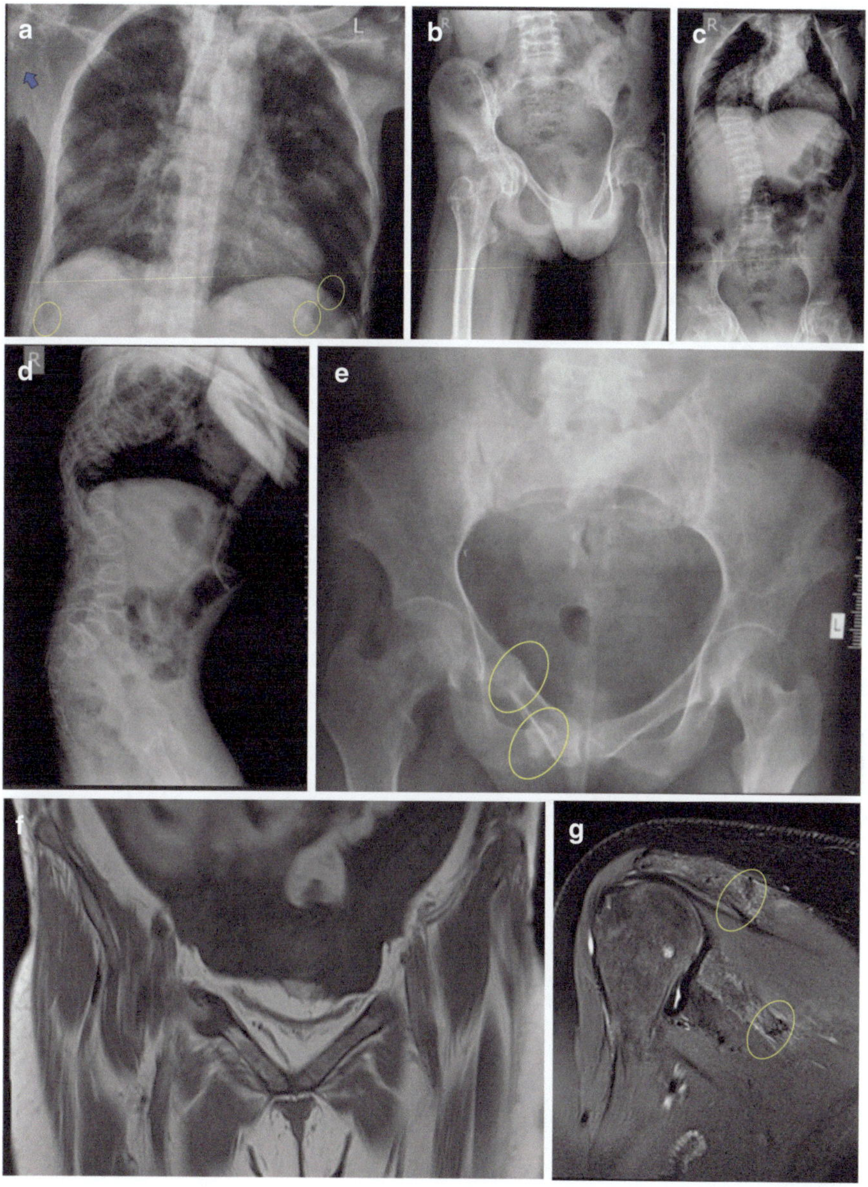

Fig. 15.5 (**a**) Chest radiograph of an adult with osteomalacia shows coarsened trabeculae and Looser's zones in axillary border of right scapula (arrow) and in bilateral ribs (oval). (**b**) Radiographs of the pelvis and hips in osteomalacia show fuzzy coarsened trabeculae. There is bilateral protrusion acetabuli with triradiate pelvis. Bowing of left femoral shaft is prominent. (**c**, **d**) Frontal and lateral radiographs of the dorsolumbar spine show kyphoscoliosis with biconcave vertebral end plates and bilateral protrusion acetabuli. (**e**, **f**) Looser's zone: on the frontal radiograph, Looser's zones are seen in the right superior and inferior pubic rami (oval). On coronal T1-weighted MR (**f**), the Looser's zone shows intermediate signal intensity. (**g**) Coronal T2-weighted MRI of the shoulder in another patient shows the linear hypointense Looser's zones (oval) in the base of acromion and axillary border of scapula with surrounding oedema

In severe cases, bony deformity with kyphoscoliosis occurs with biconcave (cod-fish) vertebrae.

Triradiate pelvis—a characteristic shape of the pelvis with medial protrusion of acetabuli into pelvic cavity and basilar invagination of the skull occurs.

Secondary hyperparathyroidism may occur with cortical tunnelling and erosions and resorptions mimicking hyperparathyroidism [1].

15.1.4 Tumour-Induced Osteomalacia

Certain tumours known as phosphaturic mesenchymal tumour (mixed connective tissue variant) can produce FGF23 (fibroblast growth factor 23), a hormone which inhibits renal phosphate reabsorption and 1α-hydroxylation of vitamin D in the kidney.

This causes renal phosphate wasting. Phosphate levels are low with low or inappropriately normal levels of 1,25-dihydroxy vitamin D and elevated FGF23. Serum calcium level is normal with raised alkaline phosphatase.

Clinical manifestations—patients present with long standing symptoms of weakness, bone pain and fractures with loss of height and a generalised debilitated status. Children with *tumour-induced osteomalacia* (TIO) may present with rickets.

The tumours are generally very small and difficult to localise and require functional imaging with PET and anatomical localisation with CT or MRI. The tumour can also be superficially located and appear as a new lump or bump. While most tumours are benign, malignancy with metastasis is known to occur [10].

Other tumour and tumour-like conditions associated with TIRO:

Haemangiopericytoma (Fig. 15.6a, b)
Haemangioma
Ossifying fibroma
Osteoblastoma
Fibrous dysplasia
Neurofibromatosis
Gorham's massive osteolysis

TIO also occurs as a paraneoplastic syndrome in prostate cancer, oat cell cancer and haematologic malignancies [10].

Imaging

18-Fluorodeoxyglucose positron emission tomography is sensitive but not specific as metabolically active foci (areas of fracture) may give false-positive results.

11Indium octreotide scintigraphy, with single-photon emission CT is also used as many of these tumours have somatostatin receptors.

68Ga-DOTANOC PET/CT utilises a modified octreotide to increase sensitivity with specificity [10].

Imaging should be done from head to toe.

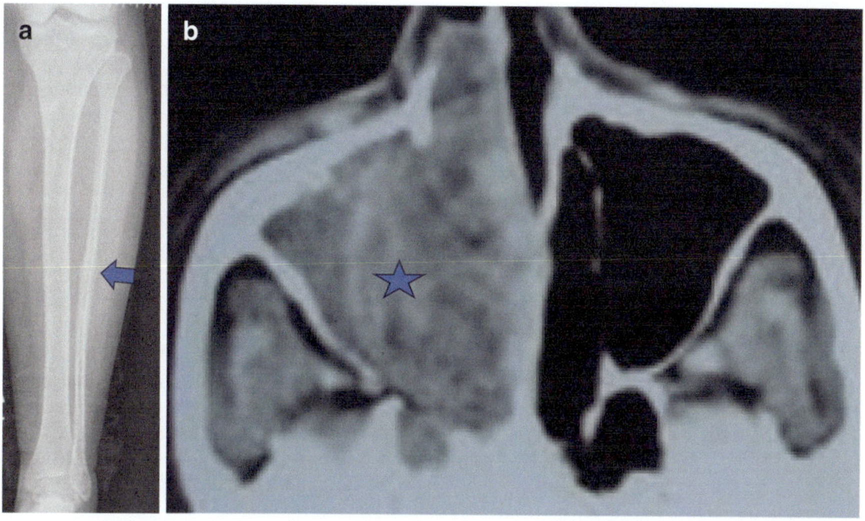

Fig. 15.6 A 30-year-old female with osteomalacia and swelling of the left cheek. Frontal radiograph of the left leg shows a Looser's zone in the mid shaft of fibula–an atypical site (arrow in **a**). Contrast-enhanced axial CT of the paranasal sinus shows a heterogeneously enhancing expansile mass in the right maxillary sinus extending into the right nasal cavity. On biopsy, it was a hemangiopericytoma (∗)

Whole-body MRI has also been attempted for localisation of these tumours.

15.1.5 Hypervitaminosis

Acute vitamin D toxicity presents with symptoms of hypercalcaemia with confusion, irritability, nausea, vomiting and even stupor and coma.

Chronic toxicity can have imaging manifestations.

Imaging findings of chronic hypervitaminosis in children include dense metaphyseal bands with alternating areas of lucency. Cortical thickening with osteopenia or osteosclerosis can occur. Calcification of falx and tentorium cerebelli can occur. Adults have osteopenia.

In both adults and children, visceral and soft tissue calcification can occur in the kidneys, in the periarticular soft tissues, muscles and vessels, mimicking tumoural calcinosis.

References

1. Adams JE. Chapter 54 – radiology of rickets and osteomalacia, In: Feldman D, editor. Vitamin D (Fourth Edition). Academic Press; 2018. p 975–1006. ISBN 9780128099650. https://doi.org/10.1016/B978-0-12-809965-0.00054-9. (https://www.sciencedirect.com/science/article/pii/B9780128099650000549)
2. Chanchlani R, Nemer P, Sinha R, Nemer L, Krishnappa V, Sochett E, Safadi F, Raina R. An overview of rickets in children. Kidney Int Rep. 2020;5(7):980–90. https://doi.org/10.1016/j.ekir.2020.03.025. PMID: 32647755; PMCID: PMC7335963
3. Meyers AL, Marquart MJ. Pediatric physeal injuries overview. [Updated 2022 Jul 25]. In: StatPearls [Internet]. Treasure Island: StatPearls; 2024. Available from: https://www.ncbi.nlm.nih.gov/books/NBK560546/
4. Salerno AE, States LJ, Kaplan BS, Chapter 15 – Rickets. In: Kaplan BS, Meyers KEC, Bell LM, editors. The requisites in pediatrics, pediatric nephrology and urology, Mosby, 2004, p 122–30. ISBN 9780323018418. https://doi.org/10.1016/B978-0-323-01841-8.50021-4. https://www.sciencedirect.com/science/article/pii/B9780323018418500214
5. Coley BD. Caffey's pediatric diagnostic imaging. E-Book. Elsevier Health Sciences; 2013.
6. Slovis TL. In 12th Edition Caffeys Coley BD. Caffey's pediatric diagnostic imaging E-Book. Elsevier Health Sciences;2018. p 223–32
7. Thacher TD, Fischer PR, Pettifor JM, Lawson JO, Manaster BJ, Reading JC. Radiographic scoring method for the assessment of the severity of nutritional rickets. J Trop Pediatr. 2000;46:132–9.
8. Thacher TD, Pettifor JM, Tebben PJ, Creo AL, Skrinar A, Mao M, Chen CY, Chang T, San Martin J, Carpenter TO. Rickets severity predicts clinical outcomes in children with X-linked hypophosphatemia: utility of the radiographic rickets severity score. Bone. 2019;122:76–81. https://doi.org/10.1016/j.bone.2019.02.010. Epub 2019 Feb 14
9. Chang CY, Rosenthal DI, Mitchell DM, Handa A, Kattapuram SV, Huang AJ. Imaging findings of metabolic bone disease. RadioGraphics. 2016;36(6):1871–87. https://doi.org/10.1148/rg.2016160004.
10. Chong WH, Molinolo AA, Chen CC, Collins MT. Tumor-induced osteomalacia. Endocr Relat Cancer. 2011;18(3):R53–77. https://doi.org/10.1530/ERC-11-0006. PMID: 21490240; PMCID: PMC3433741

Chapter 16
Vitamin D and Malignancies

Amit Bahl and Venkata Krishna Vamsi Gade

Abstract In addition to its metabolic functions, recent research highlights vitamin D's wide range of effects, including its anticancer properties. Calcitriol, the active form of vitamin D3, affects gene expression by interacting with VDRs, leading to cellular differentiation, apoptosis, and inhibition of tumor-related processes like inflammation, angiogenesis, and metastasis. Vitamin D's impact on specific cancers, such as colorectal, breast, and prostate cancer, involves the regulation of molecular pathways that control cell proliferation and hormone sensitivity. In colorectal cancer, calcitriol inhibits the WNT-beta catenin signaling pathway, while in breast cancer, it downregulates aromatase expression. Studies in animal models have shown that vitamin D deficiency may promote tumor growth, and supplementation could hinder cancer progression. However, human-level population studies have been inconclusive. Further research is required to identify the at-risk population that would most benefit from vitamin D.

Keywords Vitamin D · Cancer · Carcinogenesis

A. Bahl (✉)
Regional Cancer Center, Post Graduate Institute of Medical Education & Research (PGIMER), Chandigarh, India

V. K. V. Gade
Department of Radiotherapy & Oncology, Regional Cancer Center, Post Graduate Institute of Medical Education & Research (PGIMER), Chandigarh, India

V. Kumar, S. Patel (eds.), *Vitamin D: A Comprehensive Guide for Medical Professionals*, https://doi.org/10.1007/978-981-96-6491-7_16

Vitamin D_3 also called cholecalciferol is a fat-soluble vitamin and a prohormone. Cholesterol derivatives located on the skin surface are converted to cholecalciferol when exposed to UV radiation. Cholecalciferol then undergoes two-step enzyme-mediated hydroxylation in the hepatic and renal parenchyma to form 1,25-dihydroxy-vitamin D_3 (known as calcitriol). Calcitriol primarily regulates phosphate and calcium metabolism and maintains bone mineral density. 25-Hydroxy-vitamin D_3 is the freely circulating component that is used as a surrogate for the vitamin D levels in the body [1, 2]. In addition to its metabolic functions, recent research has revealed that vitamin D performs a wide range of functions through its actions on vitamin D receptors (VDRs) located on almost all the cells in the human body [3]. Deficiency of this micronutrient is associated with a variety of conditions like psoriasis, alopecia areata, autoimmune diseases, asthma, multiple sclerosis, and increased susceptibility to granulomatous diseases [4–9]. There has been an increasing scrutiny in its part in carcinogenesis.

Vitamin D is said to have antineoplastic effects mainly by its action on VDRs located on nucleated cells causing them to dimerize with retinoid X receptor (RXR). This leads to its nuclear translocation where it directly affects gene expression by attaching with vitamin D-responsive elements (VDRE) located on the chromosomes. Vitamin D streamlines cellular differentiation by regulating expression of factors such as PSA, prostate differentiation factor, casein, and E cadherin. Vitamin D promotes apoptosis of cells with defective DNA by upregulating BAX expression and downregulating BCL-2 expression. It limits the abnormal proliferation of cells by downregulating cyclins, CDKs, RB, and myc factors [10]. It inhibits tumor-associated inflammation by downregulating NF-kappa B, PG, and COX2 expression [11]. It inhibits tumor angiogenesis by downregulating HIF1, VEGF, and IL-8 expression [12]. It hinders the metastasis and invasion of metastatic subclones by downregulating the MMP9 and plasminogen activator genes [13].

In addition to these broad antineoplastic effects, vitamin D affects the pathogenesis of various solid tumors by acting on specific molecular targets. The WNT-beta catenin signaling pathway is the predominant pathway for sporadic colorectal cancer as part of the adenoma carcinoma sequence. By its action on VDREs, calcitriol inhibits the transcriptional activity of beta catenin [14]. Calcitriol also increases the expression of DICKKOPF1 and DICKKOPF4, which are inhibitors of WNT pathway [15]. Calcitriol also regulates the p38 and MAPK-MSK pathway to upregulate E-cadherin expression reducing colorectal cancer development and progression [11].

Vitamin D has antineoplastic effects in patients of postmenopausal hormone receptor-positive breast cancer via downregulating aromatase expression in cancer cells and adipose tissue. By downregulating PGE2, it further reduces aromatase production. It also inhibits ER expression on breast cancer cells making them less sensitive to estrogen. The various effects this molecule has on proliferation and differentiation of neoplastic cells may contribute to its protective effects in breast cancer [16]. Hence the vitamin D and calcium supplements given along with aromatase inhibitors given in this group of patients may have anticancer effects in addition to its bone mineral function. It has a role in preventing the development of castration resistance in patients of prostate cancer by increasing androgen-stimulated PSA expression and by upregulating PSA expression. It also promotes the differentiation of progenitor cancer cells into androgen receptor expressive cells. It has antineoplastic effects by increase in growth inhibitor AS3 that are induced by androgens. Calcitriol is also involved in controlling the expression of genes regulating the catabolism of androgens [17, 18].

The link between vitamin D deficiency and cancer incidence has been the subject of various preclinical and human studies. Animal studies predominantly performed on murine models showed that the deficiency of this vitamin promoted tumor development and growth. Studies mainly performed in murine models of prostate cancer and colorectal cancer carcinogenesis showed that dietary supplementation of vitamin D alone or along with calcium was found to hinder carcinogenesis. Certain studies also indicate that standard rodent feeds containing levels of vitamin D sufficient for bone health may not be sufficient to exert an antineoplastic effect. Levels higher than those sufficient for bone health seem to be required to inhibit tumor growth [19–21]. Preclinical studies have shown that expression of the CYP24A1 in some tumor models results in the development of cellular resistance to calcitriol hampering its antineoplastic effects. Potential combination of these supplements with inhibitors of cytochrome enzyme like ketoconazole has been suggested to overcome the resistance [22].

Vitamin D deficiency has been postulated to play a role in development of bony metastasis and in the development of primary bone cancers. Patients with malignant bone tumors have been shown to suffer from vitamin D deficiency or insufficiency. Calcipotriol has shown to inhibit development and metastasis of osteosarcomas in animal models [23].

The IOM expert committee defines vitamin D deficiency as 25-hydroxvitamin D levels of less than 20 ng per mL while mandating a daily intake of 600 IU for age less than 70 years [21]. A daily intake of 800 IU is suggested for people aged over 70 years. The Endocrine Society recommended a cutoff of 30 ng per mL while suggesting a higher requirement of 1000–2000 IU/day [21]. These recommendations are aimed at the maintenance of bone mineral health. The ideal dose needed for antineoplastic effect is still under investigation. A study by Jung Lee et al. showed that up to a 40% reduction in the risk of colon cancer occurred in those with higher vitamin D levels as compared to those with lower levels [24].

A recent study showed there was an inverse correlation in postmenopausal women between the risk of breast cancer development and vitamin D levels but only after a threshold of 27 ng per ml [25]. A pooling project consortium of rare cancers found no general association between vitamin D levels and cancer incidence across various sites [26]. A phase 3 randomized trial (VITAL) studied the found that while vitamin D did not significantly reduce cancer incidence, it showed a decrease in the overall cancer mortality. Vitamin D did not affect cardiovascular disease incidence. Meta-analyses incorporating the VITAL trial also found that while vitamin D did not reduce cancer incidence, it may reduce cancer mortality [27].

Vitamin D is a fat-soluble vitamin that has various antineoplastic effects at the cellular level. Preclinical animal models show that vitamin D deficiency may be associated with cancer development and tumor growth. Dietary addition of vitamin D in murine models also showed a reduction of cancer incidence and progression. However, human population-level studies investigating the association between cancer incidence and vitamin D deficiency have been inconclusive and at times contradictory. A large phase 3 randomized study showed that vitamin D supplementation reduced overall cancer mortality without affecting the cancer incidence. Further research is needed to recognize populations that are most likely to benefit from the antineoplastic effects of vitamin D supplementation.

References

1. Prosser DE, Jones G. Enzymes involved in the activation and inactivation of vitamin D. Trends Biochem Sci. 2004;29:664–73.
2. Omdahl JL, Morris HA, May BK. Hydroxylase enzymes of the vitamin D pathway: expression, function, and regulation. Annu Rev Nutr. 2002;22:139–66.
3. Henry HL, Norman AW. Vitamin D: metabolism and biological actions. Annu Rev Nutr. 1984;4:493–520.
4. Gade VKV, Mony A, Munisamy M, et al. An investigation of vitamin D status in alopecia areata. Clin Exp Med. 2018;18:577–84.
5. Barrea L, Savanelli MC, Di Somma C, et al. Vitamin D and its role in psoriasis: an overview of the dermatologist and nutritionist. Rev Endocr Metab Disord. 2017;18:195–205.
6. The impact of Vitamin D deficiency on asthma, allergic rhini.... J Family Community Med [Internet]. [cited 2024 Mar 1] Available from: https://journals.lww.com/jfcm/fulltext/2014/21030/the_impact_of_vitamin_d_deficiency_on_asthma,.2.aspx
7. Sintzel MB, Rametta M, Reder AT. Vitamin D and multiple sclerosis: a comprehensive review. Neurol Ther. 2017;7:59–85.
8. Talat N, Perry S, Parsonnet J, et al. Vitamin D deficiency and tuberculosis progression. Emerg Infect Dis J [Internet]. 2010;16(5). CDC [cited 2024 Mar 1]. Available from: https://wwwnc.cdc.gov/eid/article/16/5/09-1693_article
9. Yang C-Y, Leung PSC, Adamopoulos IE, et al. The implication of vitamin D and autoimmunity: a comprehensive review. Clin Rev Allergy Immunol. 2013;45:217–26.
10. Samuel S, Sitrin MD. Vitamin D's role in cell proliferation and differentiation. Nutr Rev. 2008;66:S116–24.
11. Molecules | Free Full-Text | Vitamin D signaling in inflammation and cancer: molecular mechanisms and therapeutic implications [Internet] [cited 2024 Mar 1] Available from: https://www.mdpi.com/1420-3049/25/14/3219
12. 1α,25-Dihydroxyvitamin D3 Inhibits angiogenesis in vitro and in vivo | Circulation Research [Internet] [cited 2024 Mar 1] Available from: https://www.ahajournals.org/doi/10.1161/01.RES.87.3.214
13. Timms PM, Mannan N, Hitman GA, et al. Circulating MMP9, vitamin D and variation in the TIMP-1 response with VDR genotype: mechanisms for inflammatory damage in chronic disorders? QJM Int J Med. 2002;95:787–96.
14. González-Sancho JM, Larriba MJ, Muñoz A. Chapter 13—vitamin D and Wnt/β-Catenin signaling [Internet]. In: Feldman D, Pike JW, Adams JS, editors. Vitamin D (Third Edition). San Diego: Academic Press; 2011. p. 235–50 [cited 2024 Mar 1]. Available from: https://www.sciencedirect.com/science/article/pii/B9780123819789100137
15. Aguilera O, Peña C, García JM, et al. The Wnt antagonist DICKKOPF-1 gene is induced by 1alpha,25-dihydroxyvitamin D3 associated to the differentiation of human colon cancer cells. Carcinogenesis. 2007;28:1877–84.

16. Vitamin D: an essential adjuvant therapeutic agent in breast cancer – Romany H Thabet, Adel A Gomaa, Laila M Matalqah, Erin M Shalaby, 2022 [Internet] [cited 2024 Mar 1] Available from: https://journals.sagepub.com/doi/10.1177/03000605221113800?icid=int.sj-full-text.similar-articles.4
17. Trump DL, Aragon-Ching JB. Vitamin D in prostate cancer. Asian J Androl. 2018;20:244–52.
18. Li H, Stampfer MJ, Hollis JBW, et al. A prospective study of plasma vitamin D metabolites, vitamin D receptor polymorphisms, and prostate cancer. PLoS Med. 2007;4:e103.
19. Hummel DM, Thiem U, Höbaus J, et al. Prevention of preneoplastic lesions by dietary vitamin D in a mouse model of colorectal carcinogenesis. J Steroid Biochem Mol Biol. 2013;136:284–8.
20. Newmark HL, Yang K, Kurihara N, et al. Western-style diet-induced colonic tumors and their modulation by calcium and vitamin D in C57Bl/6 mice: a preclinical model for human sporadic colon cancer. Carcinogenesis. 2009;30:88–92.
21. Ray R, Banks M, Abuzahra H, et al. Effect of dietary vitamin D and calcium on the growth of androgen-insensitive human prostate tumor in a murine model. Anticancer Res. 2012;32:727–31.
22. Osanai M, Lee G-H. CYP24A1-induced vitamin D insufficiency promotes breast cancer growth. Oncol Rep. 2016;36:2755–62.
23. Horas K, van Herck U, Maier GS, et al. Does vitamin D deficiency predict tumour malignancy in patients with bone tumours? Data from a multi-center cohort analysis. J Bone Oncol. 2020;25:100329.
24. Lee JE, Li H, Chan AT, et al. Circulating levels of vitamin D and colon and rectal cancer: the physicians' health study and a meta-analysis of prospective studies. Cancer Prev Res (Phila Pa). 2011;4:735–43.
25. Bauer SR, Hankinson SE, Bertone-Johnson ER, et al. Plasma vitamin D levels, menopause, and risk of breast cancer: dose-response meta-analysis of prospective studies. Medicine (Baltimore). 2013;92:123.
26. Helzlsouer KJ, VDPP Steering Committee. Overview of the Cohort Consortium vitamin D pooling project of rarer cancers. Am J Epidemiol. 2010;172:4–9.
27. Manson JE, Bassuk SS, Buring JE. Principal results of the vitamin D and OmegA-3 TriaL (VITAL) and updated meta-analyses of relevant vitamin D trials. J Steroid Biochem Mol Biol. 2020;198:105522.

Chapter 17
Vitamin D in Rheumatological Conditions

Rajesh Kumar, Adrish Biswas, and Krushna Chandra Behera

Abstract Vitamin D, a fat-soluble vitamin, needed for calcium absorption and well known for its role in maintaining bone health, has garnered substantial interest in recent times for its conjectural role in the pathogenesis and immunomodulation in autoimmune rheumatic diseases (AIRDs). The fact that vitamin D receptors are present on a vast multitude of cells including macrophages, dendritic cells, B cells and T cells in the body is in itself indicative of its multifarious role in immune regulation. There have been studies that clearly suggest vitamin D curtails antigen-presenting cells (APC) maturation, augments regulatory T cells and decreases pro-inflammatory cytokines, thereby establishing immune tolerance.

This immunomodulatory function of vitamin D has been implicated in multiple AIRDs like rheumatoid arthritis (RA), systemic lupus erythematous (SLE), Sjogren's syndrome, spondyloarthropathies, vasculitis and others, and its deficiency has been shown to be positively linked to both occurrence and severity of AIRDs. In addition to influencing adaptive immunity by inhibiting Th1 mediated cytokine storm, shifting response from Th1 to Th2, vitamin D also plays a role in altering the innate immune response by increasing production of antimicrobial peptides, promoting phagocytosis and regulating the Toll-like receptors (TLRs).

The current treatment guidelines regarding management of vitamin D deficiency in adults involve achieving serum 25(OH)D levels of at least 30 ng/mL through supplementation, with higher doses recommended for individuals in high-latitude regions or with hyper-pigmented skin.

R. Kumar (✉)
Department of General Medicine & Rheumatological Services, AIIMS Deoghar, Deoghar, Jharkhand, India

A. Biswas · K. C. Behera
Department of General Medicine, AIIMS Deoghar, Deoghar, Jharkhand, India

V. Kumar, S. Patel (eds.), *Vitamin D: A Comprehensive Guide for Medical Professionals*, https://doi.org/10.1007/978-981-96-6491-7_17

Vitamin D has shown promise in helping mitigate the burden of AIRDs in both occurrence and severity aspect. However, further clinical and randomised control trials are needed to lay down a uniform and definite evidence in treatment protocol.

Keywords Vitamin D · Immunity · Osteoarthritis · Rheumatic diseases · Rheumatoid arthritis · SLE · Sjogren's syndrome

17.1 Introduction

Vitamin D belongs to a family of fat-soluble vitamins responsible for intestinal absorption of calcium and other elements with fascinating immunomodulatory and immunoregulatory functions [1, 2]. Its deficiency is aptly labelled as 'pandemic' because of its relationship to development of number of chronic diseases. Owing to this, the relationship of vitamin D to autoimmune rheumatic disorders has been an entirely new avenue for study and research.

Ergocalciferol (D2) and cholecalciferol (D3) are two major forms of vitamin D in the body. The role of vitamin D in the pathogenesis of autoimmune disorders is multifaceted. The pleiotropic action of vitamin D on immunomodulation of the body is based on the presence of vitamin D receptors (VDRs) on various cell types, and this has been recognised worldwide [2]. Amongst all understood mechanisms, one of the most crucial ones seems to be the curbing action of vitamin D on the maturation of antigen-presenting cells (APCs) and decrease in the level of pro-inflammatory cytokines [2]. Vitamin D plays a crucial role in attuning both innate and adaptive immune responses. Ongoing studies and data suggest the beneficial role of supplementation of vitamin D in autoimmune diseases. Autoimmune rheumatic diseases (AIRDs) comprise a spectrum of diseases notably rheumatoid arthritis, systemic lupus erythematosus, systemic sclerosis, Sjogren's syndrome, spondyloarthropathies (SpA), vasculitis and others [1, 2]. Recent updates have revealed an interplay of gamut of inflammatory cytokines. Vitamin D has been associated with immunoregulatory, antiproliferative effect in these diseases, and its deficiency has been linked to augmentation of frequency and severity of disease flares [2]. The findings also suggest the deficiency of vitamin D as a causative factor of autoimmune disorders in the first place [1, 2].

It is also to be understood that clinical trials have mixed results till date, for some studies revealed no relationship between vitamin D and autoimmune diseases. The differences in the study outcomes may be due to differences in study design, baseline variations of levels of vitamin D of participants, differences of vitamin D levels used for treatment and various confounding factors. So, it is of prime importance to note that this mere association does not prove causation, and further studies will be required to arrive at a definite conclusion—one of the major reasons for this discrepancy may be the fact that the role of vitamin D in immune modulation is still not fully understood and needs further discussion.

17.2 Vitamin D and Immunity

Vit. D imposes control of both the innate and the adaptive immune responses of the body and the function of which has been at the epicenter of research [1, 2].

17.2.1 Vitamin D Impact on Dendritic Cells

Vit. D inhibits the maturation of dendritic cells (DCs)—this decreases the ability of DCs to present antigens to T cells, thus decreasing the overall immune response and promoting tolerance to self-antigens. It promotes IL 10 secretion (anti-inflammatory) and inhibits pro-inflammatory cytokines like IL 12, thereby promoting DC apoptosis. Vit. D decreases co-stimulatory molecules CD80 and CD86, thereby decreasing full activation of T cell. The mechanism mentioned above induces peripheral tolerance, decreases pro-inflammatory cytokines and increases regulatory T cell production, i.e. has an overall tolerogenic action.

17.2.2 Vitamin D Effect on APCs/MHC

Vitamin D inhibits expression of MHC II cells on antigen-presenting cells, thereby decreasing formation of autoreactive T cells, as antigen-presenting cells are unable to present antigen to CD4 T cells.

17.2.3 Vitamin D Effect on Vitamin D Receptor

Vitamin D receptor (VDR) response—Activation of VDR promotes anti-inflammatory cytokine CD10 and downregulates pro-inflammatory cytokines such as IL-2, IL-7 and TNF-alpha leading to tolerogenic phenotype. VDR activation acts as a transcription modulator which alters expression of multiple genes in such a way that it decreases autoimmunity in rheumatic diseases. Vit. D expresses VDR in many cells including APCs, T cells, B cells and other cells leading to increase in 1-25$(OH)_2$D.

17.2.4 Vit. D and Innate Immune Response

Innate immunity is regulated by 1-25$(OH)_2$D at multiple barriers of body like skin, intestine, lungs. Vitamin D influences the control of Toll-like receptors (TLRs) in the body which are needed for pattern recognition amongst pathogens, and thus, its activation. 1-25$(OH)_2$D has an facilitatory role in phagocytosis of microbes by macrophages, stimulating the overall production of immune cells like neutrophils, macrophages and immunological peptides like Defensins which are important mediators of innate immune responses of the body. Vit. D downregulates TLR 2 and TLR 4 in monocytes preventing excessive immune responses to PAMPs and DAMPs. 1-25$(OH)_2$D increases the production of defensins particularly beta-defensins and cathelicidin antimicrobial peptide (CAMP), which are involved in the modulation of immune response. Vitamin D and defensins play a significant role in skin protection and healing in autoimmune diseases like psoriasis.

17.2.5 Vit. D and Adaptive Immune Response

Vit. D promotes maturation of T_{regs} which are needed for immune tolerance [1]. Vit. D also inhibits the production of pro-inflammatory T helper cells (Th Cells) like Th_1 and Th_{17} thereby shifting towards Th_2, clearly indicating a beneficial role. Vit. D prevents proliferation of memory B cells which are important in preventing persistence of autoreactive memory B cells in autoimmune conditions. 1-25$(OH)_2$D can also curtail maturation of B cells into plasma cells, thus regulating the immune response. It inhibits the differentiation of naive T cells into Th_1 cells which is characterized by production of IFN-gamma and decreased MHC II expression, thereby reducing the risk of autoimmunity. Vitamin D also reduces the production of pro-inflammatory cytokines like TNF-alpha. One finer point to note would be the presence of enzyme 1 alpha Hydroxylase in APCs, which is the same enzyme that converts vitamin D to its active form—this also is a strong predictor of how vitamin D is involved in the autoimmune disease process. Summary of actions of vitamin D is highlighted in Fig. 17.1.

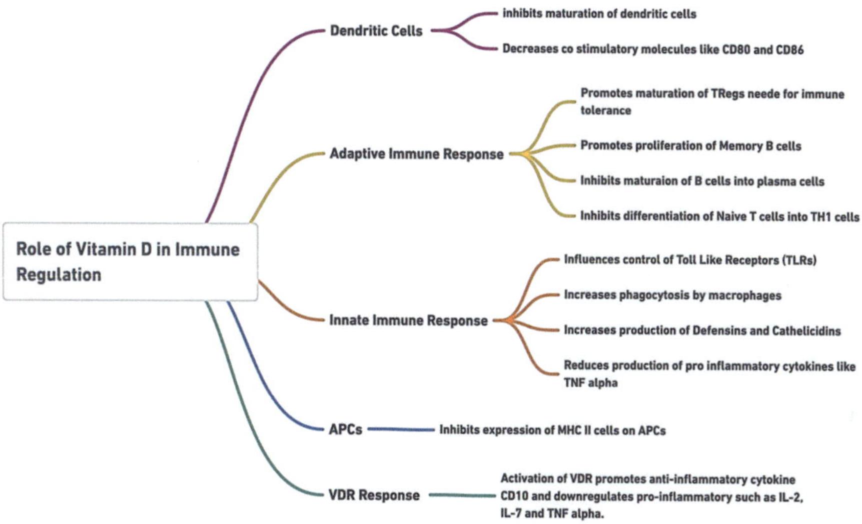

Fig. 17.1 Vitamin D role in innate and adaptive immunity

17.3 Vitamin D Deficiencies in Autoimmune Disorders

Autoimmune disorders often involve the liver and the kidneys—both essential organs in the metabolic pathway of Vit. D for conversion into its active form. Glucocorticoids often used in the treatment of autoimmune disorders interfere with the metabolism of Vit. D and increase Vit. D excretion through the kidneys.

Autoimmune diseases are often associated with cutaneous manifestations, and patients are on drugs like hydroxychloroquine, thereby limiting their exposure to sun which decreases the synthesis of vitamin D in the skin.

Autoimmune disorders often involve the mucosa of the gastrointestinal tract—this may cause a decrease in the absorption of Vit. D. These patients often have genetic variations or polymorphisms affecting the (VDRs) in the body causing less formation of 1,25 $(OH)_2D$. There are myriad of causes of low Vit. D in AIRDs; important ones are mentioned in Fig. 17.2.

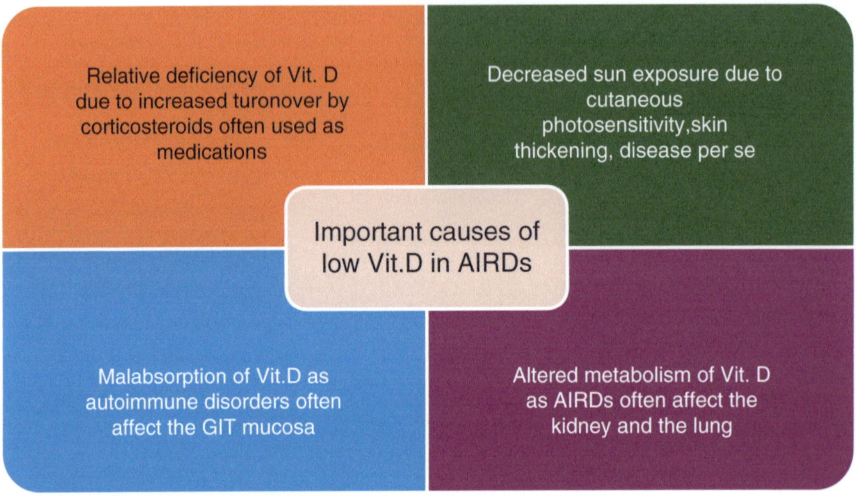

Fig. 17.2 Summary of causes of Vit. D deficiency/insufficiency in AIRDs

17.4 Vit. D in Autoimmune Rhematic Diseases

There have been clinical trials that support the role of Vit. D in prevention and treatment of AIRDs and degenerative diseases, but due to the presence of confounding factors of low Vit. D in these conditions like less exposure to sunlight due to various causes, drugs, outdoor going limitations, its acceptance requires support large data-based trials in various settings. Following are the description of role of Vit. D in various AIRDs.

17.4.1 Rheumatoid Arthritis

Rheumatoid arthritis (RA) is an autoinflammatory disorder mainly having polyarthritis which is symmetrical in nature with global prevalence of 0.46% having extraarticular complications. Though originally it was caused by a T_h1 immune response, recent studies suggest increased T_h17 and T_h22 activities and improper T_{reg} functioning contributing to its pathogenesis. Vit. D is believed to play a role in regulating RA pathogenesis and severity of the disease by suppressing Th1 and Th17 proliferation and activity, enhancing T_{reg} activity and suppressing B lymphocytes. RA disease activity is assessed by DAS 28 score which incorporates tender joints, swollen joints, ESR/CRP and visual analogue score. Observational studies by Lin et al. [3] and COMORA [4] study have shown an association between higher vitamin D status and lower incidence and severity, i.e. DAS 28 score of RA, low CRP. Negative association of Vit. D was found with VAS as in study by Higgins et al. [5, 6]. In another clinical trial by Gopinath et al., patients who were given Vit. D along with DMARDs and calcium had less pain. Observational evidence suggests that Vit. D intake to optimum levels reduces the development of RA but is not substantiated by clinical trials. Moderate evidence suggests that Vit. D supplementation may help alleviate disease activity in RA patients, but further large-scale randomised clinical trials are needed before recommending Vit. D as an adjunctive treatment for RA.

17.4.2 Systemic Lupus Erythematous

Systemic lupus erythematous (SLE) has prevalence of 0.3–23.2 cases per 100,000 individuals per year globally with diverse clinical manifestations, including inflammatory polyarthritis, multisystemic involvement and constitutional systems. Although the exact etiology of SLE remains unclear, genetic predisposition and environmental factors such as smoking are proposed to be important risk factors. Genetic variations in the VDR gene polymorphism, especially Smi and FOKI, may also influence SLE risk [1, 2]. Patients with SLE exhibit immune dysregulation, characterized by imbalances in immune cell activities and production of autoantibodies. Impaired immunity like variable T_h1, T_h2 and increased T_h17, T_h23 decreased T_{reg}. Vit. D signaling pathway alleviates autoinflammation and has immunomodulatory effects of innate immunity. Low levels of vitamin D, specifically 25(OH)D, have been associated with SLE occurrence and has been shown to be related with disease severity [7–16]. While studies suggest that vitamin D deficiency is linked to SLE disease activity, evidence on the preventive effects of vitamin D intake against SLE development is lacking. However, vitamin D supplementation has shown promising results in improving disease activity in SLE-SLEDAI, particularly in reducing fatigue severity, anti-ds DNA positivity, renal activity, photosensitivity and inducing remission [17]. Further research is necessary to determine whether maintaining vitamin D status or increasing intake can reduce the risk of development of SLE.

17.4.3 Spondyloarthropathies

SpA are a group of autoimmune diseases that have genetic predispositions (HLA B27) and clinical characteristics of inflammatory low back pain, asymmetrical peripheral arthritis, enthesitis, skin lesions and uveitis [1, 2]. This group of diseases include ankylosing spondylitis (AS), reactive arthritis, psoriatic arthritis, inflammatory bowel disease and undifferentiated spondyloarthropathy and involves formation of inflammatory cytokines causing bone formation and resorption. Vit. D plays a pivotal role in modulating the disease severity of SpA by affecting immune cells and gut microbiota.

Vitamin D low levels have been observed in patients with psoriatic arthritis, ankylosing spondylitis and inflammatory bowel disease [18–20]. DBP (GC) gene polymorphism, which is involved in manifestations of peripheral joint and eye involvement, is negatively correlated to Vit. D [21]. Lange et al. reported inverse relation of Vit. D with AS activity-BASDAI [22]. However, the effect of Vit. D supplementation in patients with SpA needs further studies.

17.4.4 Psoriatic Arthritis

Psoriasis involves skin with or without axial and/or appendicular skeletal manifestations. Gamonal et al. reported that hypovitaminosis was associated with plaque psoriasis and arthritis and was associated with increased disease activity (PASI) [23].

17.4.5 Sjogren's Syndrome

It is an autoimmune disease having glandular and extra glandular manifestations. Agmon-Levin N et al. reported that Vit. D supplementation was beneficial in decreasing presence of lymphoma and peripheral neuropathy in Sjogren's syndrome. Erten et al. reported low Vit. D levels in cohort of Sjogren's syndrome [24].

17.4.6 Systemic Sclerosis

It is a multisystemic disease causing binding of skin, Raynaud's phenomenon, interstitial lung disease, GI and cardiac involvement. Vit. D suppresses fibroblasts decreasing the TGF-β and hence leading to decreased skin thickening as reported by Slominski AT et al. [25]. Some studies reported about decreased synthesis of Vit. D due to skin thickening and increased pulmonary, gastrointestinal, cardiac involvement associated with hypovitaminosis D [26]. Groseanu et al. in his study revealed the deficiency of Vit. D with scleroderma renal crisis, a renal involvement in systemic sclerosis [27].

17.4.7 Gout

Hyperuricemia leads to precipitation of monosodium urate crystals in joints leading to joint pain and swelling. Hyperuricemia is also associated with metabolic syndrome. Vit. D deficiency and insufficiency are associated with mild increase in serum uric acid levels, with a dose-dependent relationship. The mechanism behind this effect is thought to be the suppression of parathyroid hormone, which down-regulates ABCG2 transporter in kidneys, thereby decreasing the renal clearance of uric acid. Though, there is some relation between vitamin D and uric acid. In population-based study, Vit. D status was not related to gout.

17.4.8 Other AIRDs

Idiopathic inflammatory myopathies (IIM) like dermatomyositis, polymyositis, and vasculitis are associated with decreased serum 25(OH)D levels as suggested by multiple observation studies [28]. Some patients having chronic widespread pain (CWS)–fibromyalgia associated with osteomalacia have been relieved by Vit. D supplementation [28]. However, conclusive evidence of role Vit. D in causation of these diseases is lacking.

17.5 Osteoarthritis

Osteoarthritis (OA) is the most prevalent degenerative joint disease-causing degeneration of articular cartilage found mostly in adults of more than 65 years with joint injury, ageing, obesity, and genetics being important risk factors. Serum 25(OH)D low levels have been linked with the presence of severity of knee OA. Vitamin D may influence OA development and progression by affecting bone quality, reducing inflammation, and improving skeletal muscle function. Studies also suggest vitamin D addition can alleviate pain and improve physical performance in patients with osteoarthritis of knee joint, improving WOMAC function. However, there is no effect on reversal of disease process or tibial cartilage volume [29, 30] (Table 17.1).

Table 17.1 Summary of studies highlighting the association between Vit. D and autoimmune AIRDs

Disease	Study details (year and lead author)	Conclusion
RA	COMORA Study [2]	"Serum level of Vit. D 25 (OH) is inversely related to DAS 28 disease activity" [2]
	Guan et al. 2020 [31]	"Supplementation of vitamin D caused a significant improvement in the DAS 28" [31]
	Nguyen et al. 2020 [32]	"Vitamin D intake caused only a mild reduction in RA flares defined by the DAS28 of >3.2" [1, 32]
	Hajjaj-Hassouni et al. 2017 [4]	"Pain relief when Vit. D was added to patients DMARDs and calcium" [4]
	Merlino et al. 2004 [33]	"Women in the highest tertile of Vit. D intake had decreased risk of 33% with those on lower tertile" [1, 33]
SLE	Zheng et al. 2019 [34]	"Vitamin D3 intake caused a reduction in the fatigue scores in patients with SLE" [1, 34]
	Sun et al. 2019 [35]	"Vit. D3 supplementation leads to decrease SLE activity-SEDAI" [35]
	Sairenji et al. 2008 [19]	"Vit. D has no temporal association with SLE" [19]
Sjogren's syndrome	Liu et al. 2021 [36]	"No difference of Vit. D related to Sjogren's syndrome was reported in meta-analysis" [36]
OA	Sanghi et al. 2013 [29]	"Vitamin D3 addition caused a significant reduction in VAS and WOMAC in population-based data" [29]
	Jin et al. 2016 [30]	"Vitamin D3 caused a significant improvement in WOMAC function but no difference in tibial cartilage volume" [1]

17.6 Treatment of Vitamin D Deficiency—Recommendation in Adults

The Endocrinal Society Clinical Practice guidelines recommendation for achieving serum 25(OH) level of at least 30 ng/l to have optimum benefits with minimal stimulation of secondary hyperparathyroidism. Vit. D insufficiency and deficiency is considered when serum 25(OH) levels are <30 ng/L and < 20 ng/l, respectively. The therapy mentioned in Table 17.2 Vit. D3 is preferred over Vit. D2 as it is involved in achieving faster normal levels [38].

However, there has been suggestion of higher doses of 4000–6000 IUs to maintain serum 25(OH) level of at least 40–60 ng/l to have a beneficial effect of Vit. D on AIRDs [37].

People living in high latitude, hyperpigmentation of skin, in winter season, require oral supplementation of Vit. D as sun exposure is not sufficient for desired requirements.

Table 17.2 Recommended vitamin D daily requirement and treatment of vitamin D deficiency—ESCP

Age	Daily need of Vit. D	Treatment of vitamin D deficiency
18–70 years	1500–2000 IUs (10,000 IUs is the upper limit/day)	"50,000 IU/wk. or 2000 IU/d of vitamin D3 is needed for at least 6–8 weeks to achieve serum 25(OH)D levels of >30 ng/mL (75 nmol/L) Followed by maintenance therapy of 600–1000 IU/d" [1, 37]
>70 years	1500–2000 IUs (10,000 IUs is the upper limit/day)	"50,000 IU/wk. or 2000 IU/d of vitamin D3 is needed for at least 6–8 weeks to achieve serum 25(OH)D levels of >30 ng/mL (75 nmol/L) followed by 600–1000 IU/d as maintenance therapy" [1, 37]

17.7 Conclusion

Vit. D has multimodal action not only in maintaining healthy functioning of bones and muscle but has regulatory activity on innate and adaptive immunity that is beneficial in prevention, mitigation of severity and alleviation of signs and symptoms of most of the rheumatic diseases when used as an add-on therapy. The benefits of treatment with Vit. D have been reported in many studies though recommendations are lacking. The possible benefits in AIRDs particularly in RA and SLE as supported in literatures cannot be ignored, and larger sample-based studies are required for the proven benefit for recommendation. Serum 25(OH) levels of 30 ng/l are required for proven benefit along with suggestion of oral Vit. D intake and optimum sunlight exposure with Vit. D-rich foods.

References

1. Athanassiou L, Kostoglou-Athanassiou I, Koutsilieris M, Shoenfeld Y. Vitamin D and autoimmune rheumatic diseases. Biomol Ther. 2023;13(4):709. https://doi.org/10.3390/biom13040709. PMID: 37189455; PMCID: PMC1013588
2. Charoenngam N. Vitamin D and rheumatic diseases: a review of clinical evidence. Int J Mol Sci. 2021;22(19):10659. https://doi.org/10.3390/ijms221910659. PMID: 34639000; PMCID: PMC850887
3. Lin J, Liu J, Davies ML, Chen W. Serum vitamin D level and rheumatoid arthritis disease activity: review and metanalysis. PLoS One. 2016;11:e0146351.
4. Hajjaj-Hassouni N, Mawani N, Allali F, Rkain H, Hassouni K, Hmamouchi I, Dougados M. Evaluation of vitamin D status in rheumatoid arthritis and its association with disease activity across 15 countries: "the COMORA study". Int J Rheumatol. 2017;2017:5491676. https://doi.org/10.1155/2017/5491676. Epub 2017 Jun 1. PMID: 28656048; PMCID: PMC5471553
5. Higgins MJ, Mackie SL, Thalayasingam N, Bingham SJ, Hamilton J, Kelly CA. The effect of vitamin D levels on the assessment of disease activity in rheumatoid arthritis. Clin Rheumatol. 2013;32(6):863–7. https://doi.org/10.1007/s10067-013-2174-x. Epub 2013 Jan 23. PMID: 23340834

6. Gopinath K, Danda D. Supplementation of 1,25 dihydroxy vitamin D3 in patients with treatment naive early rheumatoid arthritis: a randomised controlled trial. Int J Rheum Dis. 2011;14(4):332–9. https://doi.org/10.1111/j.1756-185X.2011.01684.x. PMID: 22004229
7. Amital H, Szekanecz Z, Szücs G, Dankó K, Nagy E, Csépány T, Kiss E, Rovensky J, Tuchynova A, Kozakova D, Doria A, Corocher N, Agmon-Levin N, Barak V, Orbach H, Zandman-Goddard G, Shoenfeld Y. Serum concentrations of 25-OH vitamin D in patients with systemic lupus erythematosus (SLE) are inversely related to disease activity: is it time to routinely supplement patients with SLE with vitamin D? Ann Rheum Dis. 2010;69(6):1155–7. https://doi.org/10.1136/ard.2009.120329. Epub 2010 May 3. PMID: 20439290
8. Watad A, Neumann SG, Soriano A, Amital H, Shoenfeld Y. Vitamin D and systemic lupus erythematosus: myth or reality? Isr Med Assoc J. 2016;18(3–4):177–82. PMID: 27228639
9. Mok CC, Birmingham DJ, Leung HW, Hebert LA, Song H, Rovin BH. Vitamin D levels in Chinese patients with systemic lupus erythematosus: relationship with disease activity, vascular risk factors and atherosclerosis. Rheumatology (Oxford). 2012;51(4):644–52. https://doi.org/10.1093/rheumatology/ker212. Epub 2011 Jun 29. PMID: 21719424
10. Shoenfeld Y, Giacomelli R, Azrielant S, Berardicurti O, Reynolds JA, Bruce IN. Vitamin D and systemic lupus erythematosus – the hype and the hope. Autoimmun Rev. 2018;17(1):19–23. https://doi.org/10.1016/j.autrev.2017.11.004. Epub 2017 Nov 3. PMID: 29108830
11. Perricone C, Agmon-Levin N, Colafrancesco S, Shoenfeld Y. Vitamins and systemic lupus erythematosus: to D or not to D. Expert Rev Clin Immunol. 2013;9(5):397–9. https://doi.org/10.1586/eci.13.23. PMID: 23634733
12. Islam MA, Khandker SS, Alam SS, Kotyla P, Hassan R. Vitamin D status in patients with systemic lupus erythematosus (SLE): a systematic review and meta-analysis. Autoimmun Rev. 2019;18(11):102392. https://doi.org/10.1016/j.autrev.2019.102392. Epub 2019 Sep 11. PMID: 31520805
13. Mak A. The impact of vitamin D on the Immunopathophysiology, disease activity, and extra-musculoskeletal manifestations of systemic lupus erythematosus. Int J Mol Sci. 2018;19(8):2355. https://doi.org/10.3390/ijms19082355. PMID: 30103428; PMCID: PMC6121378
14. Mok CC. Vitamin D and systemic lupus erythematosus: an update. Expert Rev Clin Immunol. 2013;9(5):453–63. https://doi.org/10.1586/eci.13.19. PMID: 23634739
15. Stagi S, Rigante D. Vitamin D and juvenile systemic lupus erythematosus: lights, shadows and still unresolved issues. Autoimmun Rev. 2018;17(3):290–300. https://doi.org/10.1016/j.autrev.2018.01.004. Epub 2018 Jan 31. PMID: 29353100
16. Yap KS, Morand EF. Vitamin D and systemic lupus erythematosus: continued evolution. Int J Rheum Dis. 2015;18(2):242–9. https://doi.org/10.1111/1756-185X.12489. Epub 2014 Dec 19. PMID: 25522756
17. Kamen DL, Cooper GS, Bouali H, Shaftman SR, Hollis BW, Gilkeson GS. Vitamin D deficiency in systemic lupus erythematosus. Autoimmun Rev. 2006;5(2):114–7. https://doi.org/10.1016/j.autrev.2005.05.009. Epub 2005 Jun 21. PMID: 16431339
18. Hausmann J, Kubesch A, Amiri M, Filmann N, Blumenstein I. Vitamin D deficiency is associated with increased disease activity in patients with inflammatory bowel disease. J Clin Med. 2019;8(9):1319. https://doi.org/10.3390/jcm8091319. PMID: 31461996; PMCID: PMC6780251
19. Sairenji T, Collins KL, Evans DV. An update on inflammatory bowel disease. Prim Care. 2017;44(4):673–92. https://doi.org/10.1016/j.pop.2017.07.010. Epub 2017 Oct 5. PMID: 29132528
20. de Souza HS, Fiocchi C. Immunopathogenesis of IBD: current state of the art. Nat Rev Gastroenterol Hepatol. 2016;13(1):13–27. https://doi.org/10.1038/nrgastro.2015.186. Epub 2015 Dec 2. PMID: 2662755

21. Costenbader KH, Feskanich D, Holmes M, Karlson EW, Benito-Garcia E. Vitamin D intake and risks of systemic lupus erythematosus and rheumatoid arthritis in women. Ann Rheum Dis. 2008;67(4):530–5. https://doi.org/10.1136/ard.2007.072736. Epub 2007 Jul 31. PMID: 17666449; PMCID: PMC2717608
22. Lange U, Teichmann J, Strunk J, Müller-Ladner U, Schmidt KL. Association of 1.25 vitamin D3 deficiency, disease activity and low bone mass in ankylosing spondylitis. Osteoporos Int. 2005;16(12):1999–2004. https://doi.org/10.1007/s00198-005-1990-5. Epub 2005 Sept 20. PMID: 16172800
23. Theodoridis X, Grammatikopoulou MG, Stamouli EM, Talimtzi P, Pagkalidou E, Zafiriou E, Haidich AB, Bogdanos DP. Effectiveness of oral vitamin D supplementation in lessening disease severity among patients with psoriasis: a systematic review and meta-analysis of randomized controlled trials. Nutrition. 2021;82:111024. https://doi.org/10.1016/j.nut.2020.111024. Epub 2020 Sept 18. PMID: 33183899
24. Erten Ş, Şahin A, Altunoğlu A, Gemcioğlu E, Koca C. Comparison of plasma vitamin D levels in patients with Sjögren's syndrome and healthy subjects. Int J Rheum Dis. 2015;18(1):70–5. https://doi.org/10.1111/1756-185X.12298. Epub 2014 Jan 28. PMID: 24467766
25. Slominski AT, Li W, Bhattacharya SK, Smith RA, Johnson PL, Chen J, Nelson KE, Tuckey RC, Miller D, Jiao Y, Gu W, Postlethwaite AE. Vitamin D analogs 17,20S(OH)2pD and 17,20R(OH)2pD are noncalcemic and exhibit antifibrotic activity. J Invest Dermatol. 2011;131(5):1167–9. https://doi.org/10.1038/jid.2010.425. Epub 2011 Jan 13. PMID: 21228816; PMCID: PMC3078214
26. Corrado A, Colia R, Mele A, Di Bello V, Trotta A, Neve A, Cantatore FP. Relationship between body mass composition, bone mineral density, skin fibrosis and 25(OH) vitamin D serum levels in systemic sclerosis. PLoS One. 2015;10(9):e0137912. https://doi.org/10.1371/journal.pone.0137912. Erratum in: PLoS One 2015;10(11):e0142748. PMID: 26375284; PMCID: PMC4572708
27. Groseanu L, Bojinca V, Gudu T, Saulescu I, Predeteanu D, Balanescu A, Berghea F, Opris D, Borangiu A, Constantinescu C, et al. Low vitamin D status in systemic sclerosis and the impact on disease phenotype. Eur J Rheumatol. 2016;3:50–5.
28. Häuser W, Perrot S, Sommer C, Shir Y, Fitzcharles MA. Diagnostic confounders of chronic widespread pain: not always fibromyalgia. Pain Rep. 2017;2(3):e598. https://doi.org/10.1097/PR9.0000000000000598. PMID: 29392213; PMCID: PMC5741304
29. Sanghi D, Mishra A, Sharma AC, Singh A, Natu SM, Agarwal S, Srivastava RN. Does vitamin D improve osteoarthritis of the knee: a randomized controlled pilot trial. Clin Orthop Relat Res. 2013;471(11):3556–62. https://doi.org/10.1007/s11999-013-3201-6. Epub 2013 Aug 1. PMID: 23904246; PMCID: PMC3792285
30. Jin X, Jones G, Cicuttini F, et al. Effect of vitamin D supplementation on tibial cartilage volume and knee pain among patients with symptomatic knee osteoarthritis: a randomized clinical trial. JAMA. 2016;315(10):1005–13. https://doi.org/10.1001/jama.2016.1961.
31. Guan Y, Hao Y, Guan Y, Bu H, Wang H. The effect of vitamin D supplementation on rheumatoid arthritis patients: a systematic review and meta-analysis. Front Med (Lausanne). 2020;30(7):596007. https://doi.org/10.3389/fmed.2020.596007. PMID: 33195358; PMCID: PMC7661491
32. Nguyen Y, Sigaux J, Letarouilly JG, Sanchez P, Czernichow S, Flipo RM, Soubrier M, Semerano L, Seror R, Sellam J, Daïen C. Efficacy of oral vitamin supplementation in inflammatory rheumatic disorders: a systematic review and meta-analysis of randomized controlled trials. Nutrients. 2020;13(1):107. https://doi.org/10.3390/nu13010107. PMID: 33396851; PMCID: PMC7824121
33. Merlino LA, Curtis J, Mikuls TR, Cerhan JR, Criswell LA, Saag KG. Iowa women's health study. Vitamin D intake is inversely associated with rheumatoid arthritis: results from the Iowa women's health study. Arthritis Rheum. 2004;50(1):72–7. https://doi.org/10.1002/art.11434. PMID: 14730601

34. Zheng R, Gonzalez A, Yue J, Wu X, Qiu M, Gui L, Zhu S, Huang L. Efficacy and safety of vitamin D supplementation in patients with systemic lupus erythematosus: a meta-analysis of randomized controlled trials. Am J Med Sci. 2019;358(2):104–14. https://doi.org/10.1016/j.amjms.2019.04.020. Epub 2019 Apr 26. PMID: 31331447
35. Sun J, Zhang S, Liu JS, Gui M, Zhang H. Expression of vitamin D receptor in renal tissue of lupus nephritis and its association with renal injury activity. Lupus. 2019;28(3):290–4. https://doi.org/10.1177/0961203319826704. Epub 2019 Jan 28. PMID: 30691345
36. Liu Y, Li C, Wang S, Guo J, Guo J, Fu J, Ren L, An Y, He J, Li Z. Human umbilical cord mesenchymal stem cells confer potent immunosuppressive effects in Sjögren's syndrome by inducing regulatory T cells. Mod Rheumatol. 2021;31(1):186–96. https://doi.org/10.1080/14397595.2019.1707996. Epub 2020 Jan 7. PMID: 31859545
37. Dowdy JC, Sayre RM, Holick MF. Holick's rule and vitamin D from sunlight. J Steroid Biochem Mol Biol. 2010;121(1–2):328–30. https://doi.org/10.1016/j.jsbmb.2010.04.002. Epub 2010 Apr 14. PMID: 20398766
38. Hossein-nezhad A, Holick MF. Vitamin D for health: a global perspective. Mayo Clin Proc. 2013;88(7):720–55. https://doi.org/10.1016/j.mayocp.2013.05.011. Epub 2013 Jun 18. PMID: 23790560; PMCID: PMC3761874

Chapter 18
Effects of Vitamin D: Anatomical and Physiological Perspective

Chiman Kumari, Richa Gupta, and Deepy Zohmangaihi

Abstract Vitamin D is a fat-soluble secosteroid and helps in calcium homeostasis in the body. Vitamin D receptor is present in different types of cells within the body, and through this receptor, vitamin D influences several intracellular pathways, thus exerting its metabolic function. Apart from the musculoskeletal effect, vitamin D has been attributed to participate in cellular functions of several non-skeletal systems. Due to its influence on several systems within the body, it helps to maintain the proper balance required for a healthy physiological support. Deficiency in this micronutrient may be a risk factor for several interconnected physiological imbalances. In this chapter, we focus on the source, metabolism and effect of vitamin D in the human body with special reference to its influence on clinical aspect.

Keywords Homeostasis · Vitamin D · VDR · Vitamin D-binding protein

Abbreviations

VDBP	Vitamin D-binding protein
VDR	Vitamin D receptor.
RXR	Retinoid X receptor.
VDRE	Vitamin D receptor elements.
RANKL	Receptor activator of NF-κB ligand.

C. Kumari (✉) · R. Gupta
Department of Anatomy, PGIMER, Chandigarh, India

D. Zohmangaihi
Department of Biochemistry, PGIMER, Chandigarh, India

V. Kumar, S. Patel (eds.), *Vitamin D: A Comprehensive Guide for Medical Professionals*, https://doi.org/10.1007/978-981-96-6491-7_18

RANK	Receptor activator of NF-κB.
m-CSF	Macrophage colony-stimulating factors.
TRPV6	Transient receptor potential vanilloid type 6.
PMCA	Plasma membrane calcium ATPase.
NCX	Sodium-calcium exchanger.
NaPi-IIb	Sodium-phosphate co-transporter type IIb.
PTH	Parathyroid hormone.
ACE	Angiotensin-converting enzyme.
TNF-α	Tumor necrosis factor α.
IFN-γ	Interferon-gamma.
APC	Antigen-presenting cell.

Learning Outcomes
At the end of the chapter, the learner will be able to appreciate

- Sources, structure and metabolism of Vitamin D
- Effect of Vitamin D on the proper functioning of the body
- Risks of Vitamin D deficiency
- Vitamin D toxicity

18.1 Introduction

The fat-soluble secosteroid, vitamin D, helps in many biological functions of the body, apart from the absorption of calcium, phosphate and magnesium via intestinal wall. Its calcemic action and role in biochemical processes is of utmost importance for a proper and balanced functioning of the body. This valuable vitamin exists in two forms, ergocalciferol (D2) and cholecalciferol (D3), both of which are converted to the bioactive form (1,25-dihydroxymetabolites) during the process of metabolism. While ergocalciferol is a plant-based source, cholecalciferol is an animal-based source. The bioactive form is recognized by the respective receptors present in different tissues of the body to carry out the necessary functions.

18.2 Vitamin D: Structure

The structure of active form of vitamin D resembles that of steroid hormone (Fig. 18.1). The structures of the raw forms of vitamin D (i.e. D2 and D3) are almost similar except the presence of different side chains (Fig. 18.1a, b). The differences in the side chains influence the binding of the molecules to the vitamin D-binding protein (VDBP). These molecules need to bind to the VDBP to control the gene expression within the cell.

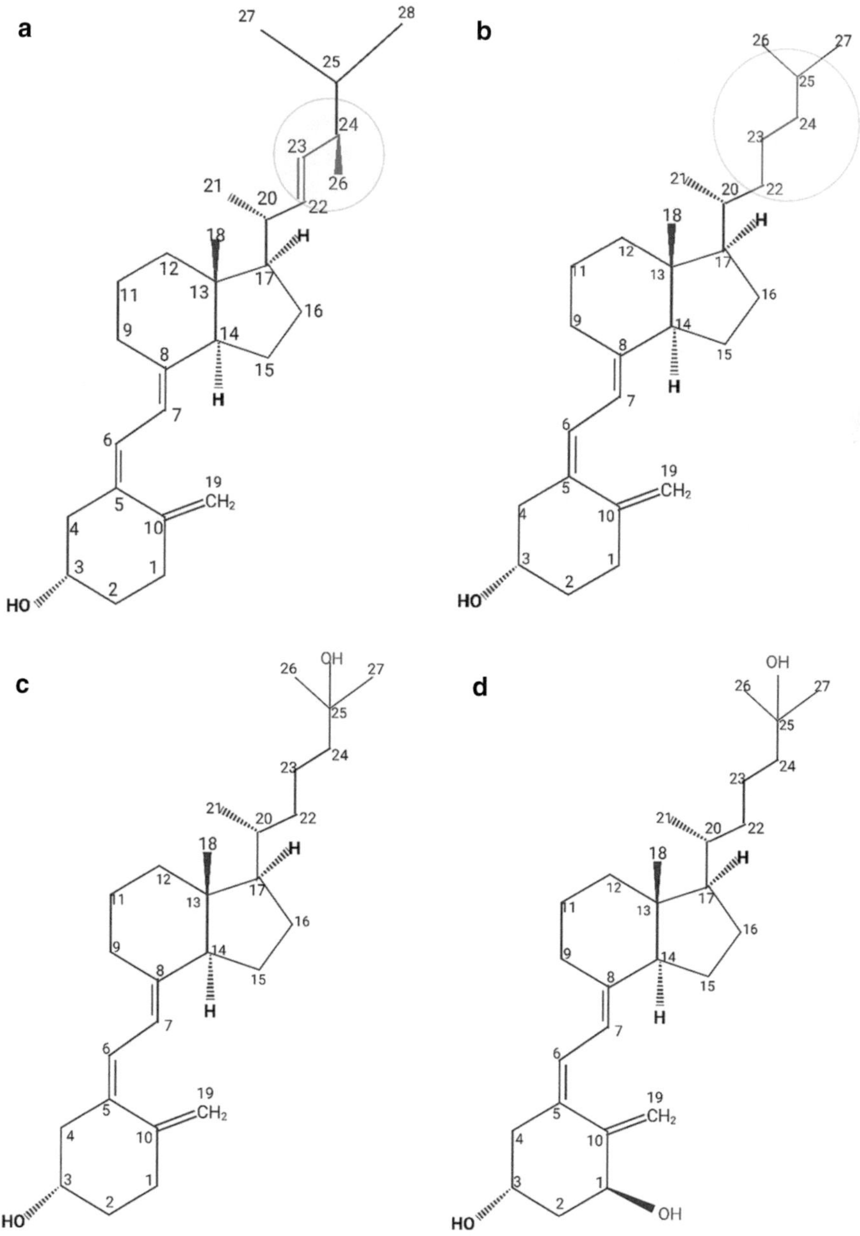

Fig. 18.1 (**a**) Ergocalciferol, (**b**) cholecalciferol, (**c**) calcidiol. There is hydroxylation in C25 position, (**d**) calcitriol, there is hydroxylation in C25 followed by C1 position. The differences in side chain alter the binding region of the molecules to vitamin D-binding protein (VDBP)

18.3 Vitamin D: Source, Metabolism (Fig. 18.2)

While there are some dietary sources of vitamin D, they are relatively limited, and for many individuals, diet alone may not provide sufficient amounts of this essential nutrient. Natural food sources of vitamin D include fatty fish such as salmon, mackerel and tuna. A rich source of vitamin D, cod liver oil, has been traditionally used as a dietary supplement. Egg yolks contain small amount of vitamin D, particularly if the chickens have been raised on vitamin D-enriched feed. Some foods are fortified with vitamin D, including milk, fruit juice, breakfast cereals and certain types of yogurt and margarine. Plant source of vitamin D includes mushrooms, which contain ergocalciferol. Besides having plant and animal sources, endogenous production of vitamin D occurs in the epidermis of the skin with the help of ultraviolet B (UVB) radiation (290–315 nm) from the sun. UVB breaks the B-ring of 7-dehydrocholesterol (cholesterol precursor) to form pre-D3 [1] which further undergoes photoisomerisation and gets converted to a precursor secosterol which then rapidly forms vitamin D3 (cholecalciferol). Vitamin D3 binds to VDBP and reaches the liver via the blood. The vitamin D-binding protein is a 58 kDa glycoprotein which is synthesized in the liver. In the liver, with the help of CYP 24A1 and CYP2R1(a cytochrome p450 enzyme), cholecalciferol gets converted to 25 hydroxy vitamin D_3 (25(OH)D3, calcidiol). 25OHD_3 is converted to active 1α, 25-dihydroxy vitamin D_3 (1,25$(OH)_2$D3, calcitriol) in the kidney cells, with the help of 25(OH)D

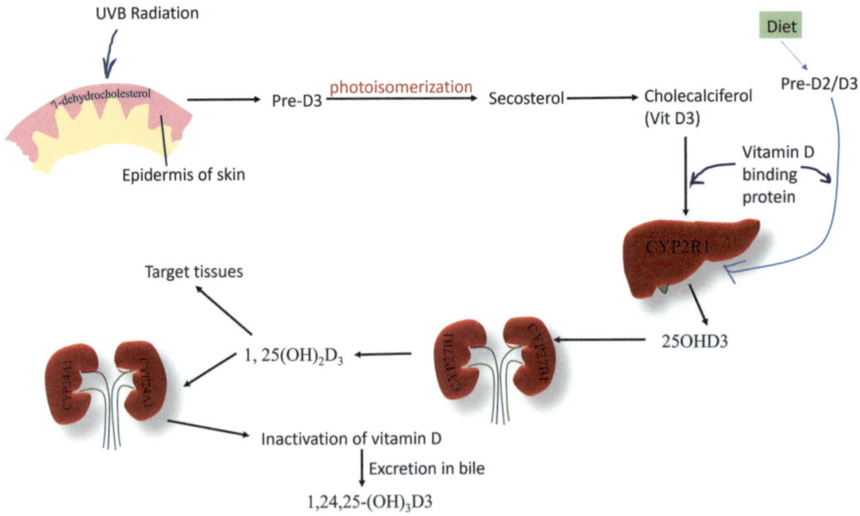

Fig. 18.2 The inactive form of vitamin D obtained either from diet or from the epidermis is carried by the VDBP to liver where, with the help of CYP2R1, it gets converted to calcidiol (25OHD3). Calcidiol is again carried by the VDBP to the kidneys and gets converted into the active metabolite, calcitriol (1,25$(OH)_2$D3) with the help of enzyme CY24A1 present in the kidneys. The active form is carried by the VDBP to the target tissues to perform its variable functions. It is later inactivated and excreted mainly via the kidneys. *VDBP* vitamin D-binding protein

1α hydroxylase (CYP27B1), and this process is under the control of parathyroid hormone (PTH) as well as calcium phosphate and fibroblast growth factor [2]. Another enzyme, renal 24-hydroxylase (CYP24A1), is responsible for inactivation of vitamin D (both 25(OH)D and 1,25$(OH)_2$D3) which later gets excreted as 1,24,25-$(OH)_3$D3 [2, 3]. With ageing, the ability of the kidneys to synthesize the activating enzymes decrease, which leads to decreased activity of vitamin D in various systems of the body. The active form is also produced in immune cells (monocytes and macrophages), keratinocytes and osteoblasts [1, 4]. While this extrarenal production of vitamin D is controlled by various cytokines, vitamin D itself has a role in production of cytokines. Since calcitriol is lipid soluble, it easily gets transported to the cell via the plasma membrane and carries out its action through its receptor [5].

A serum level of 25OHD3, which is the major circulating form of vitamin D, is usually measured to know the vitamin D status. However, most of the function of vitamin D in the body is through 1,25$(OH)_2$D3 [1]. The reason behind measuring $25OHD_3$ to determine the serum level of vitamin D3 is that serum half-life of $25OHD_3$ is more compared to that of 1,25$(OH)_2D_3$ [6].

18.4 Mechanism of Action of Vitamin D

Most of the biological actions of vitamin D are exerted via the nuclear–vitamin D receptors (VDRs) [7]. This receptor belongs to transcription factor superfamily and is distributed widely in different tissues of the body apart from the classic targets of vitamin D. VDR belongs to nuclear receptor family. The vitamin D-binding protein delivers the calcitriol to the target cells. Since vitamin D is a lipid-soluble molecule, it can easily pass through the plasma membrane. Within the cell, 1,25$(OH)_2$D3 binds to the VDR and then the whole complex binds to another receptor, retinoid X receptor (RXR) (Fig. 18.3). This complex then regulates various gene expressions through vitamin D receptor elements (VDREs), which are special DNA sequences within the nucleus of a cell [1]. Apart from its action on the DNA within the nucleus, vitamin D also exerts some non-genomic functions through cytosolic VDR by interaction with second messenger system involving ion channels, kinases and phosphatases [8]. While genomic action requires longer time to show the physiological function, the non-genomic functions occur within minutes as the non-genomic mechanism does not require the involvement of transcription and translation of the proteins [9].

Non-genomic action of vitamin D takes place through vitamin D receptor present in the cell membrane (VDR_{mem}). The action takes place by amplification of the signals through second messengers. This function through the VDR_{mem} is observed to be faster than that occurring through the nuclear VDR [10].

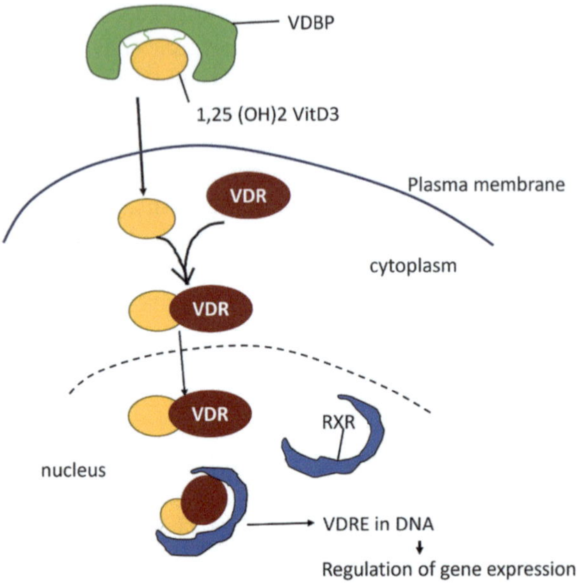

Fig. 18.3 The VDBP carries the active form of vitamin D to different regions of the body. The cells which harbor the VDR receive the active form. Both the active vitamin D and the VDR present in the cell form complex, which then later pass to the nucleus, bind with RXR, and the whole complex then regulates gene expression through the VDRE. *VDBP* vitamin D-binding protein, *VDR* vitamin D receptor, *RXR* retinoid X receptor, *VDRE* vitamin D receptor element

18.5 Vitamin D: Its Action in Cellular Function and Tissues

18.5.1 Cellular Function

The effects of vitamin D on cellular proliferation and growth have been studied mostly in cell lines. It has been observed that vitamin D and its analogue restricts cellular hyperplasia through various signaling pathways and by influencing the cellular growth cycle and apoptosis [11]. This vitamin has its target mostly on the G0-G1 phase and even in later stages of cell cycle, arresting the further ongoing process [12], thus preventing hyperproliferation. This function is carried out indirectly through modulation of the cyclins and related kinases taking part during the process of cell cycle.

The process of cell differentiation is also influenced by vitamin D. Both normal and tumor cells are influenced by the differentiating role of vitamin D. Through the genomic as well as nongenomic action, this vitamin has the capacity to direct the cellular differentiation. Accumulating evidence suggest that vitamin D has anticancer properties by which it has the ability to halt the progression of cancer through inhibition of proliferation and invasiveness. However, certain cancer cells have the ability to dysregulate the vitamin D metabolism, thus reducing this antitumorigenic effect [13, 14].

18.5.2 Musculoskeletal System

Formation of bone requires presence of calcium and phosphate, which is again dependent on presence of vitamin D for their absorption (Fig. 18.4). Hence, along with the necessary associated environment, the action of vitamin D is required for proper bone density in the body. It has been observed that calcium with the help of vitamin D maintains the bone mineral density [15]. The role of vitamin D in homeostasis of calcium and phosphorus through intestinal absorption, renal excretion as well as reabsorption has an effect in moderating the activities of osteoblasts and osteoclasts, which possess the VDR and the enzymes for metabolising vitamin D [1]. The principal role of vitamin D in osteogenesis lies in the fact that it facilitates absorption of calcium and phosphate through the intestinal wall.

1,25 $(OH)_2D3$ increases the number as well as the activities of osteoclasts by modulating the proliferation and differentiation of osteoclast progenitors, thus promoting bone resorption [16]. This osteoclastogenic activity of vitamin D takes place by its influence on the osteoblasts (Fig. 18.4). Under the influence of vitamin D, osteoblasts produce membrane-associated receptor activator of NF-κB ligand (RANKL), which is responsible for activation of receptor activator of NF-κB (RANK) in the osteoclasts. This RANK activation along with production of macrophage colony-stimulating factors (m-CSF) by the osteoblast leads to differentiation and proliferation of the precursor cells to osteoclasts (Fig. 18.4) [1, 17].

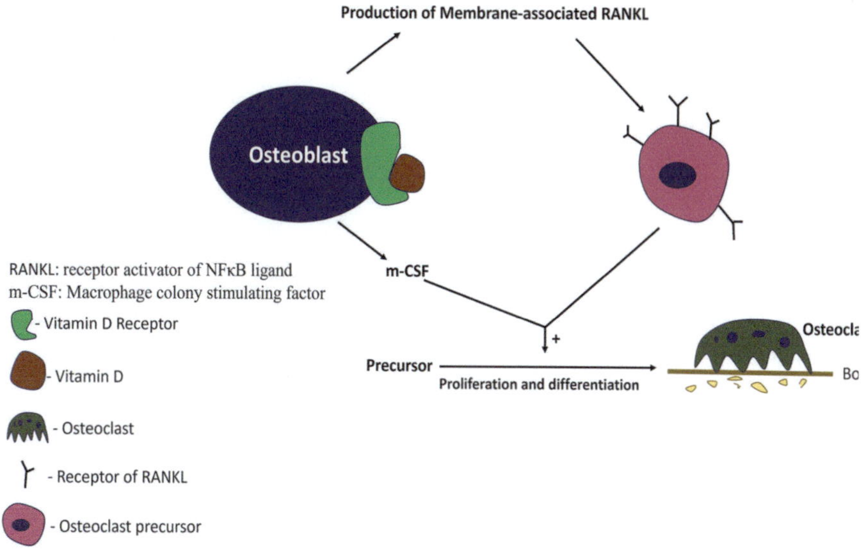

Fig. 18.4 Vitamin D through its receptor promotes the osteoblast to produce receptor activator of NF-κB ligand (RANKL), which shows its effect by binding to its receptor (RANK) present in the osteoclast precursor. Osteoblasts also produce m-CSF which along with the RANKL-RANK activation promotes the process of osteoclastogenesis

The primary role of vitamin D and its receptors in muscles is still under debate. However, vitamin D deficiency and myopathy have often been associated. It is observed that hypovitaminosis leads to oxidative stress causing mitochondrial dysfunction and muscular atrophy [18]. Ca^{2+} being an important component for muscle contractility, it has been observed that vitamin D plays an important role in Ca^{2+} mobilisation [19] and thus in muscle contractility. Again, VDBP has affinity for actin filaments, and in vitro studies show that skeletal muscles have the ability to store 25-hydroxyvitamin D (25OHD) through the VDBP. In this way, skeletal muscle may be responsible to maintain bioavailability of vitamin D in the body [20].

Osteomalacia may cause musculoskeletal pain. Vitamin D deficiency causes demineralisation of the bones due to which the gelatin-like collagen matrix may get hydrated and is expanded. This leads to an outward pressure on the highly innervated periosteal tissue. The nociceptive fibres are stimulated leading to throbbing pain [21].

18.5.3 Calcium-Phosphate Homeostasis

The calcium and phosphate obtained from food are absorbed from the small intestine (Fig. 18.4) and are transported to the target tissues by the blood. These two minerals are also reabsorbed in the renal tubules, and a homeostasis is maintained. This intestinal absorption takes place through the action of calcitriol and VDR along with the regulation of parathyroid hormone. Calcium is absorbed passively through paracellular pathway while the transcellular pathway requires the activity of $1,25(OH)_2D3$ for the passage. There are calcium transporters in the apical as well as basal sides of the enterocytes, which are regulated by this active form of vitamin D. So, any dysregulation in the vitamin D or its receptors may hamper the calcium absorption in the intestine [22]. In the apical part of the brush border intestinal epithelial cells, presence of transient receptor potential vanilloid type 6 (TRPV6) channel facilitates the entry of calcium from the lumen (Fig. 18.5). Within the cell, a calcium-binding protein (Calbidin D9K) facilitates the transport towards the basal surface of the enterocytes. This transcellular transport is dependent on $1,25(OH)_2D3$. The plasma membrane calcium ATPase 1b (PMCA1b) and sodium-calcium exchanger (NCX1) located towards the basolateral domain of the enterocytes extrudes the calcium out which is then taken up into the blood [23]. The paracellular transport needs passage through the tight junctions and is also regulated by $1,25(OH)_2D3$. The phosphate also gets absorbed from the lumen of the small intestine with the help of a sodium-phosphate co-transporter type IIb (NaPi-IIb). The production of this NaPi-IIb protein is dependent on vitamin D.

An imbalance in calcium and phosphate in the body is compensated by regulating the reabsorption of calcium and phosphate in the kidney tubules as well as resorption of bone and is modulated by vitamin D with the help of parathyroid hormone. This process requires the activity of parathyroid gland which gets feedback from the bodily concentration of calcium and phosphate whether to release

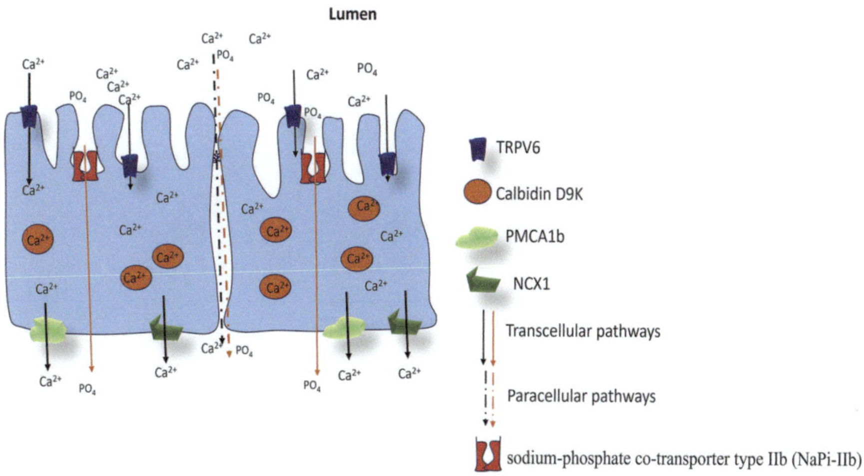

Fig. 18.5 Both the transcellular and paracellular calcium-phosphate influenced by active vitamin D metabolite (1,25(OH)$_2$D3). Calbidin D9K which is a calcium transporter is regulated by 1,25(OH)$_2$D3. Besides PMCA1b and the molecules that have influence on the tight junction of the cells are also modulated by (1,25(OH)$_2$D3). The gene responsible for transcription of NaPi-IIb is modulated by vitamin D

parathyroid hormone or not. A decrease in calcium concentration or an increase in phosphate concentration stimulates the parathyroid gland to secrete PTH. This leads to increase in absorption of calcium from the renal tubules and activation of renal vitamin D. At the same time, the phosphate gets excreted from the renal tubules. Simultaneously, resorption of bone also takes place under the influence of PTH. All this collectively bring the calcium and phosphate levels to normal again. An increase in calcium will then stimulate the thyroid gland responsible for secretion of calcitonin which blocks the activity of osteoclasts, and the chance of hypercalcaemia due to the process of resorption of bone is thus guarded. Low vitamin D status triggers secondary hyperparathyroidism, increases bone loss and leads to muscle weakness [24] (Fig. 18.6).

18.5.4 Cardiovascular System

Vitamin D receptors (VDRs) are present in the cardiomyocytes. Since vitamin D influences the intracellular content of calcium in the cells, it has a role in the contractility of the cardiomyocytes [25]. Anti-hypertensive effect of vitamin D can be attributed to its influence on the activity of renin-angiotensin-aldosterone system (RAAS). In vitro studies demonstrate that vitamin D has the capacity to inhibit transcription of renin gene [26]. Renin secreted from the renal juxtaglomerular cells helps in conversion of angiotensinogen to angiotensin I, which is then converted to

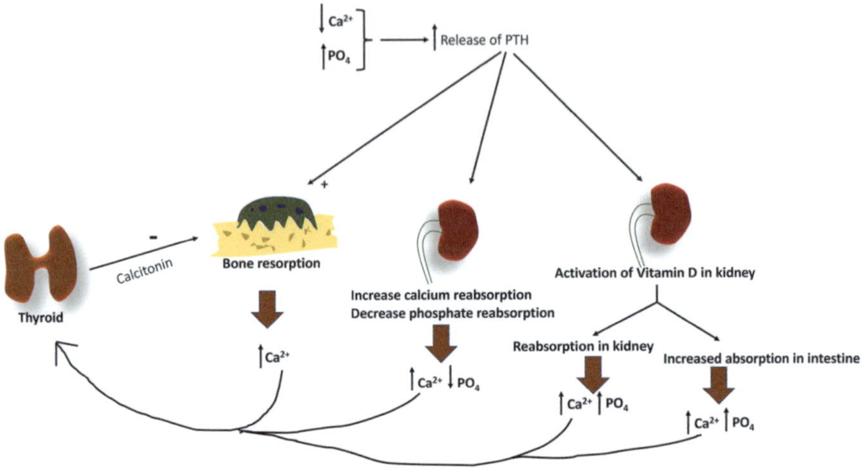

Fig. 18.6 Serum increase in calcium (Ca^{2+}) and phosphate (PO_4) stimulates release of parathyroid hormone (PTH). PTH influences bone resorption which releases Ca^{2+}. In the renal tubules, PTH stimulates Ca^{2+} and PO_4 reabsorption. PTH activates the renal vitamin D which then facilitates intestinal absorption of Ca^{2+} and PO_4. Hypercalcaemia is again prevented by feedback mechanism by which thyroid gland release calcitonin which blocks the bone resorption. Also, vitamin D promotes reabsorption of PO_4 which counteracts decrease in PO_4 caused by excretion and decrease in absorption due to PTH

angiotensin II by angiotensin-converting enzyme. The conversion of angiotensinogen to angiotensin I is modulated by vitamin D (Fig. 18.7). It was observed that people with deficient vitamin D level had higher activity of renin and in turn higher level of angiotensin II [27]. Converging evidence suggest that through inhibition of RAAS and fibroblast proliferation, it has a role in preventing cardiomyopathy [28]. Through its anti-inflammatory and anti-fibrotic properties, this vitamin tends to protect the heart from the post ischaemic risk factors, atrial fibrillations and heart failure [25].

18.5.5 Inflammation and Immunomodulation

It has been observed that the active metabolite of vitamin D has a role in immunomodulation in the tissues [29]. This action of vitamin D is possible through the receptors present in the immune cells. Presence of enzyme 25-hydroxyl vitamin D_3-1α-hydroxylase (CYP27B1), one of the key enzymes of vitamin D metabolism, along with VDR [29] in immune cells mainly in monocytes, macrophages, lymphocytes and dendritic cells, reflects the influence of vitamin D in regulation of immune and anti-inflammatory responses in the tissues (Fig. 18.8). These actions take place in an autocrine and paracrine manner and harmonise the innate as well as adaptive immunity either through activation or suppression of immunomodulatory molecules

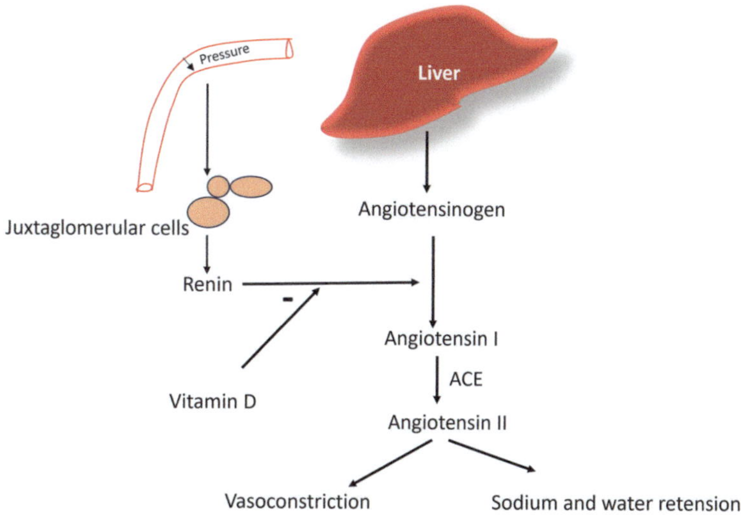

Fig. 18.7 A decrease in pressure in the renal arteriole causes secretion of renin by the juxtaglomerular cells. The angiotensinogen released from the liver gets converted to angiotensin I which again gets converted to angiotensin II with the help of angiotensin-converting enzyme (ACE). Angiotensin II causes increase in blood pressure by vasoconstriction and retention of sodium and water. Vitamin D checks the conversion of angiotensinogen to angiotensin I and thus modulates the blood pressure

Fig. 18.8 Vitamin D has a role in immune and inflammatory regulators of the tissues thus maintaining a homeostasis in the micro-environment. *APC* antigen-presenting cells

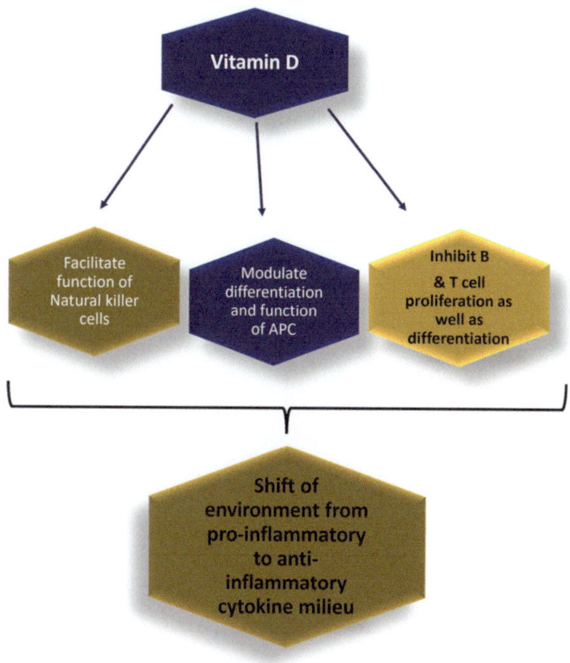

[30, 31]. Any dysregulation of this harmony may lead to autoimmune disorders and also increases the susceptibility to infections [32]. The anti-inflammatory action of 1,25(OH)$_2$D3 comes into play through the decreased production of inflammatory cytokines which include pro-inflammatory interleukins like interleukin (IL)-2, IL-12, IL-17 and IL-21, along with tumor necrosis factor α (TNF-α) and interferon-gamma (IFN-γ) [33]. 25(OH)D3 level has been found to be inversely related to the inflammatory markers [34, 35].

18.5.6 Nervous System

Accumulating evidence suggest that apart from the endocrine role of vitamin D, it also contributes intricately in development of central nervous system. The topographical location of VDR can be traced in several regions of adult central nervous system, which include the neocortex as well as the archicortex of the cerebrum along with the deeper nuclei and diencephalon [36]. Immunostaining and in situ hybridisation studies in rat central nervous system localised the VDR protein in both the nucleus and cytoplasm of neurons as well as glial cells [37]. The localisation of this protein in different functional areas of the central nervous system indicates its multifold functional aspects involving the nervous system. This vitamin has been attributed to modulate the synaptic plasticity of the brain and influence the cognitive functions [38]. Converging evidence suggest that decrease in this essential vitamin in the nervous system may stimulate a pro-inflammatory environment which may be one of the causes of mental illness and diverge disorders of the nervous system including dementia, Alzheimer's disease, schizophrenia, autism and multiple sclerosis [39–41]. The neuroprotective role of vitamin D has been attributed to its anti-oxidant, pro-differentiating, anti-inflammatory and immunomodulatory properties [41, 42]. In vitro studies have demonstrated the essential role of this vitamin in neurogenesis, thus contributing in the development of the nervous system [43]. No doubt, vitamin D has been rightly called a potent neurosteroid hormone [44].

Emerging evidence suggest the importance of vitamin D in improvement of neuropathy due to various pathologies [45]. It helps in recovery after nerve injury by assisting in myelination [46]. It has been observed in experimental studies that this secosteroid promotes differentiation of neural stem cells into glial cells and promotes myelination (Fig. 18.9) [47]. Clinical trial with vitamin D supplementation in patients with chronic low back pain having low level of vitamin D has shown improvement in the pain and associated disability of the patients [48].

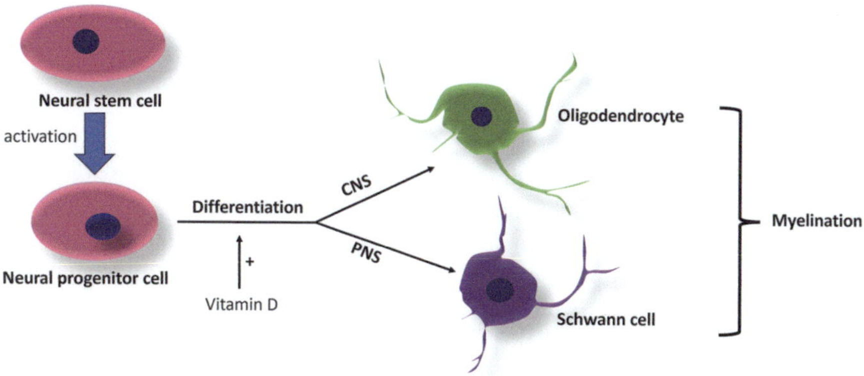

Fig. 18.9 Vitamin D promotes differentiation of neuronal stem cells to form glial cells which helps in myelination of the nerves. *CNS* central nervous system, *PNS* peripheral nervous system

18.6 Vitamin D Deficiency

Several factors may contribute to vitamin D deficiency. Besides dietary deficiency, there may be decreased absorption or increased excretion leading to low level of serum vitamin D. Again, decreased amount of sun exposure may also reduce the endogenous production of vitamin D. Melanin, the pigment responsible for skin color acts as a natural sunscreen. Darker skin contains more melanin, which reduces the skin's ability to produce vitamin D in response to sunlight. UVB intensity varies with latitude. People living in higher latitudes, especially during winter months, may receive insufficient UVB exposure for vitamin D synthesis due to the angle of the sun and shorter daylight hours. In regions with distinct seasons, vitamin D production may be limited during winter months when sunlight exposure is reduced. Older adults have reduced capacity for vitamin D synthesis compared to younger individuals due to changes in skin composition. Consequently, many people rely on dietary sources or supplements to ensure adequate vitamin D intake, especially in regions or populations with limited sunlight exposure or other risk factors for deficiency. However, moderate sun exposure remains an essential contributor to overall vitamin D status for most individuals. Balancing sunlight exposure for vitamin D synthesis with skin cancer prevention is a key consideration, emphasising the importance of individualized recommendations based on factors like age, skin type and geographical location [49, 50]. Sometimes the action of this vitamin is hampered even if the serum content is normal due to a defect in the structure or the receptor responsible for carrying out the function. Polymorphisms in vitamin D-binding proteins may affect their binding affinity for vitamin D and could be directly related to a risk of vitamin D deficiency [51].

In children, deficiency of this vitamin causes rickets, while in adults, it causes osteopenia, osteoporosis and osteomalacia which cause fragile bone prone to fracture. Muscle weakness may also be attributed to vitamin D deficiency leading to impaired balance of the body.

Autoimmune diseases such as Hashimoto's thyroiditis, Addison's disease, insulin-dependent diabetes mellitus (IDDM), multiple sclerosis, inflammatory bowel disease (IBD), systemic lupus erythematosus (SLE) and rheumatoid arthritis [52] tend to occur more in vitamin D-deficient status.

Cardiovascular health also to some extent directly and indirectly depends on adequate functioning of vitamin D. By its anti-fibrotic and anti-proliferative property, it also has a role in modulating fibrosis and cancerous growths in different tissues of the body.

18.7 Vitamin D Toxicity

Toxicity due to vitamin D is comparatively rare. However, it may occur due to increased availability of the active metabolite (1,25$(OH)_2$D3) in the body due to increased exogenous intake of vitamin D or an abnormal metabolic pathway. Many a times comorbidities may be a reason to produce excessive endogenous 1,25$(OH)_2$D3. This hypervitaminosis, which is considered to be more than 150 ng/ml of serum 25(OH)D, leads to hypercalcaemia with serious consequences, which include polyuria leading to dehydration, polydipsia, recurrent vomiting with subsequent confusion and apathy. Endogenous production occurs in certain lymphomas, in people who are hypersensitive to vitamin D and certain genetic disorders causing mutational changes [53, 54].

18.8 Conclusion

Apart from the musculoskeletal effect, vitamin D has been attributed to participate in cellular functions of several non-skeletal systems. In view of its widespread effect through its receptor, this fat-soluble vitamin has potential to maintain a healthy functioning body. Hence maintaining its adequate level in serum as well as tissues is of utmost necessity.

References

1. Bikle DD. Vitamin D and bone. Curr Osteoporos Rep. 2012;10(2):151–9. https://doi.org/10.1007/s11914-012-0098-z. PMID: 22544628; PMCID: PMC3688475

2. Bikle DD. Vitamin D: production, metabolism and mechanisms of action. In: Feingold KR, Anawalt B, Blackman MR, et al., editors. Endotext. South Dartmouth (MA): MDText.com, Inc; 2021.
3. Jenkinson C. The vitamin D metabolome: an update on analysis and function. Cell Biochem Funct. 2019;37(6):408–23.
4. Christakos S, Ajibade DV, Dhawan P, Fechner AJ, Mady LJ. Vitamin D. Metabolism. 2010;39:243–53.
5. Faye PA, Poumeaud F, Miressi F, et al. Focus on 1,25-dihydroxyvitamin D3 in the peripheral nervous system. Front Neurosci. 2019;13:348. https://doi.org/10.3389/fnins.2019.00348. Published 2019 Apr 12
6. Kumar R. The metabolism and mechanism of action of 1, 25-dihydroxyvitamin D3. Kidney Int. 1986;30(6):793–803.
7. Kato S. The function of vitamin D receptor in vitamin D action. J Biochem. 2000;127(5):717–22. https://doi.org/10.1093/oxfordjournals.jbchem.a022662.
8. Hii CS, Ferrante A. The non-genomic actions of vitamin D. Nutrients. 2016;8(3):135.
9. Zmijewski MA, Carlberg C. Vitamin D receptor (s): in the nucleus but also at membranes? Exp Dermatol. 2020;29(9):876–84.
10. Huhtakangas JA, Olivera CJ, Bishop JE, Zanello LP, Norman AW. The vitamin D receptor is present in caveolae-enriched plasma membranes and binds 1α, 25 (OH) 2-vitamin D3 in vivo and in vitro. Mol Endocrinol. 2004;18(11):2660–71.
11. Samuel S, Sitrin MD. Vitamin D's role in cell proliferation and differentiation. Nutr Rev. 2008;66(suppl_2):S116–24.
12. Umar M, Sastry KS, Chouchane AI. Role of vitamin D beyond the skeletal function: a review of the molecular and clinical studies. Int J Mol Sci. 2018;19(6):1618. https://doi.org/10.3390/ijms19061618. PMID: 29849001; PMCID: PMC6032242
13. Ma Y, Trump DL, Johnson CS. Vitamin D in combination cancer treatment. J Cancer. 2010;1:101.
14. Jeon SM, Shin EA. Exploring vitamin D metabolism and function in cancer. Exp Mol Med. 2018;50(4):1–4.
15. Muresan GC, Hedesiu M, Lucaciu O, Boca S, Petrescu N. Effect of vitamin D on bone regeneration: a review. Medicina. 2022;58(10):1337.
16. Suda T, Takahashi N, Abe E. Role of vitamin D in bone resorption. J Cell Biochem. 1992;49(1):53–8.
17. Singh PP, van der Kraan AG, Xu J, Gillespie MT, Quinn JM. Membrane-bound receptor activator of NFκB ligand (RANKL) activity displayed by osteoblasts is differentially regulated by osteolytic factors. Biochem Biophys Res Commun. 2012;422(1):48–53.
18. Dzik KP, Kaczor JJ. Mechanisms of vitamin D on skeletal muscle function: oxidative stress, energy metabolism and anabolic state. Eur J Appl Physiol. 2019;119(4):825–39. https://doi.org/10.1007/s00421-019-04104-x. Epub 2019 Mar 4. PMID: 30830277; PMCID: PMC6422984
19. Boland RL. VDR activation of intracellular signaling pathways in skeletal muscle. Mol Cell Endocrinol. 2011;347(1–2):11–6.
20. Abboud M, Puglisi DA, Davies BN, Rybchyn M, Whitehead NP, Brock KE, Cole L, Gordon-Thomson C, Fraser DR, Mason RS. Evidence for a specific uptake and retention mechanism for 25-hydroxyvitamin D (25OHD) in skeletal muscle cells. Endocrinology. 2013;154(9):3022–30.
21. Shinchuk L, Holick MF. Vitamin D and rehabilitation: improving functional outcomes. Nutr Clin Pract. 2007;22(3):297–304.
22. Khazai N, Judd SE, Tangpricha V. Calcium and vitamin D: skeletal and extraskeletal health. Curr Rheumatol Rep. 2008;10(2):110–7. https://doi.org/10.1007/s11926-008-0020-y.
23. Pu F, Chen N, Xue S. Calcium intake, calcium homeostasis and health. Food Sci Human Wellness. 2016;5(1):8–16.
24. Mendes MM, Botelho PB, Ribeiro H. Vitamin D and musculoskeletal health: outstanding aspects to be considered in the light of current evidence. Endocr Connect. 2022;11(10):e210596.

25. Cosentino N, Campodonico J, Milazzo V, De Metrio M, Brambilla M, Camera M, Marenzi G. Vitamin D and cardiovascular disease: current evidence and future perspectives. Nutrients. 2021;13(10):3603.
26. Yuan W, Pan W, Kong J, Zheng W, Szeto FL, Wong KE, Cohen R, Klopot A, Zhang Z, Li YC. 1,25-dihydroxyvitamin D3 suppresses renin gene transcription by blocking the activity of the cyclic AMP response element in the renin gene promoter. J Biol Chem. 2007;282:29821–30.
27. Ajabshir S, Asif A, Nayer A. The effects of vitamin D on the renin-angiotensin system. J Nephropathol. 2014;3(2):41–3. https://doi.org/10.12860/jnp.2014.09.
28. Freundlich M, Li YC, Quiroz Y, Bravo Y, Seeherunvong W, Faul C, Weisinger JR, Rodriguez-Iturbe B. Paricalcitol downregulates myocardial renin–angiotensin and fibroblast growth factor expression and attenuates cardiac hypertrophy in uremic rats. Am J Hypertens. 2014;27(5):720–6.
29. Brennan A, Katz DR, Nunn JD, Barker S, Hewison M, Fraher LJ, O'Riordan JL. Dendritic cells from human tissues express receptors for the immunoregulatory vitamin D3 metabolite, dihydroxycholecalciferol. Immunology. 1987;61(4):457–61. PMID: 2832307; PMCID: PMC1453440
30. Hewison M. An update on vitamin D and human immunity. Clin Endocrinol. 2012;76(3):315–25.
31. Khammissa RA, Fourie J, Motswaledi MH, Ballyram R, Lemmer J, Feller L. The biological activities of vitamin D and its receptor in relation to calcium and bone homeostasis, cancer, immune and cardiovascular systems, skin biology, and oral health. Biomed Res Int. 2018;22:2018.
32. Aranow C. Vitamin D and the immune system. J Investig Med. 2011;59(6):881–6. https://doi.org/10.2310/JIM.0b013e31821b8755.
33. Cannell JJ, Grant WB, Holick MF. Vitamin D and inflammation Dermato-endocrinology. 2014;6(1):e983401.
34. Amer M, Qayyum R. Relation between serum 25-hydroxyvitamin D and C-reactive protein in asymptomatic adults (from the continuous National Health and Nutrition Examination Survey 2001 to 2006). Am J Cardiol. 2012;109(2):226–30.
35. Bellia A, Garcovich C, D'Adamo M, Lombardo M, Tesauro M, Donadel G, Gentileschi P, Lauro D, Federici M, Lauro R, Sbraccia P. Serum 25-hydroxyvitamin D levels are inversely associated with systemic inflammation in severe obese subjects. Intern Emerg Med. 2013;8:33–40.
36. DeLuca GC, Kimball SM, Kolasinski J, Ramagopalan SV, Ebers GC. The role of vitamin D in nervous system health and disease. Neuropathol Appl Neurobiol. 2013;39(5):458–84.
37. Prüfer K, Veenstra TD, Jirikowski GF, Kumar R. Distribution of 1, 25-dihydroxyvitamin D3 receptor immunoreactivity in the rat brain and spinal cord. J Chem Neuroanat. 1999;16(2):135–45.
38. Mayne PE, Burne TH. Vitamin D in synaptic plasticity, cognitive function, and neuropsychiatric illness. Trends Neurosci. 2019;42(4):293–306.
39. Littlejohns TJ, Henley WE, Lang IA, Annweiler C, Beauchet O, Chaves PH, Fried L, Kestenbaum BR, Kuller LH, Langa KM, Lopez OL. Vitamin D and the risk of dementia and Alzheimer disease. Neurology. 2014;83(10):920–8.
40. Chiang M, Natarajan R, Fan X. Vitamin D in schizophrenia: a clinical review. BMJ Ment Health. 2016;19(1):6–9.
41. Sangha A, Quon M, Pfeffer G, Orton SM. The role of vitamin D in neuroprotection in multiple sclerosis: an update. Nutrients. 2023;15(13):2978.
42. AlJohri R, AlOkail M, Haq SH. Neuroprotective role of vitamin D in primary neuronal cortical culture. eNeurologicalSci. 2018;14:43–8. https://doi.org/10.1016/j.ensci.2018.12.004. PMID: 30619951; PMCID: PMC6312860
43. Marini F, Bartoccini E, Cascianelli G, Voccoli V, Baviglia MG, Magni MV, Garcia-Gil M, Albi E. Effect of 1α, 25-dihydroxyvitamin D3 in embryonic hippocampal cells. Hippocampus. 2010;20(6):696–705.

44. Kalueff AV, Tuohimaa P. Neurosteroid hormone vitamin D and its utility in clinical nutrition. Curr Opin Clin Nutr Metab Care. 2007;10(1):12–9.
45. Shehab D, Al-Jarallah K, Abdella N, Mojiminiyi OA, Al MH. Prospective evaluation of the effect of short-term oral vitamin D supplementation on peripheral neuropathy in type 2 diabetes mellitus. Med Princ Pract. 2015;24(3):250–6.
46. Chabas JF, Stephan D, Marqueste T, Garcia S, Lavaut MN, Nguyen C, Legre R, Khrestchatisky M, Decherchi P, Feron F. Cholecalciferol (vitamin D3) improves myelination and recovery after nerve injury. PLoS One. 2013;8(5):e65034.
47. Gomez-Pinedo U, Cuevas JA, Benito-Martín MS, et al. Vitamin D increases remyelination by promoting oligodendrocyte lineage differentiation. Brain Behav. 2020;10(1):e01498. https://doi.org/10.1002/brb3.1498.
48. Ghai B, Bansal D, Kanukula R, Gudala K, Sachdeva N, Dhatt SS, Kumar V. Vitamin D supplementation in patients with chronic low back pain: an open label, single arm clinical trial. Pain Physician. 2017;20(1):E99.
49. Holick MF. Environmental factors that influence the cutaneous production of vitamin D. Am J Clin Nutr. 1995;61:638e45. https://doi.org/10.1093/ajcn/61.3.638S.
50. Terushkin V, Bender A, Psaty EL, Engelsen O, Wang SQ, Halpern AC. Estimated equivalency of vitamin D production from natural sun exposure versus oral vitamin D supplementation across seasons at two US latitudes. J Am Acad Dermatol. 2010;62:929.e1. https://doi.org/10.1016/j.jaad.2009.07.028.929.e1-929.e9.
51. Trehan N, Afonso L, Levine DL, Levy PD. Vitamin D deficiency, supplementation, and cardiovascular health. Crit Pathw Cardiol. 2017;16:109e18.
52. Marques CD, Dantas AT, Fragoso TS, Duarte ÂL. The importance of vitamin D levels in autoimmune diseases. Rev Bras Reumatol. 2010;50:67–80.
53. Tebben PJ, Singh RJ, Kumar R. Vitamin D-mediated hypercalcemia: mechanisms, diagnosis, and treatment. Endocr Rev. 2016;37(5):521–47.
54. Marcinowska-Suchowierska E, Kupisz-Urbańska M, Łukaszkiewicz J, Płudowski P, Jones G. Vitamin D toxicity: a clinical perspective. Front Endocrinol (Lausanne). 2018;9:550. https://doi.org/10.3389/fendo.2018.00550. Published 2018 Sept 20

Chapter 19
Vitamin D: Research and Emerging Trends

Mehar Dhillon, Parth Bansal, and Vishal Kumar

Abstract Although vitamin D is well known for its critical role in calcium and phosphorous metabolism and maintaining bone health, recent research has revealed its broader physiological functions, with vitamin D receptors found in nearly every cell of the body. This chapter explores emerging trends in the extraskeletal roles of vitamin D, focusing on its impact on respiratory disorders, diabetes mellitus, neurodegenerative and immune disorders, malignancies and cardiovascular diseases. Additionally, the chapter delves into innovative drug delivery systems for vitamin D supplementation. While much of this research is still in its early stages, the promising results are driving further studies and have the potential to translate into real-world therapeutic applications for various chronic diseases.

Keywords Vitamin D · Extraskeletal functions · Chronic diseases · Supplementation · Drug delivery systems

Learning Objectives

1. Research over the past decade that has shed light on the multifaceted role of vitamin D in various processes throughout the body.
2. Innovations and development in the field of drug delivery methods for vitamin D.
3. Innovations and development in supplementation and potential therapeutic uses of vitamin D.

M. Dhillon · P. Bansal (✉) · V. Kumar
Department of Orthopaedics, PGIMER, Chandigarh, India

V. Kumar, S. Patel (eds.), *Vitamin D: A Comprehensive Guide for Medical Professionals*, https://doi.org/10.1007/978-981-96-6491-7_19

The role of vitamin D in maintaining bone health and in the regulation of calcium and phosphorous metabolism in the body is well established and has already been discussed extensively elsewhere. But research over the past few years, especially in the last two decades, has revealed that vitamin D receptors (VDRs) are present in nearly every cell of the body, and it is widely believed that all cells are targets for vitamin D at some stage in their life cycle [1]. In a review on extraskeletal actions of vitamin D published in 2016, the author highlighted the increasing body of research and evidence regarding vitamin D with 67,463 results on a PubMed search; this has since increased to 104,602 results in March 2024 [2]. As a result, there is extensive research ongoing to ascertain the role of vitamin D in various physiological and pathological processes, and to determine whether interventions can be planned based on any of these findings. While a complete detailed review is beyond the scope of this chapter, we will aim to provide a brief overview of the emerging roles and research trends and hope to educate our readers about future therapeutic potential of vitamin D.

Broadly speaking, the research is divided into two directions. The majority of studies are focused on the role of vitamin D in various chronic diseases and disorders, while there are some studies that aim to investigate the therapeutic dosing and delivery methods for vitamin D.

19.1 Respiratory Disorders

Vitamin D deficiency has been shown to be highly prevalent in chronic airway inflammation, often correlating with increased disease severity. Previous data has revealed the role of vitamin D in modulating abnormal immune responses and restricting bacterial and viral colonisation in the lungs. There is substantial evidence to suggest the role of vitamin D in childhood asthma, chronic obstructive pulmonary diseases, lung cancer and COVID-19 [3].

Vitamin D deficiency has been correlated with increased risk of asthma and other respiratory symptoms [4, 5]. There is a debate whether vitamin D deficiency is a modifiable risk factor for asthma or a consequence of the disease itself [3]. Studies on the role of vitamin D supplementation in asthmatics have shown conflicting evidence with a recent metanalysis has depicted that vitamin D supplementation reduces the rate of asthma exacerbation, along with reduced need of systemic corticosteroids [6]. Deficiency of vitamin D in early childhood or even in mothers during the prenatal period has been associated with development of chronic lung diseases, and this has led to the investigation of vitamin D supplementation in early childhood to reduce the risk of such diseases.

Similar to asthma, deficiency of vitamin D has also been noted in chronic obstructive pulmonary disease (COPD). Research in this area is difficult due to the possibility of multiple reasons for vitamin D deficiency including hospitalisation, poor physiological status and old age. While a plethora of literature is available, there is no conclusive evidence to suggest therapeutic dosing of vitamin D. While the role of vitamin D dosing is debated, its role as a modulator for TNFα which itself has an essential role in pathogenesis is not debated [7]. Vitamin D has shown promising results in patients with cystic fibrosis with a role in decreasing severity and exacerbations [8, 9].

In the last few years, vitamin D has also been shown to be implicated in the occurrence and severity of COVID-19. Sartini et al. [10] concluded that there was a significant protective role of vitamin D in the incidence of COVID 19 infection and in rates of ICU admission. In addition to its protective effect with respect to COVID-19, polymorphisms of VDRs have been shown to be associated with increased risk of severe COVID-19 [11].

19.2 Diabetes Mellitus

Recent studies have reported increasing data on vitamin D supplementation in diabetic complications like peripheral neuropathy, diabetic foot ulcers and in cardiovascular autonomic neuropathy in patients diagnosed with type 2 diabetes. Vitamin D supplementation in diabetic neuropathy has reported to slow down the progressions and provide relief in neural damage [12, 13]. Vitamin D deficiency in pregnancy has also been reported to increase risk of type 1 diabetes in neonates. Furthermore, heat deficiency has been reported to be associated with painful neuropathy in diabetic patients [14]. The role of vitamin D in diabetes as receptor for advanced glycation end products (RAGE) has recently gained popularity regarding its etiology. Its role in type 1 and 2 diabetes is supported by the discovery of 1 alpha hydroxylase expression in pancreatic cells and immune cells [15].

19.3 Neurodegenerative and Immune Disorders

Vitamin D has been recently regarded as an immunomodulator and a neurosteroid. Recent studies have reported debatable reports on vitamin D in neurological, psychiatric and autoimmune diseases. There is a possible role of vitamin D being used as a biomarker for these conditions. Vitamin D is expressed as a receptor in in various brain structures like hippocampus, hypothalamus, substantia nigra and the thalamus. It also regulates the expression of neurotrophins and neural growth factors. Recent research is regarding potential role in treatment, prevention and further progression of these degenerative conditions [16].

Recent meta-analysis combining six randomised controlled trials (RCTs) on administration of vitamin D in multiple sclerosis depicted no benefits of the supplementation in the disease, contradicting some of the recent data [17]. There is still debatable data on the supplementation of vitamin D in neurodegenerative disorders like Parkinson's and Alzheimer's disease. Whereas some authors report no benefits of supplementation as well as no correlation between low serum levels and these diseases, other data suggests that vitamin D supplementation alleviates postural balance and strength and thus helps in slowing regression [18, 19].

Some authors have reported the administration of vitamin D in juvenile systemic lupus erythematosus (SLE) patients in reducing fatigue, and further supplementation decreases autoantibodies, decreasing the TH1 andTH17 and reducing fatigue [20]. UV light is a major reason for flare-up of SLE, and sun avoidance further increases vitamin D deficiency, thus requiring supplementation. Another meta-analysis reported the beneficial role of vitamin D in irritable bowel syndrome in improving the symptom severity score [21].

19.4 Malignancy

Previous data has reported that low levels of vitamin D have been associated with increased risk of severity of symptoms in cancer as well as metastatic potential. Medrano et al. reported that vitamin supplementation is significantly associated with increased survival and decreased risk of relapse in myeloid conditions and not lymphoid conditions in transplant recipients, advocating the possible correlation between vitamin D and immune regulation of tumours [22].

Studies also report the use of vitamin D and its analogues as adjuvants in radiotherapy. It has also depicted benefits in treatment of benign lesions like uterine fibroids derived from smooth muscle cells, keeping in mind the antiproliferative and antifibrotic properties of the vitamin [23–25]. The vitamin D not only possesses biological activity but its precursors could also inhibit proliferation of melanoma cells and can be used as a radio-sensitising agent in treatment [26].

19.5 Cardiac Disease

While there has been extensive research on the role of vitamin D in the prevention and causation of cardiovascular disease, there have been no large-scale trials which have confirmed such an association [2]. Severe deficiency of vitamin D has been associated with congestive heart failure and cardiomyopathy in children [27]. Additionally, there have been multiple studies which have highlighted the role of vitamin D in myocardial infarction, coronary artery disease and association with decreased cardiac function—but these need to be confirmed by well-designed large-scale trials [28–30].

19.6 Reproductive Health

Vitamin D has been known to improve reproductive health due to its antiproliferative action. Matias et al. reported reduced pre-eclampsia-related inflammasome by supplementation [31]. Supplementation can also improve insulin receptiveness by reducing chances of gestational diabetes and associated complications. Recent data supports vitamin D as a predictor in success of in vitro fertilisation and supplementation improves pregnancy outcomes [32]. Low vitamin D has been reported as a risk factor for postpartum haemorrhage and increased likelihood of low haemoglobin, and thus the levels should be corrected along the course of pregnancy to avoid complications like postpartum haemorrhage [33]. Shan et al. reported lower serum levels were associated with hyperandrogenaemia in patients with polycystic ovary syndrome [34]. Thus, vitamin D has growing trend as a supplementary adjuvant in maintaining reproductive health.

19.7 Drug Dosing and Delivery

The daily recommendations for vitamin D supplementation range from 400 to 800 IU per day; however, recent reports based on the benefits of vitamin D have proposed a dose of 2000 IU in the form of daily supplementation. Higher intakes are required from different ethnicity especially in dark skinned individuals, and these are also reported keeping in mind a safe guideline to prevent toxicity [35]. In 2023 an RCT documented a safety dose of 2000 IU per day for long periods in vitamin D-deficient individuals depicting no toxicity signs [36]. A recent meta-analysis of 80 RCTs reported the supplementation of vitamin D reduced the all-cause mortality with an odds ratio of 0.95 comparing to a placebo group emphasising the role of the vitamin in all-purpose healthcare and life expectancy.

Multiple novel formulation techniques have been explored for improving the drug delivery and the bio-availability of vitamin D. Fat-soluble nature of vitamin D reduces its ability to be used in aqueous solutions and thus increases the need to formulate strategies for better drug delivery and action [36]. The various drug delivery formulations include solid lipid nanoparticles (SLNs), nanoemulsions, self-emulsifying drug delivery system (SEDDS), nanoparticles and microdispersions amongst others. These have been investigated in the treatment of pulmonary infections, or to target specific cells in diabetic patients [37, 38].

There is ongoing debate as to whether vitamin D should be treated only as a supplement, eventually used in prophylactics, or if it could be also considered in the therapy of multiple disorders and future research is required to corroborate this data.

19.8 Vitamin D Research

For any therapeutic or prophylactic application of vitamin D to be implemented, high level of evidence is needed before changes can be made to pre-existing guidelines and policy. The difficulty in conducting high volume trials and analysis with vitamin D are many. First and foremost, it is difficult to patent anything related to vitamin D, and thus, trials will not receive support from pharmaceutical companies.

Secondly, due to widespread over-the-counter use of vitamin D and multiple other confounding factors, studies will need to involve a large number of participants to eliminate all of these, and to properly analyse the data.

Despite this, promising strides have been made with regards to both drug delivery alternatives and novel therapeutic applications of vitamin D.

RCTs with vitamin D are not easy to conduct, and most have been improperly designed and carried out. The current trends in clinical trials are concentrating on the role of vitamin D in cancers, autoimmune diseases, neurodegenerative conditions and reproductive health primarily.

References

1. Wang Y, Zhu J, DeLuca HF. Where is the vitamin D receptor? Arch Biochem Biophys. 2012;523(1):123–33. https://doi.org/10.1016/j.abb.2012.04.001. Epub 2012 Apr 6. PMID: 22503810
2. Bikle DD. Extraskeletal actions of vitamin D. Ann N Y Acad Sci. 2016;1376(1):29–52. https://doi.org/10.1111/nyas.13219. PMID: 27649525; PMCID: PMC5031366
3. Gaudet M, Plesa M, Mogas A, Jalaleddine N, Hamid Q, Al HS. Recent advances in vitamin D implications in chronic respiratory diseases. Respir Res. 2022;23(1):252. https://doi.org/10.1186/s12931-022-02147-x. PMID: 36117182; PMCID: PMC9483459
4. Han YY, Forno E, Celedón JC. Vitamin D insufficiency and asthma in a US nationwide study. J Allergy Clin Immunol Pract. 2017;5(3):790–796.e1. https://doi.org/10.1016/j.jaip.2016.10.013. Epub 2016 Nov 29. PMID: 27913247; PMCID: PMC5423854
5. Niruban SJ, Alagiakrishnan K, Beach J, Senthilselvan A. Association between vitamin D and respiratory outcomes in Canadian adolescents and adults. J Asthma. 2015;52(7):653–61. https://doi.org/10.3109/02770903.2015.1004339. Epub 2015 Jun 1. PMID: 25563060
6. Jolliffe DA, Greenberg L, Hooper RL, Griffiths CJ, Camargo CA Jr, Kerley CP, Jensen ME, Mauger D, Stelmach I, Urashima M, Martineau AR. Vitamin D supplementation to prevent asthma exacerbations: a systematic review and meta-analysis of individual participant data. Lancet Respir Med. 2017;5(11):881–90. https://doi.org/10.1016/S2213-2600(17)30306-5. Epub 2017 Oct 3. Erratum in: Lancet Respir Med. 2018 Jun;6(6):e27. PMID: 28986128; PMCID: PMC5693329
7. Ilyas M, Agussalim A, Megawati M, Massi N, Djaharuddin I, Bakri S, As'ad S, Arief M, Bahar B, Seweng A, Prihantono P. Relationship between vitamin D level and serum TNF-α concentration on the severity of chronic obstructive pulmonary disease. Open Access Maced J Med Sci. 2019;7(14):2298–304. https://doi.org/10.3889/oamjms.2019.663. PMID: 31592277; PMCID: PMC6765078
8. Abu-Fraiha Y, Elyashar-Earon H, Shoseyov D, Cohen-Cymberknoh M, Armoni S, Kerem E, Wilschanski M. Increasing vitamin D serum levels is associated with reduced pulmonary exacerbations in patients with cystic fibrosis. J Pediatr Gastroenterol Nutr. 2019;68(1):110–5. https://doi.org/10.1097/MPG.0000000000002126. PMID: 30095576
9. Pincikova T, Paquin-Proulx D, Sandberg JK, Flodström-Tullberg M, Hjelte L. Clinical impact of vitamin D treatment in cystic fibrosis: a pilot randomized, controlled trial. Eur J Clin Nutr. 2017;71(2):203–5. https://doi.org/10.1038/ejcn.2016.259. Epub 2016 Dec 14. PMID: 27966575
10. Sartini M, Del Puente F, Oliva M, Carbone A, Bobbio N, Schinca E, Giribone L, Cristina ML. Preventive vitamin D supplementation and risk for COVID-19 infection: a systematic review and meta-analysis. Nutrients. 2024;16(5):679. https://doi.org/10.3390/nu16050679. PMID: 38474807; PMCID: PMC10935157

11. Tentolouris N, Achilla C, Anastasiou IA, Eleftheriadou I, Tentolouris A, Basoulis D, Kosta O, Lambropoulos A, Yavropoulou MP, Chatzikyriakidou A, Jude EB. The association of vitamin D receptor polymorphisms with COVID-19 severity. Nutrients. 2024;16(5):727. https://doi.org/10.3390/nu16050727. PMID: 38474855; PMCID: PMC10935253
12. Papandreou D, Hamid ZT. The role of vitamin D in diabetes and cardiovascular disease: an updated review of the literature. Dis Markers. 2015;2015:1–15.
13. Putz Z, Tordai D, Hajdú N, Vági OE, Kempler M, Békeffy M, Körei AE, Istenes I, Horváth V, Stoian AP, Rizzo M, Papanas N, Kempler P. Vitamin D in the prevention and treatment of diabetic neuropathy. Clin Ther. 2022;44(5):813–23. https://doi.org/10.1016/j.clinthera.2022.03.012. Epub 2022 Apr 12. PMID: 3542852
14. Bilezikian JP, et al. Mechanisms in endocrinology: vitamin D and COVID-19. Eur J Endocrinol. 2020;183:R133–47.
15. Akhter A, Alouffi S, Shahab U, Akasha R, Fazal-Ur-Rehman M, Ghoniem ME, Ahmad N, Kaur K, Pandey RP, Alshammari A, Akhter F, Ahmad S. Vitamin D supplementation modulates glycated hemoglobin (HBA1c) in diabetes mellitus. Arch Biochem Biophys. 2024;753:109911. https://doi.org/10.1016/j.abb.2024.109911. Epub 2024 Jan 26. PMID: 38280562
16. Zmijewski MA. Vitamin D and human health. Int J Mol Sci. 2019;20(1):145.
17. Doosti-Irani A, Tamtaji OR, Mansournia MA, Ghayour-Mobarhan M, Ferns G, Daneshvar Kakhaki R, Rezaei Shahmirzadi A, Asemi Z. The effects of vitamin D supplementation on expanded disability status scale in people with multiple sclerosis: a critical, systematic review and metaanalysis of randomized controlled trials. Clin Neurol Neurosurg. 2019;187:105564.
18. Rutjes AW, Denton DA, Di Nisio M, Chong LY, Abraham RP, Al-Assaf AS, Anderson JL, Malik MA, Vernooij RW, Martínez G, et al. Vitamin and mineral supplementation for maintaining cognitive function in cognitively healthy people in mid and late life. Cochrane Database Syst Rev. 2018;12:CD011906.
19. Han JH, Ginde AA, Brown SM, Baughman A, Collar EM, Ely EW, Gong MN, Hope AA, Hou PC, Hough CL, Iwashyna TJ, Jackson JC, Khan A, Orun OM, Patel MB, Raman R, Rice TW, Ringwood N, Semler MW, Shapiro NI, Talmor DS, Self WH, Vitamin D to Improve Outcomes by Leveraging Early Treatment Network Investigators. Effect of early high-dose vitamin D3 repletion on cognitive outcomes in critically Ill adults. Chest. 2021;160(3):909–18. https://doi.org/10.1016/j.chest.2021.03.046. Epub 2021 Apr 2. PMID: 33819472; PMCID: PMC8449000
20. Bivona G, Gambino CM, Lo Sasso B, Scazzone C, Giglio RV, Agnello L, Ciaccio M. Serum vitamin D as a biomarker in autoimmune, psychiatric and neurodegenerative diseases. Diagnostics. 2022;12(1):130.
21. Yan C, Hu C, Chen X, Jia X, Zhu Z, Ye D, Wu Y, Guo R, Jiang M. Vitamin D improves irritable bowel syndrome symptoms: a meta-analysis. Heliyon. 2023;9(6):e16437. https://doi.org/10.1016/j.heliyon.2023.e16437. PMID: 37260904; PMCID: PMC10227324
22. Liu W, Zhang L, Xu H-J, Li Y, Hu C-M, Yang J-Y, Sun M-Y. The anti-inflammatory effects of vitamin D in tumorigenesis. Int J Mol Sci. 2018;19:2736.
23. Śniegocka M, Podgórska E, Płonka PM, Elas M, Romanowska-Dixon B, Szczygieł M, Żmijewski MA, Cichorek M, Markiewicz A, Brożyna AA, Słominski AT, Urbańska K, Transplantable Melanomas in Hamsters and Gerbils as Models for Human Melanoma. Sensitization in melanoma radiotherapy-from animal models to clinical trials. Int J Mol Sci. 2018;19(4):1048. https://doi.org/10.3390/ijms19041048. PMID: 29614755; PMCID: PMC5979283
24. Podgorska E, Drzal A, Matuszak Z, Swakon J, Slominski A, Elas M, Urbanska K. Calcitriol and calcidiol can sensitize melanoma cells to low⁻LET proton beam irradiation. Int J Mol Sci. 2018;19(8):2236. https://doi.org/10.3390/ijms19082236. PMID: 30065179; PMCID: PMC6122082
25. Ciebiera M, Włodarczyk M, Ciebiera M, Zaręba K, Łukaszuk K, Jakiel G. Vitamin D and uterine fibroids-review of the literature and novel concepts. Int J Mol Sci. 2018;19(7):2051. https://doi.org/10.3390/ijms19072051. PMID: 30011902; PMCID: PMC6073230

26. Wasiewicz T, Piotrowska A, Wierzbicka J, Slominski AT, Zmijewski MA. Antiproliferative activity of non-calcemic vitamin D analogs on human melanoma lines in relation to VDR and PDIA3 receptors. Int J Mol Sci. 2018;19(9):2583. https://doi.org/10.3390/ijms19092583. PMID: 30200275; PMCID: PMC6163194
27. Uysal S, Kalayci AG, Baysal K. Cardiac functions in children with vitamin D deficiency rickets. Pediatr Cardiol. 1999;20(4):283–6. https://doi.org/10.1007/s002469900464. PMID: 10368454
28. Aleksova A, Janjusevic M, Zhou XNO, Zandonà L, Chicco A, Stenner E, Beltrami AP, D'Errico S, Sinagra G, Marketou M, Fluca AL, Zwas DR. Persistence of vitamin D deficiency among Italian patients with acute myocardial infarction. Nutr Metab Cardiovasc Dis. 2024;S0939-4753(24):00078–4. https://doi.org/10.1016/j.numecd.2024.02.007. Epub ahead of print. PMID: 38494368
29. Gerede Uludag DM, Aydogan BI, Tan TS, Acıbuca A, Turan N, Emral R, Dincer I, Erol C. Evaluation of the relationship between vitamin D deficiency and subclinical cardiac dysfunction using 2D/3D strain echocardiography in healthy people. Kardiologiia. 2024;64(2):73–9. https://doi.org/10.18087/cardio.2024.2.n2331. PMID: 38462807
30. Bakthavatchalam R, Bakthavatchalam S, Chandran I, Gaur A, Natarajaboopathy R, Geetha J, Reddy KS, Sindhura G, Varatharajan S. Association of vitamin D with risk factors for coronary artery disease. Maedica (Bucur). 2023;18(4):563–70. https://doi.org/10.26574/maedica.2023.18.4.563. PMID: 38348080; PMCID: PMC10859193
31. Matias ML, Romao-Veiga M, Ribeiro VR, Nunes PR, Gomes VJ, Devides AC, et al. Progesterone and vitamin D downregulate the activation of the NLRP1/NLRP3 inflammasomes and TLR4-MyD88-NF-κB pathway in monocytes from pregnant women with preeclampsia. J Reprod Immunol. 2021;144:103286. https://doi.org/10.1016/j.jri.2021.103286.
32. Zhao J, Liu S, Wang Y, Wang P, Qu D, Liu M, et al. Vitamin D improves in vitro fertilization outcomes in infertile women with polycystic ovary syndrome and insulin resistance. Minerva Med. 2019;110(3):199–208. https://doi.org/10.23736/S0026-4806.18.05946-3.
33. Li WJ, Chen KH, Huang LW, Tsai YL, Seow KM. Low maternal serum 25-hydroxyvitamin D concentration is associated with postpartum hemorrhage: a retrospective observational study. Front Endocrinol (Lausanne). 2022;13:816480. https://doi.org/10.3389/fendo.2022.816480. PMID: 35370939; PMCID: PMC8968120
34. Shan C, Zhu YC, Yu J, Zhang Y, Wang YY, Lu N, Cai J, Liu W, Tao T. Low serum 25-Hydroxyvitamin D levels are associated with hyperandrogenemia in polycystic ovary syndrome: a cross-sectional study. Front Endocrinol (Lausanne). 2022;13:894935. https://doi.org/10.3389/fendo.2022.894935. PMID: 35586624; PMCID: PMC9108253
35. Adebayo FA, Itkonen ST, Ohman T, Kiely M, Cashman KD, Lamberg-Allardt C, On Behalf Of The Odin Consortium. Safety of vitamin D food fortification and supplementation: evidence from randomized controlled trials and observational studies. Food Secur. 2021;10:3065.
36. Pludowski P, Grant WB, Karras SN, Zittermann A, Pilz S. Vitamin D supplementation: a review of the evidence arguing for a daily dose of 2000 international units (50 μg) of vitamin D for adults in the general population. Nutrients. 2024;16(3):391. https://doi.org/10.3390/nu16030391. PMID: 38337676; PMCID: PMC10857599
37. Castoldi A, Herr C, Niederstraßer J, Labouta HI, Melero A, Gordon S, Schneider-Daum N, Bals R, Lehr CM. Calcifediol-loaded liposomes for local treatment of pulmonary bacterial infections. Eur J Pharm Biopharm. 2017;118:62–7. https://doi.org/10.1016/j.ejpb.2016.11.026. Epub 2016 Nov 22. PMID: 27888144
38. Cristelo C, Sá AF, Lúcio M, Sarmento B, Gama FM. Vitamin D loaded into lipid nanoparticles shows insulinotropic effect in INS-1E cells. Eur J Pharm Sci. 2024;196:106758. https://doi.org/10.1016/j.ejps.2024.106758. Epub ahead of print. PMID: 38570054

Chapter 20
Patient Education, Communication & Counselling: How Important Is It?

Mehar Dhillon and Sandeep Patel

Abstract The following is a patient information **factsheet** which emphasizes the frequently asked questions regarding vitamin D and further targets to educate the patient regarding its importance.

Keywords Vitamin D · Patient factsheet · Rickets · Osteoporosis · Vitamin D hypervitaminosis

20.1 Patient Education, Communication, and Counseling: How Important Is It?

The following is a patient information **factsheet** which emphasizes the frequently asked questions regarding vitamin D and further targets to educate the patient regarding its importance.

M. Dhillon · S. Patel (✉)
Department of Orthopaedics, PGIMER, Chandigarh, India

V. Kumar, S. Patel (eds.), *Vitamin D: A Comprehensive Guide for Medical Professionals*, https://doi.org/10.1007/978-981-96-6491-7_20

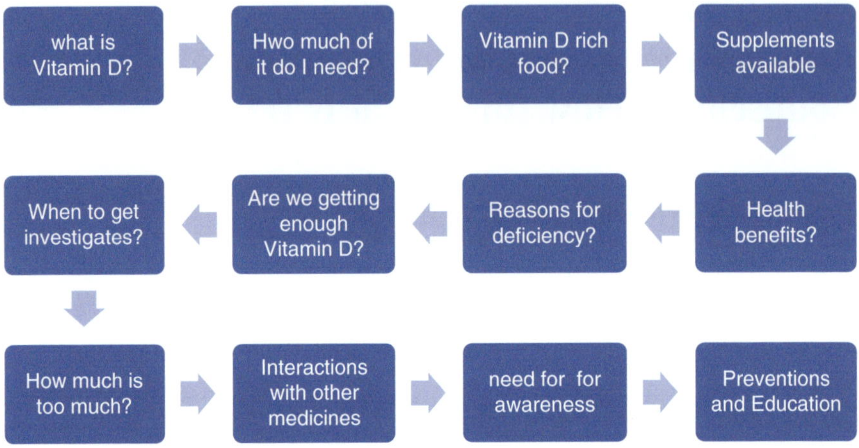

20.2 What Is Vitamin D?

It has nutritive value and is required to maintain good health. It is stored in our body's tissue containing fat and is required to conduct many functions in the body.

- Absorbance of calcium and phosphorous from the intestines, helping in building adequate bone health.
- It suppresses parathyroid hormone causing resorption of bones, thus preventing breakdown of bones.
- Vitamin D, along with calcium, helps in maintaining bone health and preventing osteoporosis. It is a condition that causes thinning and weakness of bones causing breakdown and fractures.

20.3 Burden of the Disease

Vitamin D-related ailments are categorized into serum values of 25-hydroxy and are categorized by the Indian Academy of Pediatrics [1].

- Deficiency: <12 ng/mL
- Insufficiency: 12–20 ng/mL
- Sufficiency: >20 ng/mL
- Toxicity: serum 25OHD >100 ng/mL with hypercalcemia and/or hypercalciuria

The global prevalence of the deficiency (<30 nmol/) was found to be as 15.7% between the year 2000 and 2022 with increased prevalence especially in people of Eastern Mediterranean region with low income [2].

In 2023, the **burden of the disease** in India was found to be [3]:

1–4 years: 14%
5–9 years: 18%
10–19 years: 24%

20.4 How Much Vitamin D Do I Need?

The amount of vitamin D required depends on the age of the individual.

Daily recommended amounts are found to be:

Birth to 12 months: 10 mcg (400 IU)
Children 1–13 years: 15 mcg (600 IU)
Teens 14–18 years: 15 mcg (600 IU)
Adults 19–70 years: 15 mcg (600 IU)
Pregnant and breastfeeding women: 15 mcg (600 IU)
Adults 71 years and older: 20 mcg (800 IU)

based on the Indian Academy of Pediatrics [1, 3, 4].

20.5 What Foods Provide Vitamin D?

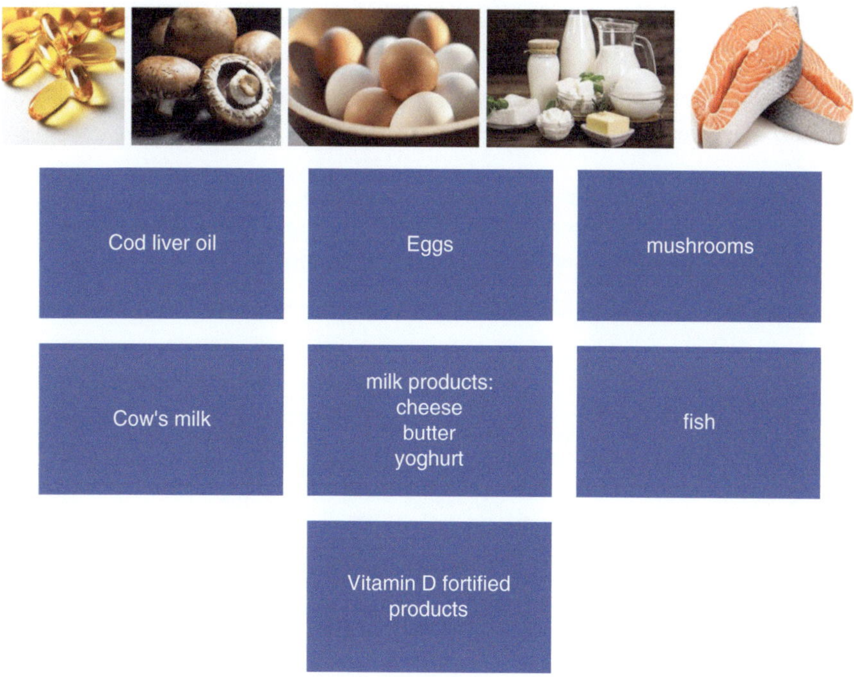

Beef and **egg yolks** provide a large amount of natural source of the vitamin [5]. Cheese along with mushrooms also provides a little vitamin D [4, 5].

Eggs Eggs provide a large amount of vitamin D as a natural source. 82% of the daily requirement of vitamin D can be attained by eating two eggs. The yolk contains 37 IU of it. The egg yolk also contains large amount of the nutrient; thus, whole eggs should be eaten and not just egg whites.

Mushrooms Mushrooms are rich in nonanimal sources of vitamin D2, D3, and D4. They synthesize vitamin D on being exposed to the ultraviolet light of the sun. By including them in your diet at least four times a week, you can achieve the body's requirement.

Cow's Milk (Pasteurized) It is rich in the nutrient but may vary among different brands. Most production companies add the vitamin during processing to reach an adequate nutrition value. Milk, however, does not contain the vitamin but is a high source of calcium. Calcium and vitamin D specifically work well together and help in attaining maximum bone strength with adequate calcium absorption.

Milk-Based Products

- **Cheese:** It does have some amount of the vitamin. Among all types of cheeses, **ricotta and cheddar** are widely rich in the nutrient.
- **Yoghurt:** Like milk, yoghurt can also be fortified with nutrients and thus not only contains calcium but also vitamins, which in combination provide adequate bone health.
- **Butter:** It is rich **antioxidants** and is also a good source of vitamin D. 11% of the daily requirement of the nutrient can be achieved by consuming one tablespoon of this.

Fortified Foods Many products in the market are fortified with macro- and micronutrients like cereals to make the diet rich in fiber and multiple vitamins.

Fish Salmon is an excellent animal source of vitamin D and is not only important for bone health but also provides good nutrition to the muscles. 66% of the daily need of nutrient can be attained by consuming 100 g of salmon.

The table below shows vitamin D amounts in some foods:

Products for consumption	Serving	IU of vitamin D
Cod liver oil	1 tablespoon	1360
Cooked salmon	3 ounces	500–800
Canned tuna fish	3 ounces	70–230
Shrimp	4 ounces	165
Fortified milk	1 cup	100–120
Fortified yoghurt	1 cup	50–100
Fortified orange juice	1 cup	100
Fortified cereal	3/4–1 cup	40–100
Cheese	1 ounce	5–20

20.6 Can I Get Vitamin D from the Sun?

The primary source of vitamin D among people is sunlight. The human body produces vitamin D when our bare skin is under the direct exposure of sunlight. However, due to the presence of clouds, formation of smog, increasing age, and dark-colored skin tone, there is reduction of formation of the nutrient [6].

Sun Exposure
- 5–30 min of sun exposure
- 10 am–3 pm
- Two times each week is adequate for the human body to produce vitamin D.

Sunscreen usually inhibits the body's ability to make vitamin D [7, 8]. Vitamin D deficiency among Indians is also attributed to decreased production of vitamin D as a consequence of comparatively darker skin despite adequate exposure to sunlight [6].

20.7 What Kinds of Dietary Supplements Are Available?

Vitamin D is found in all types of supplements. It can be found in isolated form or even combined with other nutrients.

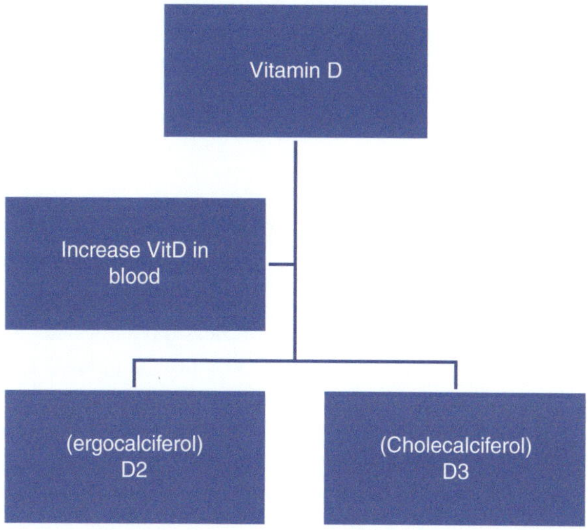

D3 might raise more content in the blood and for a longer while than D2.

Vitamin D is a fat-soluble nutrient, and thus, its absorption increases after fatty meals.

20.8 What Happens If I Don't Get Enough Vitamin D?

The deficiency of the nutrient causes the following:

Rickets: a condition in children, where the bones become soft, weak, deformed, and painful.

In teens and adults, vitamin D deficiency causes rickets.

Osteomalacia: A condition among adults that causes pain in bones and muscular weakness. Long-term shortages of vitamin D and calcium cause your bones to become fragile and break more easily; this condition is called osteomalacia.

Osteoporosis: Long-term deficiency of the nutrient causes increased fragility of the bones causing multiple fractures.

Healthy muscles are the stepping stones for adequate bone health. Deficiency of the vitamin can lead to pain in the muscles.

20.9 What Are Some Effects of Vitamin D on Our Health?

Consumption of required amounts of the vitamin and calcium from daily dietary intake of supplements also provides adequate bone health and prevents conditions like osteoporosis and muscle pain.

Heart Disease Vitamin D is essential for providing health blood vessels and leads to maintaining adequate blood pressure.

Some studies show that vitamin D supplements might help reduce blood cholesterol levels and high blood pressure—two of the main risk factors for heart disease.

Multiple Sclerosis People living closer to the equator are exposed to more sunlight and have higher amount of vitamin D levels. Multiple sclerosis (MS) is a disease affecting the nerves that carry messages from the brain to whole body. Recent studies have found link between low vitamin blood levels and the increased risk to develop the condition. However, researchers have not been able to find any correlation between consumption of the vitamin supplements and prevention of MS.

Type 2 Diabetes

Vitamin D aids in regulating the body's blood sugar levels. However, research has shown that consuming vitamin D-rich supplements does not prevent prediabetics from becoming diabetics.

20.10 The Common Reasons for Deficiency of Vitamin D

Lack of vitamin D-rich diet and inadequate exposure to the sun: Adults who have less access to sunlight develop deficiency of this nutrient, especially individuals with skin pigmentation. In addition to this, increased age, dark skin color as well as use of sunscreen also reduce the synthesis of the vitamin.

Reduction in absorption from the intestines: There are certain conditions that develop, due to which there is hindered fat absorption. This leads to inadequate absorption of the vitamin from intestines, as the nutrient is fat soluble. Celiac disease, Crohn's disease, and cystic fibrosis are some examples causing fat malabsorption. Certain surgeries causing removal of a part of the body's intestine also lead to such deficiency.

Reduced processing in patients suffering from kidney or liver disease: The liver and kidney produce certain enzymes that change vitamin D from the skin or the diet to its biologically active form. People with chronic kidney and liver disease have more chances of developing deficiency due to the lack of enzymes leading to the formation of the active form.

There are certain familial or acquired (noninherited) conditions that effect the enzymes in the liver or kidney and prevent formation of the biologically active form of the vitamin.

20.11 The Primary Reasons for Low Levels Among Indians

India is a tropical country with adequate sun exposure. Most of the Indian population reside in areas providing optimum exposure to sunlight throughout. However, there is a high incidence of vitamin D deficiency in India as [9]:

- Dark skin tone
- Sunscreen block
- Dietary habits of Indians (vegetarian diet)
- Decreased consumption of fortified foods
- Indian diet has increased salt content leading to hypercalciuria. Thus, decreased calcium affects bone health
- There is a large amount of caffeine in the Indian diet containing tea and coffee. The proportion of milk is very low in these drinks. There is low calcium in such beverages
- Vitamin D is stable up to 200 °C. In India, milk is boiled for several minutes before intake, and thus, calcium content may be reduced in boiled milk left for some time
- Lactose intolerance is highly prevalent, leading to low nutrient levels
- Indian diet contains large amount of phytate content. Phytate is a chelating agent, which causes reduction in certain micronutrients such as calcium leading to inadequate in absorption from the intestines

20.12 Am I at Predisposed to Deficiency?

One can develop deficiency if:

- One doesn't consume fortified foods or food rich in vitamin D.
- Consume a vegetarian diet.
- Do not eat or drink **dairy products**, including milk due to an allergy or **lactose intolerance**.
- Are over **50 years old.**
- Have certain medical problems, such as **kidney disease**, **liver disease**, or problems with your **intestinal tract or digesting fat**.
- Live somewhere with low amounts of sunlight.
- Wear sunscreen regularly.

20.13 How to Diagnose Vitamin D Deficiency

Levels of vitamin D in the body can be checked by evaluating the amount of 25-hydroxyvitamin D in the body. The following recommendations can be made [1, 3]:

- **Sufficiency**: ≥20 ng/mL, 50 nmol/L
- **Insufficiency**: 12 to <20 ng/mL, 30–50 nmol/L
- **Deficiency**: <12 ng/mL, 30 nmol/L
- **Hypervitaminosis** is defined as a levels >100 ng/mL or > 250 nmol/mL

20.14 Who Requires Testing?

- Elderly individuals or patients in a **long-term facility**
- Individuals diagnosed with osteoporosis or at-risk individuals
- Patients with low levels of calcium in blood or even phosphate levels

20.15 Can Too Much of Vitamin D Be Harmful?

Very high levels of vitamin D in your blood, greater than 375 nmol/L, can cause symptoms of vomiting, nausea, muscular weakness, pain, decreased appetite, dehydration, and increased urination and can even lead to formation of stones.

Extremely Increased Levels Can Lead to the Following

- Renal impairment
- Atrial fibrillation (abnormal beat of the heart)

One cannot get hypervitaminosis with just exposure to sunlight and is usually caused by intake of supplements.

The upper limits on daily basis are as follows:

< 6 months: 25 mcg (1000 IU)
7–12 months: 38 mcg (1500 IU)
12–36 months: 63 mcg (2500 IU)
4–8 years: 75 mcg (3000 IU)
9–18 years: 100 mcg (4000 IU)
>19 years: 100 mcg (4000 IU)
Pregnant and lactating women: 100 mcg (4000 IU)

20.16 Am I Getting Enough Vitamin D?

- **>20 ng/mL** is found to be optimal for bone health.
- **<12 ng/mL** may cause weakness of bones.
- **More than 50 ng/mL** is found to be too high.
- Breast milk alone lacks adequate amount of vitamin D.

20.17 Can I Take Too Much of the Vitamin D?

On consuming large doses of the supplement for a long time can lead to hypercalcemia, increasing the prevalence of renal calculi. It is imperative to tell the doctor regarding any underlining symptoms, which may be unusual or unexplained. The recommended dose for adults is 600 IU and 800 IU for elderly patients [10, 11].

20.18 Are There Interactions with Other Supplements or Medicines?

Medications with some interactions with the vitamin [12]:

- **Antacids** with magnesium.
- **Steroids** can reduce serum levels of the nutrient.
- The efficacy of **statins** may be reduced on taking vitamin D supplements.
- **Anti-epilepsy** medications.
- **Thiazide diuretics** can increase serum calcium levels on consuming vitamin D supplements.
- **Weight loss** medications like orlistat can decrease absorption of the vitamin.

One must inform the treating doctor regarding any other medications or supplements one is consuming before taking vitamin D supplementation.

20.19 Need for Awareness

The burden of the disease can be reduced by the following:

- **Fortification:** All types of milk, along with its products, can be easily fortified. Indian diet containing wheat flour and rice flour can also be fortified. These fortified products should be available in large quantity at affordable rates.
- **Programs** educating individuals regarding vitamin D deficiency and its complications should be performed. Education of school-going children is imperative to reinforce the importance of the nutrient at the grassroot level.
- **Vitamin D supplementation** should be readily provided at affordable rates at the peripheral health centers, along with regular affordable testing available to timely diagnose and treat patients.
- The recommended low **daily requirement** of the nutrient should be revised by ICMR.
- **Research groups** should be supported and encouraged to enforce new fortification strategies.

20.20 Patient Education

- To reinforce the need for timely effective **treatment and supplementation**. It is also imperative to educate regarding **"over supplementation."**
- Patients should be informed about the significance of **follow-up appointments** while taking high-dose vitamin D supplements to mitigate the risk of vitamin D toxicity.
- **Regular laboratory tests** to diagnose and monitor.
- Cooperation with all healthcare professionals for standardized protocol for supplementation.
- **Hypervitaminosis D** is rare and usually caused by excessive doses of vitamin D due to misuse of over-the-counter supplements or erroneous prescriptions. It can also be rarely caused by poisoning from exposure to rodenticides containing cholecalciferol. The increased amount causes increase in calcium, formation of renal stones, and renal impairment.

References

1. Gupta P, Dabas A, Seth A, et al. Indian Academy of Pediatrics Revised (2021) Guidelines on Prevention and Treatment of Vitamin D Deficiency and Rickets. Indian Pediatr. 2022;59:142–58.
2. Cui A, Zhang T, Xiao P, Fan Z, Wang H, Zhuang Y. Global and regional prevalence of vitamin D deficiency in population-based studies from 2000 to 2022: a pooled analysis of 7.9 million participants. Front Nutr. 2023;10:1070808. https://doi.org/10.3389/fnut.2023.1070808. PMID: 37006940; PMCID: PMC10064807

3. Rana G, Abraham RA, Sachdev HS, et al. Prevalence and correlates of vitamin D deficiency among children and adolescents from a national representative survey in India. Indian Pediatr. 2023;60:202.
4. Institute of Medicine, Food and Nutrition Board. Dietary reference intakes for calcium and Vitamin D. Washington, DC: National Academy Press; 2010.
5. Roseland JM, Phillips KM, Patterson KY, Pehrsson PR, Taylor CL. Vitamin D in foods: An evolution of knowledge. In: Feldman D, Pike JW, Bouillon R, Giovannucci E, Goltzman D, Hewison M, editors. Vitamin D, Volume 2: Health, disease and therapeutics. 4th ed. Elsevier; 2018. p. 41–78.
6. Brown LL, Cohen B, Tabor D, Zappala G, Maruvada P, Coates PM. The vitamin D paradox in black Americans: a systems-based approach to investigating clinical practice, research, and public health—expert panel meeting report. BMC Proc. 2018;12(Suppl 6):6.
7. Holick MF. Vitamin D deficiency. N Engl J Med. 2007;357:266–81.
8. Bouillon R. Comparative analysis of nutritional guidelines for vitamin D. Nat Rev Endocrinol. 2017;13:466–79. [PubMed abstract]
9. Selvarajan S, Gunaseelan V, Anandabaskar N, Xavier AS, Srinivasamurthy S, Kamalanathan SK, Sahoo JP. Systematic review on vitamin D level in apparently healthy Indian population and analysis of its associated factors. Indian J Endocrinol Metab. 2017;21(5):765–75. https://doi.org/10.4103/ijem.IJEM_168_17. PMID: 28989890; PMCID: PMC5628552
10. Holick MF. Vitamin D is not as toxic as was once thought: a historical and an up-to-date perspective. Mayo Clin Proc. 2015;90(5):561–4. [PubMed] [Reference list]
11. Marcinowska-Suchowierska E, Kupisz-Urbańska M, Łukaszkiewicz J, Płudowski P, Jones G. Vitamin D toxicity—a clinical perspective. Front Endocrinol (Lausanne). 2018;9:550.
12. Robien K, Oppeneer SJ, Kelly JA, Hamilton-Reeves JM. Drug-vitamin D interactions: a systematic review of the literature. Nutr Clin Pract. 2013;28:194–208.

Chapter 21
Unveiling the Truth: Dispelling Common Misconceptions About Vitamin D and Its Redressal

Amol Dubepuria, Mainak Roy, and Vishal Kumar

Abstract **Introduction:** Vitamin D deficiency is a pervasive issue affecting people worldwide, regardless of age, gender, race, or location. Despite the ability of the human body to synthesize vitamin D through sunlight exposure, various factors, including lifestyle choices, cultural norms, and socioeconomic constraints, contribute to this widespread deficiency.

Dispelling misconceptions surrounding vitamin D is crucial for promoting accurate information and optimal health practices. A balanced approach, including moderate sun exposure, dietary intake, routine testing, and personalized supplementation, is essential to address vitamin D deficiency and ensure overall health optimization.

Objectives

1. To understand the multidimensional impact of vitamin D deficiency beyond bone health.
2. To dispel misconceptions surrounding sun exposure and vitamin D synthesis.
3. To highlight the broader health implications of vitamin D deficiency.
4. To recognize the risks associated with excessive vitamin D supplementation.
5. To promote personalized approaches and routine testing for vitamin D levels.
6. To encourage balanced strategies for attaining optimal vitamin D levels.

Keywords Vitamin D deficiency · Sun exposure · Bone health · Autoimmune disorders · Cardiovascular health · Vitamin D supplementation · Public health Personalized medicine

---

A. Dubepuria (✉) · M. Roy
Department of Orthopaedics, All India Institute of Medical Sciences, Nagpur, India

V. Kumar
Department of Orthopaedics, PGIMER, Chandigarh, India

V. Kumar, S. Patel (eds.), *Vitamin D: A Comprehensive Guide for Medical Professionals*, https://doi.org/10.1007/978-981-96-6491-7_21

Learning Objectives from This Chapter

1. *Understanding vitamin D deficiency's multidimensional impact:*
   - Recognizing the scope of vitamin D deficiency beyond bone health, linking it to a range of conditions like infectious diseases, autoimmune disorders, cardiovascular problems, diabetes, and cancer.
   - Comprehending the implications of vitamin D deficiency based on epidemiological and biochemical studies to underscore its significance in overall health.
2. *Dispelling misconceptions surrounding sun exposure and vitamin D synthesis:*
   - Understanding that sole reliance on sun exposure for obtaining adequate vitamin D levels might not be feasible in certain regions due to cultural norms, urban living conditions, pollution, and other social factors.
   - Addressing misconceptions about sunscreen usage and its impact on vitamin D synthesis, emphasizing the balance between sun protection and vitamin D production.
3. *Broadening the understanding of vitamin D deficiency beyond skeletal health:*
   - Exploring the diverse health implications of vitamin D deficiency, such as its association with immunity, cardiovascular health, diabetes, and various cancers.
   - Emphasizing the need for robust evidence through globally conducted randomized controlled trials to establish the causative relationship between vitamin D deficiency and various health conditions.
4. *Recognizing risks associated with vitamin D supplementation:*
   - Understanding the potential risks of excessive vitamin D intake, including toxicity, hypercalcemia, kidney stones, and soft tissue calcification.
   - Identifying scenarios and conditions predisposing individuals to vitamin D toxicity, such as prescription errors, fortified foods, or supplements with elevated vitamin D content, and conditions causing excessive production of active vitamin D metabolites.
5. *Promoting personalized approaches and routine testing for vitamin D levels:*
   - Highlighting the importance of tailored approaches to vitamin D supplementation based on individual factors like age, health status, lifestyle, and geographical location.

(continued)

- Advocating routine testing for vitamin D levels to determine personalized supplementation needs and optimize health outcomes.

6. *Encouraging balanced strategies for attaining optimal vitamin D levels:*
    - Advocating a balanced approach encompassing moderate sun exposure, dietary sources, supplementation under healthcare guidance, and public health initiatives to combat widespread vitamin D deficiency.
    - Promoting awareness about the significance of vitamin D in health optimization while dispelling prevalent misconceptions.

These learning objectives aim to educate individuals, healthcare professionals, and policymakers about the complexities of vitamin D, its deficiency, and the necessity to rectify misconceptions while promoting strategies for attaining and maintaining adequate vitamin D levels for optimal health.

21.1 Introduction

Vitamin D deficiency is a widespread issue globally, affecting people regardless of age, gender, race, or location [1]. Despite the body's ability to produce vitamin D through sunlight exposure, deficiency is prevalent even in sunny regions like India due to various factors like lifestyle choices, cultural norms, limited dietary options, and socioeconomic constraints [2]. This deficiency does not just impact bone health but also contribute to other health issues. In India, where a significant population lacks proper nutrition, especially among socioeconomically disadvantaged groups, access to vitamin D-rich foods or supplements is limited. Studies across India indicate a deficiency in vitamin D levels among apparently healthy individuals, often diagnosed through markers like serum 25-hydroxyvitamin D levels [2].

Vitamin D metabolism involves its synthesis in the skin upon exposure to sunlight and its conversion into its active form through hydroxylations in the liver and kidneys [3]. This active form plays a critical role in calcium and phosphorus balance, crucial for bone health [4]. Deficiency leads to skeletal issues like rickets in children and osteomalacia and osteoporosis in adults.

Beyond skeletal health, vitamin D deficiency has been associated with various diseases, including infectious diseases, autoimmune disorders, cardiovascular problems, and different types of cancer [4]. However, while numerous studies suggest these associations, more robust evidence, particularly from well-designed randomized controlled trials, is necessary to establish causation.

Research indicates potential links between vitamin D deficiency and immunity [5], cardiovascular health [6], type 1 and type 2 diabetes [7], and various cancers [8]. These associations have been observed in epidemiological studies, highlighting the need for further exploration and substantiation through rigorous trials.

While vitamin D deficiency affects various aspects of health, its widespread prevalence in India necessitates urgent attention from healthcare practitioners and policymakers to address this public health concern, considering strategies like food fortification for its mitigation.

However, amid the wealth of information available, misconceptions regarding vitamin D have emerged, leading to confusion among the public. This chapter endeavors to dive into and dispel prevalent misconceptions surrounding vitamin D, shedding light on its significance, sources, benefits, deficiencies, supplementation, and addressing the fallacies associated with it.

Some common misconceptions about vitamin D, which are prevalent in our society, and few steps to curb them are highlighted below.

21.2 Misconception 1: Sun Exposure Alone Suffices for Adequate Vitamin D

In India, despite the ample sunlight available, vitamin D deficiency remains a pressing health issue. Understanding the adequate exposure needed for the skin to synthesize vitamin D is essential. Melanin, found in higher amounts in darker skin tones, acts as a natural sunblock [9]. Consequently, individuals with darker skin produce notably less vitamin D compared to those with fairer skin, like Caucasians [10].

For individuals with Indian skin tones, it is estimated that a minimum of 45 minutes of direct sun exposure to the bare face, arms, and legs is required daily to absorb the sun's ultraviolet (UV) rays, specifically in the 290–310 nm wavelength range necessary for vitamin D synthesis [11]. However, due to social and religious norms emphasizing modesty, most parts of the body, regardless of gender, are often covered, limiting sun exposure for many Indians.

The challenges exacerbate in urban areas. Major cities in India typically house a majority of people in densely populated areas, living in overcrowded multistory tenements. Such living conditions restrict direct sunlight from entering homes, depriving individuals of adequate sun exposure even indoors. Additionally, limited space for outdoor activities further diminishes opportunities for sun exposure.

Atmospheric pollution, a persistent issue in metropolitan India, also contributes to the challenge of maintaining optimal vitamin D levels. Furthermore, the use of sunscreen creams and umbrellas, while offering protection from harmful UV rays, inadvertently hinders the synthesis of vitamin D [12].

The scorching heat during the Indian summer, coupled with the desire for fairer skin complexion, leads many individuals to actively avoid sun exposure. The discomfort caused by the intense heat and societal preferences often override the need for sufficient vitamin D levels. Moreover, there is also fear of increased incidence of skin cancers due to prolonged sun exposure [13]. Consequently, the quest for vitamin D sufficiency frequently takes a backseat.

Redressal Considering these complexities, it becomes evident that solely relying on adequate sun exposure is impractical in the Indian scenario for attaining optimal vitamin D levels. Alternative strategies beyond sun exposure are crucial to address the prevalent vitamin D deficiency in the country. These might include dietary changes, fortified foods, supplements, and public health initiatives aimed at raising awareness and offering accessible solutions to combat this widespread health concern [14].

21.3 Misconception 2: All Sunscreens Completely Block Vitamin D Synthesis

Another prevalent misconception revolves around the belief that all types of sunscreens entirely inhibit the synthesis of vitamin D. Research suggests that while certain sunscreens can impede vitamin D production [12], individuals using sunscreen can still attain sufficient vitamin D levels with prolonged sun exposure. The process of generating vitamin D3 in the skin starts with exposure to UVB rays found in terrestrial sunlight, and this synthesis can occur with relatively small areas of exposed skin [15]. Daily use of sunscreen for incidental sun exposure usually involves products with low Sun Protection Factor (SPF) for UVB protection and high UVA Protection Factor (UVA-PF) for broader protection against UVA rays. However, this kind of sunscreen use is unlikely to significantly affect the production of vitamin D [15]. Studies conducted so far have consistently failed to establish a link between the regular use of sunscreen, even with SPF greater than 15, and vitamin D deficiency [12]. It is important to note that the time spent outdoors and the amount of skin exposed to the sun have shown a direct positive correlation with vitamin D levels [13]. Overall, other protective measures against sun exposure (like seeking shade, wearing protective clothing, and covering skin with long sleeves) may have a more significant impact on vitamin D levels than the use of sunscreen alone [16].

Redressal Recommending the use of sunscreens with lower SPF and allowing some unprotected sun exposure can be a pragmatic approach. Nonetheless, acknowledging other sources such as dietary intake or supplements becomes imperative, especially for individuals with restricted sun exposure due to various reasons.

21.4 Misconception 3: Vitamin D Deficiency Only Causes Bone-Related Issues

A widely held misconception is the notion that vitamin D deficiency predominantly affects bone health, leading to conditions like rickets in children and osteomalacia in adults. However, emerging research has shed light on the broader implications of vitamin D deficiency, linking it to various ailments [17]. Biochemical studies have linked vitamin D deficiency to various chronic illnesses, spanning from infectious diseases to autoimmune conditions, cardiovascular ailments, diabetes, and cancer [2]. While many epidemiological publications support the health benefits of vitamin D, most of these are association studies or small-scale randomized controlled trials (RCTs). However, stronger evidence is needed, necessitating well-designed, globally conducted RCTs to unequivocally assess the advantages of vitamin D supplementation both as a preventive measure and as adjunctive therapy.

- Immunity: Historical records show that as far back as the nineteenth century, cod liver oil, rich in vitamin D, was used to treat tuberculosis (TB) [5]. Recent studies have correlated vitamin D deficiency with increased risks of developing TB [18], otitis media [19], upper respiratory tract infections [20], and influenza and even COVID-19 [21].
- Cardiovascular health: Cardiovascular diseases (CVDs), encompassing conditions like heart failure and coronary artery disease, pose significant global health threats. Observational studies have accumulated evidence suggesting an association between CVDs and vitamin D deficiency [6]. Hypertension, diabetes, obesity, high triglyceride levels, and increased cardiovascular mortality are linked to low 25(OH)D levels (<21 ng/mL) [7]. Patients with previous or prevalent cardiovascular or cerebrovascular diseases have exhibited reduced 25(OH)D concentrations [22].
- Type 1 diabetes: Type 1 diabetes (T1D) results from the autoimmune destruction of pancreatic β cells, leading to insulin-dependent diabetes [23]. Vitamin D intake has been associated with a reduced risk of T1D, with observational studies suggesting a 30% lower risk in children receiving vitamin D supplements [24].
- Type 2 diabetes: Type 2 diabetes (T2D) involves insulin resistance (IR) [25]. Observational studies have shown an inverse relationship between 25(OH)D levels, calcium status, insulin resistance, and hyperglycemia [26]. Combined supplementation with both nutrients has demonstrated benefits in optimizing glucose levels [7].
- Cancer: Individuals residing in higher latitude regions are more prone to developing and succumbing to various cancers like colorectal, prostate, ovarian, breast, lung, and esophageal cancer [2]. Studies have revealed a heightened risk of developing and dying from these cancers when 25(OH)D levels are below 20 ng/mL, with percentages ranging from 30% to 50% for various cancer types [8].

Redressal Recognizing the multifaceted role of vitamin D in the body is crucial. Raising awareness about its wider health implications is pivotal to encourage regular testing for vitamin D levels and appropriate supplementation if a deficiency is identified.

21.5 Misconception 4: Vitamin D Supplementation Poses No Risks

There exists a misconception that Vitamin D supplementation is entirely risk-free. Contrarily, excessive intake of vitamin D can lead to hypervitaminosis D, characterized by symptoms like hypercalcemia, kidney stones, and soft tissue calcification [27].

In healthy individuals, excessive vitamin D intake, whether intentional or accidental, can lead to vitamin D toxicity. Prescription errors without regular monitoring of vitamin D levels can also result in this condition [27]. Toxicity often occurs in patients requiring high doses to manage conditions like osteoporosis, renal osteodystrophy, psoriasis, gastric bypass surgery, celiac disease, or inflammatory bowel disease [28].

Vitamin D toxicity can also arise from fortified foods or supplements containing higher-than-intended amounts of vitamin D. Instances of recalls by health authorities due to excessive vitamin D content in multivitamins and supplements have led to cases of toxicity, with reports of adverse effects particularly in children [29].

While vitamin D is naturally present in fish, meat, and dairy products, the dosage from these sources is usually insufficient to cause toxicity [30]. However, individuals on high vitamin D doses, combined with consuming significantly fortified milk unintentionally, are at an increased risk of vitamin D toxicity. Conditions leading to excessive production of 1,25(OH)2D, such as granulomatous disorders and lymphomas, can also contribute to toxicity [31]. Excessive exposure to sunlight does not cause vitamin D toxicity as the body regulates and converts vitamin D into inactive forms effectively [30].

Redressal Underlining the significance of moderate supplementation under the guidance of healthcare professionals is paramount. Regular monitoring and adjustment of dosage based on individual requirements and periodic test results are imperative to prevent adverse effects. Public awareness should be done about probable risks of over enthusiastic intake of vitamin D.

21.6 Misconception 5: Everyone Requires the Same Vitamin D Levels

Optimal vitamin D levels can vary among individuals based on several factors. While there are general guidelines for vitamin D intake, the specific requirements can differ depending on various aspects such as age, health status, geographical location, skin pigmentation, and individual medical conditions [32].

- *Age:* Different age groups may have different vitamin D needs. For instance, infants, children, and adolescents require adequate vitamin D for growth, bone development, and overall health. Older adults might need more vitamin D due to potential reduced skin synthesis and decreased dietary intake [32].

- *Geographical location and sun exposure:* Sunlight exposure is a primary way the body produces vitamin D. Individuals living in regions with limited sunlight or during seasons with less sunlight may struggle to synthesize enough vitamin D naturally. This affects their overall requirements [3].
- *Skin tone:* Melanin, the pigment responsible for skin color, can affect vitamin D synthesis. Darker skin tones have higher natural protection against harmful UV rays but may also require more time in the sun to produce sufficient vitamin D compared to lighter skin tones [9].
- *Health conditions:* Certain health conditions or medications can interfere with the absorption or utilization of vitamin D. Individuals with conditions affecting the gut's ability to absorb nutrients or those on medications that interfere with vitamin D metabolism might need higher levels of supplementation [11].
- *Individual variability:* Each person's body processes and absorbs nutrients differently. Some individuals may naturally have higher or lower levels of vitamin D even if they are exposed to similar sunlight or consume similar amounts through diet [2].

Redressal Tailoring supplementation based on factors like age, existing health conditions, lifestyle, and environmental factors is indispensable for optimizing health outcomes. Regular monitoring of vitamin D levels, especially for those at higher risk of deficiency, and adjusting intake accordingly is important to maintain optimal health. Consulting with a healthcare provider is the best approach to determine the appropriate vitamin D levels for each person based on their unique circumstances.

21.7 Misconception 6: Only Older Adults Need Vitamin D

Vitamin D is a crucial nutrient that plays a fundamental role in various bodily functions, and its necessity extends beyond just older individuals [17]. It is essential for people of all ages due to its numerous functions in the body [11]. Despite its significance, many people, regardless of age, may struggle to maintain sufficient vitamin D levels [33]. Factors such as limited sun exposure (which is one of the primary sources of vitamin D), darker skin pigmentation, certain medical conditions, and dietary restrictions can contribute to deficiencies [2].

Redressal It is important for individuals of all ages to ensure they get enough vitamin D through various means, including exposure to sunlight, consuming vitamin D-rich foods (such as fatty fish, fortified dairy products, and fortified cereals), and sometimes using supplements under the guidance of healthcare professionals. Therefore, while older adults might be more prone to vitamin D deficiency due to factors like reduced mobility and less sun exposure, ensuring adequate vitamin D intake is crucial for everyone's overall health and well-being, regardless of age. Proper awareness should be made among the common people regarding this.

21.8 Conclusion

In conclusion, dispelling misconceptions surrounding vitamin D is pivotal for disseminating accurate information and fostering optimal health practices. Acknowledging the complexities associated with vitamin D sources, absorption, deficiency implications, and supplementation is crucial. Encouraging a balanced approach that includes moderate sun exposure, dietary intake, routine testing, and personalized supplementation under healthcare guidance is essential to rectify misconceptions and ensure individuals attain and sustain adequate vitamin D levels for overall well-being and health optimization.

References

1. van Schoor NM, Lips P. Worldwide vitamin D status. Best Pract Res Clin Endocrinol Metab. 2011;25(4):671–80.
2. G R, Gupta A. Vitamin D deficiency in India: prevalence, causalities and interventions. Nutrients. 2014;6(2):729–75.
3. Holick MF. Sunlight and vitamin D for bone health and prevention of autoimmune diseases, cancers, and cardiovascular disease23. Am J Clin Nutr. 2004;80(6):1678S–88S.
4. Khazai N, Judd SE, Tangpricha V. Calcium and vitamin D: skeletal and extraskeletal health. Curr Rheumatol Rep. 2008;10(2):110–7.
5. Ismailova A, White JH. Vitamin D, infections and immunity. Rev Endocr Metab Disord. 2022;23(2):265–77.
6. de la Guía-Galipienso F, Martínez-Ferran M, Vallecillo N, Lavie CJ, Sanchis-Gomar F, Pareja-Galeano H. Vitamin D and cardiovascular health. Clin Nutr Edinb Scotl. 2021;40(5):2946–57.
7. Mitri J, Pittas AG. Vitamin D and diabetes. Endocrinol Metab Clin N Am. 2014;43(1):205–32.
8. Carlberg C, Velleuer E. Vitamin D and the risk for cancer: a molecular analysis. Biochem Pharmacol. 2022;196:114735.
9. Naik PP, Farrukh SN. Influence of ethnicities and skin color variations in different populations: a review. Skin Pharmacol Physiol. 2021;35(2):65–76.
10. Xiang F, Lucas R, de Gruijl F, Norval M. A systematic review of the influence of skin pigmentation on changes in the concentrations of vitamin D and 25-hydroxyvitamin D in plasma/serum following experimental UV irradiation. Photochem Photobiol Sci. 2015;14(12):2138–46.
11. Vitamin D in health and disease | Application of silk in medical field | Blog [Internet]. [cited 2024 Jan 4]. Available from: https://www.ssbs.edu.in/VitaminD-in-health-and-disease
12. Passeron T, Bouillon R, Callender V, Cestari T, Diepgen TL, Green AC, et al. Sunscreen photoprotection and vitamin D status. Br J Dermatol. 2019;181(5):916–31.
13. Lindqvist PG, Epstein E, Landin-Olsson M. Sun exposure - hazards and benefits. Anticancer Res. 2022;42(4):1671–7.
14. Ritu G, Gupta A. Fortification of foods with vitamin D in India. Nutrients. 2014;6(9):3601–23.
15. Springbett P, Buglass S, Young AR. Photoprotection and vitamin D status. J Photochem Photobiol B. 2010;101(2):160–8.
16. Matsuoka LY, Wortsman J, Hollis BW. Use of topical sunscreen for the evaluation of regional synthesis of vitamin D3. J Am Acad Dermatol. 1990;22(5):772–5.
17. Chang SW, Lee HC. Vitamin D and health - the missing vitamin in humans. Pediatr Neonatol. 2019;60(3):237–44.
18. Charoenngam N, Holick MF. Immunologic effects of vitamin D on human health and disease. Nutrients. 2020;12(7):2097.

19. Walker RE, Bartley J, Camargo CA, Mitchell EA. Vitamin D and otitis media. Curr Allergy Asthma Rep. 2019;19(7):33.
20. Maqbool MA, Gangadhara Somayaji KS, Nayana VG. Vitamin D - an elixir for recurrent upper respiratory tract infection. Indian J Otolaryngol Head Neck Surg. 2023;75(3):1370–5.
21. Grant WB, Lahore H, McDonnell SL, Baggerly CA, French CB, Aliano JL, et al. Evidence that vitamin D supplementation could reduce risk of influenza and COVID-19 infections and deaths. Nutrients. 2020;12(4):988.
22. Pál É, Ungvári Z, Benyó Z, Várbíró S. Role of vitamin D deficiency in the pathogenesis of cardiovascular and cerebrovascular diseases. Nutrients. 2023;15(2):334.
23. Gillespie KM. Type 1 diabetes: pathogenesis and prevention. CMAJ Can Med Assoc J. 2006;175(2):165–70.
24. Yu J, Sharma P, Girgis CM, Gunton JE. Vitamin D and Beta cells in type 1 diabetes: a systematic review. Int J Mol Sci. 2022 Jan;23(22):14434.
25. Insulin resistance: definition and consequences - PubMed [Internet]. [cited 2024 Jan 4]. Available from: https://pubmed.ncbi.nlm.nih.gov/11460565/
26. Lips P, Eekhoff M, van Schoor N, Oosterwerff M, de Jongh R, Krul-Poel Y, et al. Vitamin D and type 2 diabetes. J Steroid Biochem Mol Biol. 2017;173:280–5.
27. Kaur P, Mishra SK, Mithal A. Vitamin D toxicity resulting from overzealous correction of vitamin D deficiency. Clin Endocrinol. 2015;83(3):327–31.
28. Mouton Sclaunich H, Marchand C, Rekik A, Chenailler C, Varin R, Doucet J. A case of iatrogenic vitamin D toxicity revealed by drug reconciliation. Therapie. 2022;77(3):388–90.
29. Holmes RP, Kummerow FA. The relationship of adequate and excessive intake of vitamin D to health and disease. J Am Coll Nutr. 1983;2(2):173–99.
30. Holick MF, Vitamin D. Is not as toxic as was once thought: a historical and an up-to-date perspective. Mayo Clin Proc. 2015;90(5):561–4.
31. Hoorn EJ, Zietse R. Disorders of calcium and magnesium balance: a physiology-based approach. Pediatr Nephrol Berl Ger. 2013;28(8):1195–206.
32. Yu HJ, Kwon MJ, Woo HY, Park H. Analysis of 25-hydroxyvitamin D status according to age, gender, and seasonal variation. J Clin Lab Anal. 2016;30(6):905–11.
33. Huang SJ, Wang XH, Liu ZD, Cao WL, Han Y, Ma AG, et al. Vitamin D deficiency and the risk of tuberculosis: a meta-analysis. Drug Des Devel Ther. 2017;11:91–102.

MIX
Papier aus verantwortungsvollen Quellen
Paper from responsible sources
FSC® C105338

If you have any concerns about our products,
you can contact us on
ProductSafety@springernature.com

In case Publisher is established outside the EU,
the EU authorized representative is:
Springer Nature Customer Service Center GmbH
Europaplatz 3, 69115 Heidelberg, Germany

Printed by Libri Plureos GmbH
in Hamburg, Germany